Early Engineering Mathematics

Early Engineering Mathematics

J. O. Bird B.Sc(HONS), C.MATH, FIMA, C.ENG, MIEE, FCOLLP, FIEIE

NEWNES

Newnes
An imprint of Butterworth-Heinemann Ltd
Linacre House, Jordan Hill, Oxford OX2 8DP

ɛ A member of the Reed Elsevier plc group

OXFORD LONDON BOSTON
MUNICH NEWDELHI SINGAPORE SYDNEY
TOKYO TORONTO WELLINGTON

First published 1995

British Library Cataloguing in Publication Data

Bird, J. O.
  Early Engineering Maths
  I. Title
  510

ISBN 0 7506 2133 8

Library of Congress Cataloging in Publication Data

Bird, J. O.
  Early Engineering Mathematics: J. O. Bird.
      p.   cm.
    Includes index.
    ISBN 0 7506 2133 8
    1. Mathematics. I. Title
  QA37.2.B56
  510–dc20   94–22234
                                                 CIP

Typeset by Graphicraft Typesetters Ltd., Hong Kong
Printed and bound in Great Britain

Contents

Preface

Early Engineering Mathematics introduces and then consolidates basic mathematical principles and promotes awareness of mathematical concepts for students needing a broad base for further vocational work. The book covers specifically the **GNVQ mandatory Mathematics content** of the **Science and Mathematics for Engineering** at Intermediate Level (i.e. GNVQ 2) and **Mathematics for Engineering** at Advanced Level (i.e. GNVQ 3) in Engineering. However, it can be regarded as a basic textbook in mathematics for a much wider range of courses; for example, it will be useful for GCSE and A level revision, for technical and further education departments in Australia, for East and West Africa Examining Councils and for similar technical examining authorities in English-speaking countries worldwide.

Early Engineering Mathematics provides a lead into **Engineering Mathematics**, which provides coverage of the optional **Further Mathematics** unit at Advanced Level GNVQ in Engineering.

Each topic considered in the text is presented in a way that assumes in the reader little previous knowledge of that topic.

Theory is introduced in each chapter by a brief outline of essential definitions, formulae, laws, procedures etc. – but is kept to a minimum, for problem-solving is extensively used to establish and exemplify the theory. It is intended that readers will gain real understanding through seeing problems solved and then solving problems themselves.

The textbook contains some 600 worked problems, followed by nearly 1000 further problems (all with answers). Nearly 300 line diagrams enhance the understanding of the theory. Where at all possible the problems mirror practical situations found in engineering and science.

I would like to express my appreciation for the friendly co-operation and helpful advice given by the publishers, and to Mrs Elaine Woolley for the excellent typing of the manuscript.

'Learning by Example' is at the heart of *Early Engineering Mathematics*.

John O. Bird
Highbury College
Portsmouth

1

Revision of basic arithmetic

1.1 Arithmetic operations

Whole numbers are called **integers**. +3, +5, +72 are called positive integers: −13, −6, −51 are called **negative integers**. Between positive and negative integers is the number 0 which is neither positive nor negative.

The four basic arithmetic operators are: add (+), subtract (−), multiply (×) and divide (÷).

For addition and subtraction, when **unlike signs** are together in a calculation, the overall sign is **negative**. Thus, adding minus 4 to 3 is 3 + −4 and becomes 3 − 4 = −1. **Like signs** together give an overall **positive** sign. Thus subtracting minus 4 from 3 is 3 − −4 and becomes 3 + 4 = 7.

For multiplication and division, when the numbers have **unlike signs**, the answer is **negative**, but when the numbers have **like signs** the answer is **positive**. Thus 3 × −4 = −12, whereas −3 × −4 = +12. Similarly

$$\frac{4}{-3} = -\frac{4}{3} \quad \text{and} \quad \frac{-4}{-3} = +\frac{4}{3}$$

Problem 1   Add 27, −74, 81 and −19.

This problem is written as 27 − 74 + 81 − 19.

Adding the positive integers:	27
	81
Sum of positive integers is:	108
Adding the negative integers:	74
	19
Sum of negative integers is:	93

Taking the sum of the negative integers from the sum of the positive integers gives:

$$\begin{array}{r} 108 \\ -\ \ 93 \\ \hline 15 \end{array}$$

Thus **27 − 74 + 81 − 19 = 15**

Problem 2   Subtract 89 from 123.

This is written mathematically as 123 − 89

$$\begin{array}{r} 123 \\ -\ \ 89 \\ \hline 34 \end{array}$$

Thus **123 − 89 = 34**

Problem 3   Subtract −74 from 377.

This problem is written as 377 − −74. Like signs together give an overall positive sign, hence

$$377 - -74 = 377 + 74 \qquad \begin{array}{r} 377 \\ +\ \ 74 \\ \hline 451 \end{array}$$

Thus **377 − −74 = 451**

Problem 4   Subtract 243 from 126.

The problem is 126 − 243. When the second number is larger than the first, take the smaller number from the larger and make the result negative.

Thus $126 - 243 = -(243 - 126)$

$$
\begin{array}{r}
243 \\
- \ 126 \\
\hline
117
\end{array}
$$

Thus **126 − 243 = −117**

Problem 5   Subtract 318 from −269.

−269 − 318. The sum of the negative integers is

$$
\begin{array}{r}
269 \\
+ \ 318 \\
\hline
587
\end{array}
$$

Thus **−269 − 318 = −587**

Problem 6   Multiply 74 by 13.

This is written as 74×13

$$
\begin{array}{rl}
74 & \\
13 & \\
\hline
222 & \leftarrow 74 \times 3 \\
740 & \leftarrow 74 \times 10 \\
\hline
\end{array}
$$

Adding: 962

Thus **74 × 13 = 962**

Problem 7   Multiply 178 by −46.

When the numbers have different signs, the result will be negative. (With this in mind, the problem can now be solved by multiplying 178 by 46.)

$$
\begin{array}{r}
178 \\
46 \\
\hline
1068 \\
7120 \\
\hline
8188
\end{array}
$$

Thus $178 \times 46 = 8188$ and **178 × (−46) = −8188**

Problem 8   Divide 1043 by 7.

When dividing by the numbers 1 to 12, it is usual to use a method called **short division**.

$$
7\overline{)1\,0^3\,4^6\,3} \quad \frac{1\ 4\ 9}{}
$$

Step 1.   7 into 10 goes 1, remainder 3. Put 1 above the 0 of 1043 and carry the 3 remainder to the next digit on the right, making it 34.

Step 2.   7 into 34 goes 4, remainder 6. Put 4 above the 4 of 1043 and carry the 6 remainder to the next digit on the right, making it 63.

Step 3.   7 into 63 goes 9, remainder 0. Put 9 above the 3 of 1043.

Thus **1043 ÷ 7 = 149**

Problem 9   Divide 378 by 14.

When dividing by numbers which are larger than 12, it is usual to use a method called **long division**.

$$
\begin{array}{r}
27 \\
14\overline{)378} \\
28 \\
\hline
98 \\
98 \\
\hline
\cdot\cdot
\end{array}
$$

(2) 2×14

(4) 7×14

(1)   14 into 37 goes twice. Put 2 above the 7 of 378.

(3)   {Subtract. Bring down the 8. 14 into 98 goes 7 times. Put 7 above the 8 of 378.

(5)   Subtract.

Thus **378 ÷ 14 = 27**

Problem 10   Divide 5669 by 46.

This problem may be written as $\dfrac{5669}{46}$ or $5669 \div 46$ or 5669/46. Using the long division method shown in *Problem 9* gives:

$$
\begin{array}{r}
123 \\
46\overline{)5669} \\
46 \\
\hline
106 \\
92 \\
\hline
149 \\
138 \\
\hline
11
\end{array}
$$

As there are no more digits to bring down,

5669 ÷ 46 = 123, remainder 11 or $123\dfrac{11}{46}$

Further problems on arithmetic operations may be found in section 1.8, Problems 1 to 21, page 9.

1.2   Highest common factors and lowest common multiples

When two or more numbers are multiplied together, the individual numbers are called **factors**. Thus a factor is a number which divides into another number exactly. The

highest common factor (HCF) is the largest number which divides into two or more numbers exactly.

A **multiple** is a number which contains another number an exact number of times. The smallest number which is exactly divisible by each of two or more numbers is called the **lowest common multiple (LCM)**.

Problem 11 Determine the HCF of the numbers 12, 30 and 42.

Each number is expressed in terms of its lowest factors. This is achieved by repeatedly dividing by the prime numbers 2, 3, 5, 7, 11, 13 ... (where possible) in turn. Thus

$$12 = 2 \times 2 \times 3$$
$$30 = 2 \quad\times 3 \times 5$$
$$42 = 2 \quad\times 3 \times 7$$

The factors which are common to each of the numbers are 2 in column 1 and 3 in column 3, shown by the broken lines. Hence the **HCF is 2×3, i.e. 6**. That is, 6 is the largest number which will divide into all of the given numbers.

Problem 12 Determine the HCF of the numbers 30, 105, 210 and 1155.

Using the method shown in *Problem 11*:

$$30 = 2 \times 3 \times 5$$
$$105 = \quad 3 \times 5 \times 7$$
$$210 = 2 \times 3 \times 5 \times 7$$
$$1155 = \quad 3 \times 5 \times 7 \times 11$$

The factors which are common to each of the numbers are 3 in column 3 and 5 in column 3.

Hence **the HCF is $3 \times 5 = 15$**

Problem 13 Determine the LCM of the numbers 12, 42 and 90.

The LCM is obtained by finding the lowest factors of each of the numbers, as shown in *Problems 11* and *12* above, and then selecting the largest group of any of the factors present. Thus

$$12 = 2 \times 2 \times 3$$
$$42 = 2 \quad\times 3 \qquad\qquad \times 7$$
$$90 = 2 \quad \times 3 \times 3 \times 5$$

The largest group of any of the factors present are shown by the broken lines and are 2×2 in 12, 3×3 in 90, 5 in 90 and 7 in 42.

Hence the LCM is $2 \times 2 \times 3 \times 3 \times 5 \times 7 = 1260$, and is the smallest number which 12, 42 and 90 will all divide into exactly.

Problem 14 Determine the LCM of the numbers 150, 210, 735 and 1365.

Using the method shown in *Problem 13* above:

$$150 = 2 \times 3 \times 5 \times 5$$
$$210 = 2 \times 3 \times 5 \qquad\quad \times 7$$
$$735 = \qquad\quad 3 \times 5 \qquad\quad \times 7 \times 7$$
$$1365 = \qquad\quad 3 \times 5 \qquad\quad \times 7 \qquad\quad \times 13$$

The LCM is $2 \times 3 \times 5 \times 5 \times 7 \times 7 \times 13 = 95\,550$

Further problems on highest common factors and lowest common multiples may be found in section 1.8, Problems 22 to 27, page 10.

1.3 Order of precedence and brackets

When a particular arithmetic operation is to be performed first, the numbers and the operator(s) are placed in brackets. Thus 3 times the result of 6 minus 2 is written as $3 \times (6 - 2)$.

In arithmetic operations, the order in which operations are performed are:

(i)   to determine the values of operations contained in brackets;
(ii) multiplication and division (the word 'of' also means multiply); and
(iii) addition and subtraction.

This **order of precedence** can be remembered by the word **BODMAS**, standing for **B**rackets, **O**f, **D**ivision, **M**ultiplication, **A**ddition and **S**ubtraction, taken in that order.

The basic laws governing the use of brackets and operators are shown by the following examples:

(i) $2 + 3 = 3 + 2$, i.e. the order of numbers when adding does not matter;
(ii) $2 \times 3 = 3 \times 2$, i.e. the order of numbers when multiplying does not matter;
(iii) $2 + (3 + 4) = (2 + 3) + 4$, i.e. the use of brackets when adding does not affect the result;
(iv) $2 \times (3 \times 4) = (2 \times 3) \times 4$, i.e. the use of brackets when multiplying does not affect the result;
(v) $2 \times (3 + 4) = 2(3 + 4) = (3 + 4)2 = 2 \times 3 + 2 \times 4$, i.e. a number placed outside of a bracket indicates that the

whole contents of the bracket must be multiplied by that number;

(vi) $(2 + 3)(4 + 5) = (5)(9) = 45$, i.e. adjacent brackets indicate multiplication;

(vii) $2[3 + (4 \times 5)] = 2 [3 + 20] = 2 \times 23 = 46$, i.e. when an expression contains inner and outer brackets, the inner brackets are removed first.

Problem 15 Find the value of $6 + 4 \div (5 - 3)$.

The order of precedence of operations is remembered by the word BODMAS.

Thus
$$6 + 4 \div (5 - 3) = 6 + 4 \div 2 \quad \text{(Brackets)}$$
$$= 6 + 2 \quad \text{(Division)}$$
$$= \mathbf{8} \quad \text{(Addition)}$$

Problem 16 Determine the value of $13 - 2 \times 3 + 14 \div (2 + 5)$.

$$13 - 2 \times 3 + 14 \div (2 + 5)$$
$$= 13 - 2 \times 3 + 14 \div 7 \quad \text{(B)}$$
$$= 13 - 2 \times 3 + 2 \quad \text{(D)}$$
$$= 13 - 6 + 2 \quad \text{(M)}$$
$$= 15 - 6 \quad \text{(A)}$$
$$= \mathbf{9} \quad \text{(S)}$$

Problem 17 Evaluate $16 \div (2 + 6) + 18[3 + (4 \times 6) - 21]$.

$$16 \div (2 + 6) + 18[3 + (4 \times 6) - 21]$$
$$= 16 \div (2 + 6) + 18[3 + 24 - 21]$$
$$= 16 \div 8 + 18 \times 6 \quad \text{(B)}$$
$$= 2 + 18 \times 6 \quad \text{(D)}$$
$$= 2 + 108 \quad \text{(M)}$$
$$= \mathbf{110} \quad \text{(A)}$$

Problem 18 Find the value of $23 - 4(2 \times 7) + \dfrac{(144 \div 4)}{(14 - 8)}$.

$$23 - 4(2 \times 7) + \frac{(144 \div 4)}{(14 - 8)}$$

$$= 23 - 4 \times 14 + \frac{36}{6} \quad \text{(B)}$$
$$= 23 - 4 \times 14 + 6 \quad \text{(D)}$$
$$= 23 - 56 + 6 \quad \text{(M)}$$
$$= 29 - 56 \quad \text{(A)}$$
$$= \mathbf{-27} \quad \text{(S)}$$

Further problems on the order of precedence and brackets may be found in section 1.8, Problems 28 to 34, page 10.

1.4 Fractions

When 2 is divided by 3, it may be written as $\dfrac{2}{3}$ or 2/3. $\dfrac{2}{3}$ is called a fraction. The number above the line, i.e. 2, is called the **numerator** and the number below the line, i.e. 3, is called the **denominator**.

When the value of the numerator is less than the value of the denominator, the fraction is called a **proper fraction**; thus $\dfrac{2}{3}$ is a proper fraction. When the value of the numerator is greater than the denominator, the fraction is called an **improper fraction**. Thus $\dfrac{7}{3}$ is an improper fraction and can also be expressed as a **mixed number**, that is, an integer and a proper fraction. Thus the improper fraction $\dfrac{7}{3}$ is equal to the mixed number $2\dfrac{1}{3}$.

When a fraction is simplified by dividing the numerator and denominator by the same number, the process is called **cancelling**. Cancelling by 0 is not permissible.

Problem 19 Simplify $\dfrac{1}{3} + \dfrac{2}{7}$.

The LCM of the two denominators is 3×7, i.e. 21.

Expressing each fraction so that their denominators are 21, gives:

$$\frac{1}{3} + \frac{2}{7} = \frac{1}{3} \times \frac{7}{7} + \frac{2}{7} \times \frac{3}{3} = \frac{7}{21} + \frac{6}{21}$$

$$= \frac{7 + 6}{21} = \frac{\mathbf{13}}{\mathbf{21}}$$

Alternatively:

$$\begin{array}{cc} \text{Step (2)} & \text{Step (3)} \\ \downarrow & \downarrow \end{array}$$

$$\frac{1}{3} + \frac{2}{7} = \frac{(7 \times 1) + (3 \times 2)}{21}$$

$$\uparrow$$

$$\text{Step (1)}$$

Step 1. the LCM of the two denominators;

Step 2. for the fraction $\dfrac{1}{3}$, 3 into 21 goes 7 times, 7 × the numerator is 7 × 1;

Step 3. for the fraction $\dfrac{2}{7}$, 7 into 21 goes 3 times, 3 × the numerator is 3 × 2.

Thus $\dfrac{1}{3} + \dfrac{2}{7} = \dfrac{7+6}{21} = \dfrac{13}{21}$ as obtained previously.

Problem 20 Find the value of $3\dfrac{2}{3} - 2\dfrac{1}{6}$.

One method is to split the mixed numbers into integers and their fractional parts. Then

$$3\frac{2}{3} - 2\frac{1}{6} = \left(3 + \frac{2}{3}\right) - \left(2 + \frac{1}{6}\right)$$

$$= 3 + \frac{2}{3} - 2 - \frac{1}{6} = 1 + \frac{2}{3} - \frac{1}{6} = 1 + \frac{4}{6} - \frac{1}{6}$$

$$= 1\frac{3}{6} = 1\frac{1}{2}$$

Another method is to express the mixed numbers as improper fractions.

Since $3 = \dfrac{9}{3}$, then $3\dfrac{2}{3} = \dfrac{9}{3} + \dfrac{2}{3} = \dfrac{11}{3}$

Similarly, $2\dfrac{1}{6} = \dfrac{12}{6} + \dfrac{1}{6}\qquad = \dfrac{13}{6}$

Thus $\qquad 3\dfrac{2}{3} - 2\dfrac{1}{6}\qquad\quad = \dfrac{11}{3} - \dfrac{13}{6}$

$$= \frac{22}{6} - \frac{13}{6} = \frac{9}{6} = 1\frac{1}{2}$$

as obtained previously.

Problem 21 Evaluate $7\dfrac{1}{8} - 5\dfrac{3}{7}$.

$$7\frac{1}{8} - 5\frac{3}{7} = \left(7 + \frac{1}{8}\right) - \left(5 + \frac{3}{7}\right) = 7 + \frac{1}{8} - 5 - \frac{3}{7}$$

$$= 2 + \frac{1}{8} - \frac{3}{7} = 2 + \frac{7 \times 1 - 8 \times 3}{56}$$

$$= 2 + \frac{7 - 24}{56} = 2 + \frac{-17}{56}$$

$$= 2 - \frac{17}{56} = \frac{112}{56} - \frac{17}{56}$$

$$= \frac{112 - 17}{56} = \frac{95}{56} = 1\frac{39}{56}$$

Problem 22 Determine the value of $4\dfrac{5}{8} - 3\dfrac{1}{4} + 1\dfrac{2}{5}$.

$$4\frac{5}{8} - 3\frac{1}{4} + 1\frac{2}{5} = (4 - 3 + 1) + \left(\frac{5}{8} - \frac{1}{4} + \frac{2}{5}\right)$$

$$= 2 + \frac{5 \times 5 - 10 \times 1 + 8 \times 2}{40}$$

$$= 2 + \frac{25 - 10 + 16}{40} = 2 + \frac{31}{40} = 2\frac{31}{40}$$

Problem 23 Find the value of $\dfrac{3}{7} \times \dfrac{14}{15}$.

Dividing numerator and denominator by 3 gives:

$$\frac{{}^{1}\cancel{3}}{7} \times \frac{14}{\cancel{15}_{5}} = \frac{1 \times 14}{7 \times 5}$$

Dividing numerator and denominator by 7 gives:

$$\frac{1 \times \cancel{14}^{2}}{{}_{1}\cancel{7} \times 5} = \frac{2}{5}$$

This process of dividing both the numerator and denominator of a fraction by the same factor(s) is called **cancelling**.

Problem 24 Evaluate $1\dfrac{3}{5} \times 2\dfrac{1}{3} \times 3\dfrac{3}{7}$.

Mixed numbers **must** be expressed as improper fractions before multiplication can be performed. Thus,

$$1\frac{3}{5} \times 2\frac{1}{3} \times 3\frac{3}{7} = \left(\frac{5}{5} + \frac{3}{5}\right) \times \left(\frac{6}{3} + \frac{1}{3}\right) \times \left(\frac{21}{7} + \frac{3}{7}\right)$$

$$= \frac{8}{5} \times \frac{{}^{1}\cancel{7}}{{}_{1}\cancel{3}} \times \frac{\cancel{24}^{8}}{\cancel{7}_{1}} = \frac{8 \times 8}{5}$$

$$= \frac{64}{5} = 12\frac{4}{5}$$

Problem 25 Simplify $\dfrac{3}{7} \div \dfrac{12}{21}$.

$$\frac{3}{7} \div \frac{12}{21} = \frac{\dfrac{3}{7}}{\dfrac{12}{21}}$$

Multiplying both numerator and denominator by the reciprocal of the denominator gives:

$$\frac{\frac{3}{7}}{\frac{12}{21}} = \frac{1}{1}\frac{\overset{1}{\cancel{3}}}{\cancel{7}} \times \frac{\overset{3}{\cancel{21}}}{\cancel{12}4} = \frac{3}{4}\frac{3}{1} = \frac{3}{4}$$

This method can be remembered by the rule: invert the second fraction and change the operation from division to multiplication. Thus:

$$\frac{3}{7} \div \frac{12}{21} = \frac{\overset{1}{\cancel{3}}}{1\cancel{7}} \times \frac{\overset{3}{\cancel{21}}}{\cancel{12}4} = \frac{3}{4} \quad \text{as obtained previously.}$$

Problem 26   Find the value of $5\frac{3}{5} \div 7\frac{1}{3}$.

The mixed numbers must be expressed as improper fractions. Thus,

$$5\frac{3}{5} \div 7\frac{1}{3} = \frac{28}{5} \div \frac{22}{3}$$

$$= \frac{\overset{14}{\cancel{28}}}{5} \div \frac{3}{\cancel{22}11} = \frac{42}{55}$$

Problem 27   Simplify $\frac{1}{3} - \left(\frac{2}{5} + \frac{1}{4}\right) \div \left(\frac{3}{8} \times \frac{1}{3}\right)$.

The order of precedence of operations for problems containing fractions is the same as that for integers, i.e. remembered by BODMAS (Brackets, Of, Division, Multiplication, Addition and Subtraction). Thus,

$$\frac{1}{3} - \left(\frac{2}{5} + \frac{1}{4}\right) \div \left(\frac{3}{8} \times \frac{1}{3}\right)$$

$$= \frac{1}{3} - \frac{4 \times 2 + 5 \times 1}{20} \div \frac{\overset{1}{\cancel{3}}}{\cancel{24}8} \qquad \text{(B)}$$

$$= \frac{1}{3} - \frac{13}{\cancel{20}5} \times \frac{\overset{2}{\cancel{8}}}{1} \qquad \text{(D)}$$

$$= \frac{1}{3} - \frac{26}{5} \qquad \text{(M)}$$

$$= \frac{(5 \times 1) - (3 \times 26)}{15} \qquad \text{(S)}$$

$$= \frac{-73}{75} = -4\frac{13}{15}$$

Problem 28   Determine the value of $\frac{7}{6}$ of $\left(3\frac{1}{2} - 2\frac{1}{4}\right) + 5\frac{1}{8} \div \frac{3}{16} - \frac{1}{2}$.

$$\frac{7}{6} \text{ of } \left(3\frac{1}{2} - 2\frac{1}{4}\right) + 5\frac{1}{8} \div \frac{3}{16} - \frac{1}{2}$$

$$= \frac{7}{6} \text{ of } 1\frac{1}{4} + \frac{41}{8} \div \frac{3}{16} - \frac{1}{2} \qquad \text{(B)}$$

$$= \frac{7}{6} \times \frac{5}{4} + \frac{41}{8} \div \frac{3}{16} - \frac{1}{2} \qquad \text{(O)}$$

$$= \frac{7}{6} \times \frac{5}{4} + \frac{41}{1\cancel{8}} \times \frac{\overset{2}{\cancel{16}}}{3} - \frac{1}{2} \qquad \text{(D)}$$

$$= \frac{35}{24} + \frac{82}{3} - \frac{1}{2} \qquad \text{(M)}$$

$$= \frac{35 + 656}{24} - \frac{1}{2} \qquad \text{(A)}$$

$$= \frac{691 - 12}{24} \qquad \text{(S)}$$

$$= \frac{679}{24} = 28\frac{7}{24}$$

Further problems on fractions may be found in section 1.8, Problems 35 to 47, page 10.

1.5   Ratio and proportion

The **ratio** of one quantity to another is a fraction, and is the number of times one quantity is contained in another quantity **of the same kind**.

If one quantity is **directly proportional** to another, then as one quantity doubles, the other quantity also doubles. When a quantity is **inversely proportional** to another, then as one quantity doubles, the other quantity is halved.

Problem 29   Divide 126 in the ratio of 5 to 13.

Because the ratio is to be 5 parts to 13 parts, then the total number of parts is 5 + 13, that is 18. Then,

18 parts correspond to 126

Hence 1 part corresponds to $\frac{126}{18} = 7$, 5 parts correspond to $5 \times 7 = \mathbf{35}$ and 13 parts correspond to $13 \times 7 = \mathbf{91}$.

(Check: the parts must add up to the total 35 + 91 = **126** = the total.)

Problem 30   A piece of timber 273 cm long is cut into three pieces in the ratio of 3 to 7 to 11. Determine the lengths of the three pieces.

The total number of parts is 3 + 7 + 11, that is, 21. Hence 21 parts correspond to 273 cm

1 part corresponds to $\dfrac{273}{21}$ = 13 cm

3 parts correspond to 3 × 13 = 39 cm
7 parts correspond to 7 × 13 = 91 cm
11 parts correspond to 11 × 13 = 143 cm

i.e. **the lengths of the three pieces are 39 cm, 91 cm and 143 cm.**
(Check: 39 + 91 +143 = 273.)

Problem 31 Express 25p as a ratio of £4.25.

Working in quantities **of the same kind**, the required ratio is $\dfrac{25}{425}$, i.e. $\dfrac{1}{17}$.

That is, 25p is $\dfrac{1}{17}$th of £4.25. This may be written either as:

25 : 425 : : 1 : 17 (stated as '25 is to 425 as 1 is to 17')

or as $\dfrac{25}{425} = \dfrac{1}{17}$

Problem 32 If 3 people can complete a task in 4 hours, find how long it will take 5 people to complete the task, assuming the rate of work remains constant.

The more the number of people, the more quickly the task is done, hence inverse proportion exists.
3 people complete the task in 4 hours,
1 person takes three times as long, i.e. 4 × 3 = 12 hours,
5 people can do it in one fifth of the time that one person takes, that is $\dfrac{12}{5}$ hours or **2 hours 24 minutes.**

Further problems on ratio and proportion may be found in section 1.8, Problems 48 to 53, page 11.

1.6 Decimals

The decimal system of numbers is based on the **digits** 0 to 9. A number such as 53.17 is called a **decimal fraction**, a **decimal point** separating the integer part, i.e. 53, from the fractional part, i.e. 0.17.

A number which can be expressed exactly as a decimal fraction is called a **terminating decimal** and those which cannot be expressed exactly as a decimal fraction are called **non-terminating decimals**. Thus, $\dfrac{3}{2}$ = 1.5 is a terminating decimal, but $\dfrac{4}{3}$ = 1.333 33 . . . is a **non-terminating decimal**. 1.333 33 . . . can be written as $1.\dot{3}$, called 'one point-three recurring'.

The answer to a non-terminating decimal may be expressed in two ways, depending on the accuracy required:

(i) correct to a number of **significant figures**, that is, figures which signify something; and
(ii) correct to a number of **decimal places**, that is, the number of figures after the decimal point.

The last digit in the answer is unaltered if the next digit on the right is in the group of numbers 0, 1, 2, 3 or 4, but is increased by 1 if the next digit on the right is in the group of numbers 5, 6, 7, 8 or 9. Thus the non-terminating decimal 7.618 3 . . . becomes 7.62, correct to 3 significant figures, since the next digit on the right is 8, which is in the group of numbers 5, 6, 7, 8 or 9. Also 7.618 3 . . . becomes 7.618, correct to 3 decimal places, since the next digit on the right is 3, which is in the group of numbers 0, 1, 2, 3 or 4.

Problem 33 Evaluate 42.7 + 3.04 + 8.7 + 0.06.

The numbers are written so that the decimal points are under each other. Each column is added, starting from the right.

```
   42.7
    3.04
    8.7
    0.06
  ─────
   54.50
```

Thus **42.7 + 3.04 + 8.7 + 0.06 = 54.50**

Problem 34 Take 81.7 from 87.23.

The numbers are written with the decimal points under each other.

```
    87.23
 − 81.70
  ─────
    5.53
```

Thus **87.23 − 81.7 = 5.53**

Problem 35 Find the value of 23.4 − 17.83 − 57.6 + 32.68.

The sum of the positive decimal fractions is 23.4 + 32.68 = 56.08.

The sum of the negative decimal fractions is $17.83 + 57.6 = 75.43$.

Taking the sum of the negative decimal fractions from the sum of the positive decimal fractions gives: $56.08 - 75.43$, i.e.

$$- (75.43 - 56.08) = -19.35$$

Problem 36   Determine the value of 74.3×3.8.

When multiplying decimal fractions: (i) the numbers are multiplied as if they are integers, and (ii) the position of the decimal point in the answer is such that there are as many digits to the right of it as the sum of the digits to the right of the decimal points of the two numbers being multiplied together. Thus

(i)
```
      743
       38
    5 944
   22 290
   28 234
```

(ii) As there are $(1 + 1) = 2$ digits to the right of the decimal points of the two numbers being multiplied together, $(74.\underline{3} \times 3.\underline{8})$, then **74.3 $\times$ 3.8 = 282.34**

Problem 37   Evaluate $37.81 \div 1.7$, correct to (i) 4 significant figures and (ii) 4 decimal places.

$$37.81 \div 1.7 = \frac{37.81}{1.7}$$

The denominator is changed into an integer by multiplying by 10. The numerator is also multiplied by 10 to keep the fraction the same. Thus

$$37.81 \div 1.7 = \frac{37.81 \times 10}{1.7 \times 10} = \frac{378.1}{17}$$

The long division is similar to the long division of integers and the first four steps are as shown:

```
        22.241 17...
  17)378.100 00
     34
     ──
     38
     34
     ──
     41
     34
     ──
     70
     68
     ──
     20
```

(i) **37.81 $\div$ 1.7 = 22.24, correct to 4 significant figures,** and

(ii) **37.81 $\div$ 1.7 = 22.241 2, correct to 4 decimal places**.

Problem 38   Convert (a) 0.437 5 to a proper fraction and (b) 4.285 to a mixed number.

(a)   0.437 5 can be written as $\dfrac{0.437\ 5 \times 10\ 000}{10\ 000}$ without changing its value,

i.e. $0.437\ 5 = \dfrac{4375}{10\ 000}$

By cancelling $\dfrac{4375}{10\ 000} = \dfrac{875}{2000} = \dfrac{175}{400} = \dfrac{35}{80} = \dfrac{7}{16}$

i.e. **$0.437\ 5 = \dfrac{7}{16}$**

(b)   Similarly, $4.285 = 4\dfrac{285}{1000} = \mathbf{4\dfrac{57}{200}}$

Problem 39   Express as decimal fractions: (a) $\dfrac{9}{16}$ and

(b) $5\dfrac{7}{8}$.

(a)   To convert a proper fraction to a decimal fraction, the numerator is divided by the denominator. Division by 16 can be done by the long division method, or, more simply, by dividing by 2 and then 8:

```
     4.50                0.5 6 2 5
  2)9¹.00             8)4.5⁵0²0⁴0
```
Thus, $\dfrac{9}{16} = \mathbf{0.562\ 5}$

(b)   For mixed numbers, it is only necessary to convert the proper fraction part of the mixed number to a decimal fraction. Thus, dealing with the $\dfrac{7}{8}$ part gives:

```
     0.8 7 5
  8)7.0⁶0⁴0
```
i.e. $\dfrac{7}{8} = 0.875$ Thus $\mathbf{5\dfrac{7}{8} = 5.875}$

Further problems on decimals may be found in section 1.8, Problems 54 to 68, page 11.

1.7   Percentages

Percentages are used to give a common standard and are fractions having the number 100 as their denominators. For example, 25 per cent means $\dfrac{25}{100}$, i.e. $\dfrac{1}{4}$, and is written 25%.

Problem 40   Express as percentages: (a) 1.875 and (b) 0.012 5.

A decimal fraction is converted to a percentage by multiplying by 100. Thus,

(a) 1.875 corresponds to $1.875 \times 100\%$, i.e. **187.5%**.
(b) 0.012 5 corresponds to $0.012\ 5 \times 100\%$, i.e. **1.25%**.

Problem 41 Express as percentages:

(a) $\frac{5}{16}$ and (b) $1\frac{2}{5}$

To convert fractions to percentages, they are (i) converted to decimal fractions and (ii) multiplied by 100.

(a) By division, $\frac{5}{16} = 0.312\ 5$, hence $\frac{5}{16}$ corresponds to $0.312\ 5 \times 100\%$, i.e. **31.25%**.

(b) Similarly, $1\frac{2}{5} = 1.4$ when expressed as a decimal fraction.

Hence $1\frac{2}{5} = 1.4 \times 100\% = \mathbf{140\%}$

Problem 42 Find $12\frac{1}{2}\%$ of £378.

$12\frac{1}{2}\%$ of £378 means $\frac{12\frac{1}{2}}{100} \times 378$, since per cent means 'per hundred'.

Hence $12\frac{1}{2}\%$ of £378 $= \frac{\overset{1}{12\frac{1}{2}}}{\underset{8}{100}} \times 378 = \frac{378}{8} = \mathbf{£47.25}$

Problem 43 Express 25 minutes as a percentage of 2 hours, correct to the nearest 1%.

Working in minute units, 2 hours = 120 minutes.
Hence 25 minutes is $\frac{25}{120}$ ths of 2 hours.

By cancelling, $\frac{25}{120} = \frac{5}{24}$

Expressing $\frac{5}{24}$ as a decimal fraction gives 0.208 3.

Multiplying by 100 to convert the decimal fraction to a percentage gives:

$0.208\ \dot{3} \times 100 = 20.8\dot{3}\%$

Thus **25 minutes is 21% of 2 hours**, correct to the nearest 1%.

Problem 44 A German silver alloy consists of 60% copper, 25% zinc and 15% nickel. Determine the masses of the copper, zinc and nickel in a 3.74 kilogram block of the alloy.

By direct proportion:

100% corresponds to 3.74 kg
1% corresponds to $\frac{3.74}{120} = 0.037\ 4$ kg
60% corresponds to $60 \times 0.037\ 4 = 2.244$ kg
25% corresponds to $25 \times 0.037\ 4 = 0.935$ kg
15% corresponds to $15 \times 0.037\ 4 = 0.561$ kg.

Thus, the masses of the copper, zinc and nickel are **2.244 kg, 0.935 kg and 0.561 kg**, respectively.
(Check: $2.244 + 0.935 + 0.561 = 3.74$.)

Further problems on percentages may be found in section 1.8, Problems 69 to 75, page 12.

1.8 Further problems on basic arithmetic

Arithmetic operations

In *Problems 1* to *21*, determine the values of the expressions given:

1 $67 - 82 + 34$
 [19]
2 $124 - 273 + 481 - 398$
 [−66]
3 $927 - 114 + 182 - 183 - 247$
 [565]
4 $2417 - 487 + 2424 - 1778 - 4712$
 [−2136]
5 $-38\ 419 - 2177 + 2440 - 799 + 2834$
 [−36 121]
6 $2715 - 18\ 250 + 11\ 471 - 1529 + 113\ 274$
 [107 701]
7 $73 - 57$
 [16]
8 $813 - (-674)$
 [1487]
9 $647 - 872$
 [−225]
10 $3151 - (-2763)$
 [5914]
11 $4872 - 4683$
 [189]
12 $-23\ 148 - 47\ 724$
 [−70 872]

13 38 441 − 53 774
    [−15 333]
14 (a) 261 × 7; (b) 462 × 9
    [(a) 1827 (b) 4158]
15 (a) 783 × 11; (b) 73 × 24
    [(a) 8613 (b) 1752]
16 (a) 27 × 38; (b) 77 × 29
    [(a) 1026 (b) 2233]
17 (a) 448 × 23; (b) 143 × (−31)
    [(a) 10 304 (b) −4433]
18 (a) 288 ÷ 6; (b) 979 ÷ 11
    [(a) 48 (b) 89]
19 (a) $\dfrac{1813}{7}$; (b) $\dfrac{896}{16}$
    [(a) 259 (b) 56]
20 (a) $\dfrac{21\,432}{47}$; (b) 15 904 ÷ 56
    [(a) 456 (b) 284]
21 (a) $\dfrac{88\,738}{187}$; (b) 46 857 ÷ 79
    $\left[\text{(a) } 474\dfrac{100}{187} \text{ (b) } 593\dfrac{10}{79}\right]$

Highest common factors and lowest common multiples

In *Problems 22* to *27* find (a) the HCF and (b) the LCM of the numbers given:

22 6, 10, 14
    [(a) 2 (b) 210]
23 12, 30, 45
    [(a) 3 (b) 180]
24 10, 15, 70, 105
    [(a) 5 (b) 210]
25 90, 105, 300
    [(a) 15 (b) 6300]
26 210, 196, 910, 462
    [(a) 14 (b) 420 420]
27 196, 350, 770
    [(a) 14 (b) 53 900]

Order of precedence and brackets

Simplify the expressions given in *Problems 28* to *34*:

28 14 + 3 × 15
    [59]
29 17 − 12 ÷ 4
    [14]
30 86 + 24 ÷ (14 − 2)
    [88]
31 7(23 − 18) ÷ (12 − 5)
    [5]

32 63 − 28(14 ÷ 2) + 26
    [−107]
33 $\dfrac{112}{16} - 119 \div 17 + (3 \times 19)$
    [57]
34 $\dfrac{(50 - 14)}{3} + 7\,(16 - 7) - 7$
    [68]

Fractions

Evaluate the expressions given in *Problems 35 to 47*:

35 (a) $\dfrac{1}{2} + \dfrac{2}{5}$ (b) $\dfrac{7}{16} - \dfrac{1}{4}$
    $\left[\text{(a) } \dfrac{9}{10} \text{ (b) } \dfrac{3}{16}\right]$

36 (a) $\dfrac{2}{7} + \dfrac{3}{11}$ (b) $\dfrac{2}{9} - \dfrac{1}{7} + \dfrac{2}{3}$
    $\left[\text{(a) } \dfrac{43}{77} \text{ (b) } \dfrac{47}{63}\right]$

37 (a) $5\dfrac{3}{13} + 3\dfrac{3}{4}$ (b) $4\dfrac{5}{8} - 3\dfrac{2}{5}$
    $\left[\text{(a) } 8\dfrac{51}{52} \text{ (b) } 1\dfrac{9}{40}\right]$

38 (a) $10\dfrac{3}{7} - 8\dfrac{2}{3}$ (b) $3\dfrac{1}{4} - 4\dfrac{4}{5} + 1\dfrac{5}{6}$
    $\left[\text{(a) } 1\dfrac{16}{21} \text{ (b) } \dfrac{17}{60}\right]$

39 (a) $\dfrac{3}{4} \times \dfrac{5}{9}$ (b) $\dfrac{17}{35} \times \dfrac{15}{119}$
    $\left[\text{(a) } \dfrac{5}{12} \text{ (b) } \dfrac{3}{49}\right]$

40 (a) $\dfrac{3}{5} \times \dfrac{7}{9} \times 1\dfrac{2}{7}$ (b) $\dfrac{13}{17} \times 4\dfrac{7}{11} \times 3\dfrac{4}{39}$
    $\left[\text{(a) } \dfrac{3}{5} \text{ (b) } 11\right]$

41 (a) $\dfrac{1}{4} \times \dfrac{3}{11} \times 1\dfrac{5}{39}$ (b) $\dfrac{3}{4} \div 1\dfrac{4}{5}$
    $\left[\text{(a) } \dfrac{1}{13} \text{ (b) } \dfrac{5}{12}\right]$

42 (a) $\dfrac{3}{8} \div \dfrac{45}{64}$ (b) $1\dfrac{1}{3} \div 2\dfrac{5}{9}$
    $\left[\text{(a) } \dfrac{8}{15} \text{ (b) } \dfrac{12}{23}\right]$

43 $\dfrac{1}{3} - \dfrac{3}{4} \times \dfrac{16}{27}$

$\left[-\dfrac{1}{9}\right]$

44 $\dfrac{1}{2} + \dfrac{3}{5} \div \dfrac{9}{15} - \dfrac{1}{3}$

$\left[1\dfrac{1}{6}\right]$

45 $\dfrac{7}{15}$ of $\left(15 \times \dfrac{5}{7}\right) + \left(\dfrac{3}{4} \div \dfrac{15}{16}\right)$

$\left[5\dfrac{4}{5}\right]$

46 $\dfrac{1}{4} \times \dfrac{2}{3} - \dfrac{1}{3} \div \dfrac{3}{5} + \dfrac{2}{7}$

$\left[-\dfrac{13}{126}\right]$

47 $\left(\dfrac{2}{3} \times 1\dfrac{1}{4}\right) \div \left(\dfrac{2}{3} + \dfrac{1}{4}\right) + 1\dfrac{3}{5}$

$\left[2\dfrac{28}{55}\right]$

Ratio and proportion

48 Divide 312 in the ratio of 7 to 17.
[91 to 221]
49 Divide 621 in the ratio of 3 to 7 to 13.
[81 to 189 to 351]
50 £4.94 is to be divided between two people in the ratio of 9 to 17. Determine how much each person will receive.
[£1.71 and £3.23]
51 When mixing a quantity of paints, dyes of four different colours are used in the ratio of 7 : 3 : 19 : 5. If the mass of the first dye used is $3\dfrac{1}{2}$ g, determine the total mass of the dyes used.
[17 g]
52 It takes 21 hours for 12 men to resurface a stretch of road. Find how many men it takes to resurface a similar stretch of road in 50 hours 24 minutes, assuming the work rate remains constant.
[5]
53 It takes 3 hours 15 minutes to fly from city A to city B at a constant speed. Find how long the journey takes if

(a) the speed is $1\dfrac{1}{2}$ times that of the original speed and

(b) if the speed is three-quarters of the original speed.
[(a) 2 h 10 min (b) 4 h 20 min]

Decimals

In *Problems 54* to *60*, determine the values of the expressions given:

54 23.6 + 14.71 − 18.9 − 7.421
[11.989]
55 73.84 − 113.247 + 8.21 − 0.068
[−31.265]
56 5.73 × 4.2
[24.066]
57 3.8 × 4.1 × 0.7
[10.906]
58 374.1 × 0.006
[2.244 6]
59 421.8 ÷ 17, (a) correct to 4 significant figures and (b) correct to 3 decimal places.
[(a) 24.81 (b) 24.812]
60 $\dfrac{0.014\ 7}{2.3}$, (a) correct to 5 decimal places and (b) correct to 2 significant figures.
[(a) 0.006 39 (b) 0.006 4]
61 Convert to proper fractions:
(a) 0.65 (b) 0.84 (c) 0.012 5 (d) 0.282 and (e) 0.024
$\left[\text{(a) } \dfrac{13}{20} \text{ (b) } \dfrac{21}{25} \text{ (c) } \dfrac{1}{80} \text{ (d) } \dfrac{141}{500} \text{ (e) } \dfrac{3}{125}\right]$
62 Convert to mixed numbers:
(a) 1.82 (b) 4.275 (c) 14.125 (d) 15.35 and (e) 16.212 5
$\left[\text{(a) } 1\dfrac{41}{50} \text{ (b) } 4\dfrac{11}{40} \text{ (c) } 14\dfrac{1}{8} \text{ (d) } 15\dfrac{7}{20} \text{ (e) } 16\dfrac{17}{80}\right]$

In *Problems 63* to *68*, express as decimal fractions to the accuracy stated:

63 $\dfrac{4}{9}$, correct to 5 significant figures.
[0.444 44]
64 $\dfrac{17}{27}$, correct to 5 decimal places.
[0.629 63]
65 $1\dfrac{9}{16}$, correct to 4 significant figures.
[1.563]
66 $53\dfrac{5}{11}$, correct to 3 decimal places.
[53.455]
67 $13\dfrac{31}{37}$, correct to 2 decimal places.
[13.84]
68 $8\dfrac{9}{13}$, correct to 3 significant figures.
[8.69]

Percentages

69 Convert to percentages:
   (a) 0.057   (b) 0.374 and   (c) 1.285.
   [(a) 5.7% (b) 37.4% (c) 128.5%]

70 Express as percentages, correct to 3 significant figures:
   (a) $\frac{7}{33}$ (b) $\frac{19}{24}$ and (c) $1\frac{11}{16}$
   [(a) 21.2% (b) 79.2% (c) 169%]

71 Calculate correct to 4 significant figures:
   (a) 18% of 2758 tonnes,
   (b) 47% of 18.42 grams, and
   (c) 147% of 14.1 seconds.
   [(a) 496.4 t (b) 8.657 g (c) 20.73 s]

72 Express:
   (a) 140 kg as a percentage of 1 t,
   (b) 47 s as a percentage of 5 min, and
   (c) 13.4 cm as a percentage of 2.5 m.
   [(a) 14% (b) 15.6% (c) 5.36%]

73 A block of monel alloy consists of 70% nickel and 30% copper. If it contains 88.2 g of nickel, determine the mass of copper in the block.
   [37.8 g]

74 Two kilograms of a compound contains 30% of element A, 45% of element B and 25% of element C. Determine the masses of the three elements present.
   [A 0.6 kg, B 0.9 kg, C 0.5 kg]

75 A concrete mixture contains seven parts by volume of ballast, four parts by volume of sand and two parts by volume of cement. Determine the percentage of each of these three constituents correct to the nearest 1% and the mass of cement in a two tonne dry mix, correct to 1 significant figure.
   [54%, 31%, 15%, 0.3 t]

2

Indices and standard form

2.1 Indices

The lowest factors of 2000 are $2 \times 2 \times 2 \times 2 \times 5 \times 5 \times 5$. These factors are written as $2^4 \times 5^3$, where 2 and 5 are called **bases** and the numbers 4 and 3 are called **indices**.

When an index is an integer it is called a **power**. Thus, 2^4 is called 'two to the power of four', and has a base of 2 and an index of 4. Similarly, 5^3 is called 'five to the power of 3' and has a base of 5 and an index of 3.

Special names may be used when the indices are 2 and 3, these being called 'squared' and 'cubed', respectively. Thus 7^2 is called 'seven squared' and 9^3 is called 'nine cubed'. When no index is shown, the power is 1, i.e. 2^1 means 2.

Reciprocal

The **reciprocal** of a number is when the index is -1 and its value is given by 1 divided by the base. Thus the reciprocal of 2 is 2^{-1} and its value is $\frac{1}{2}$ or 0.5. Similarly, the reciprocal of 5 is 5^{-1} which means $\frac{1}{5}$ or 0.2.

Square root

The **square root** of a number is when the index is $\frac{1}{2}$, and the square root of 2 is written as $2^{(1/2)}$ or $\sqrt{2}$. The value of a square root is the value of the base which when multiplied by itself gives the number. Since $3 \times 3 = 9$, then $\sqrt{9} = 3$. However, $(-3) \times (-3) = 9$, so $\sqrt{9} = -3$. There are always

two answers when finding the square root of a number and this is shown by putting both a $+$ and a $-$ sign in front of the answer to a square root problem. Thus $\sqrt{9} = \pm 3$ and $4^{(1/2)} = \sqrt{4} = \pm 2$, and so on.

Laws of indices

When simplifying calculations involving indices, certain basic rules or laws can be applied, called the **laws of indices**. These are given below.

(i) When multiplying two or more numbers having the same base, the indices are added. Thus $3^2 \times 3^4 = 3^{2+4} = 3^6$.

(ii) When a number is divided by a number having the same base, the indices are subtracted. Thus

$$\frac{3^5}{3^2} = 3^{5-2} = 3^3$$

(iii) When a number which is raised to a power is raised to a further power, the indices are multiplied. Thus $(3^5)^2 = 3^{5 \times 2} = 3^{10}$.

(iv) When a number has an index of 0, its value is 1. Thus $3^0 = 1$.

(v) A number raised to a negative power is the reciprocal of that number raised to a positive power. Thus $3^{-4} = \frac{1}{3^4}$. Similarly, $\frac{1}{2^{-3}} = 2^3$.

(vi) When a number is raised to a fractional power the denominator of the fraction is the root of the number and the numerator is the power. Thus

$$8^{(2/3)} = \sqrt[3]{8^2} = (2)^2 = 4$$

and

$$25^{(1/2)} = \sqrt{25^1} = \pm 5$$

Problem 1   Evaluate: (a) $5^2 \times 5^3$, (b) $3^2 \times 3^4 \times 3$ and (c) $2 \times 2^2 \times 2^5$.

From law (i):

(a) $5^2 \times 5^3 = 5^{(2+3)} = 5^5 = 5 \times 5 \times 5 \times 5 \times 5 = \mathbf{3125}$

(b) $3^2 \times 3^4 \times 3 = 3^{(2+4+1)} = 3^7 = 3 \times 3 \times \ldots$ to 7 terms = **2187**

(c) $2 \times 2^2 \times 2^5 = 2^{(1+2+5)} = 2^8 = \mathbf{256}$

Problem 2   Find the value of: (a) $\dfrac{7^5}{7^3}$ and (b) $\dfrac{5^7}{5^4}$.

From law (ii):

(a) $\dfrac{7^5}{7^3} = 7^{(5-3)} = 7^2 = \mathbf{49}$

(b) $\dfrac{5^7}{5^4} = 5^{(7-4)} = 5^3 = \mathbf{125}$

Problem 3   Evaluate: (a) $5^2 \times 5^3 \div 5^4$ and (b) $(3 \times 3^5) \div (3^2 \times 3^3)$.

From laws (i) and (ii):

(a) $5^2 \times 5^3 \div 5^4 \qquad = \dfrac{5^2 \times 5^3}{5^4} = \dfrac{5^{(2+3)}}{5^4}$

$$= \dfrac{5^5}{5^4} \qquad = 5^{(5-4)}$$

$$= 5^1 \qquad\quad = \mathbf{5}$$

(b) $(3 \times 3^5) \div (3^2 \times 3^3) = \dfrac{3 \times 3^5}{3^2 \times 3^3} = \dfrac{3^{(1+5)}}{3^{(2+3)}}$

$$= \dfrac{3^6}{3^5} \qquad\quad = 3^{6-5}$$

$$= 3^1 \qquad\quad = \mathbf{3}$$

Problem 4   Simplify: (a) $(2^3)^4$ and (b) $(3^2)^5$, expressing the answers in index form.

From law (iii):

(a) $(2^3)^4 = 2^{3\times4} = \mathbf{2^{12}}$

(b) $(3^2)^5 = 3^{2\times5} = \mathbf{3^{10}}$

Problem 5   Evaluate:

$$\dfrac{(10^2)^3}{10^4 \times 10^2}$$

From the laws of indices:

$$\dfrac{(10^2)^3}{10^4 \times 10^2} = \dfrac{10^{(2\times3)}}{10^{(4+2)}} = \dfrac{10^6}{10^6} = 10^{6-6} = 10^0 = \mathbf{1}$$

Problem 6   Find the value of

(a) $\dfrac{2^3 \times 2^4}{2^7 \times 2^5}$ and (b) $\dfrac{(3^2)^3}{3 \times 3^9}$

From the laws of indices:

(a) $\dfrac{2^3 \times 2^4}{2^7 \times 2^5} = \dfrac{2^{(3+4)}}{2^{(7+5)}} = \dfrac{2^7}{2^{12}}$

$$= 2^{7-12} \ = 2^{-5} = \dfrac{1}{2^5} = \dfrac{1}{\mathbf{32}}$$

(b) $\dfrac{(3^2)^3}{3 \times 3^9} \ = \dfrac{3^{2\times3}}{3^{1+9}} = \dfrac{3^6}{3^{10}}$

$$= 3^{6-10} \ = 3^{-4} = \dfrac{1}{3^4} \ \dfrac{1}{\mathbf{81}}$$

Problem 7   Evaluate (a) $4^{1/2}$ (b) $16^{3/4}$ (c) $27^{2/3}$ (d) $9^{-1/2}$.

(a) $4^{1/2} = \sqrt{4} = \pm\mathbf{2}$ (b) $16^{3/4} = \sqrt[4]{16^3} = (2)^3 = \mathbf{8}$

(Note that it does not matter whether the 4th root of 16 is found first or whether 16 cubed is found first – the same answer will result.)

(c) $27^{2/3} = \sqrt[3]{27^2} = (3)^2 = \mathbf{9}$

(d) $9^{-1/2} = \dfrac{1}{9^{1/2}} = \dfrac{1}{\sqrt{9}} = \dfrac{1}{\pm3} = \pm\dfrac{1}{\mathbf{3}}$

Problem 8   Evaluate:

$$\dfrac{3^3 \times 5^7}{5^3 \times 3^4}$$

The laws of indices only apply to terms **having the same base**. Grouping terms having the same base, and then applying the laws of indices to each of the groups independently gives:

$$\dfrac{3^3 \times 5^7}{5^3 \times 3^4} = \dfrac{3^3}{3^4} \times \dfrac{5^7}{5^3}$$

$$= 3^{(3-4)} \times 5^{(7-3)} = 3^{-1} \times 5^4$$

$$= \dfrac{5^4}{3^1} = \dfrac{625}{3} = \mathbf{208\dfrac{1}{3}}$$

Problem 9 Find the value of

$$\frac{2^3 \times 3^5 \times (7^2)^2}{7^4 \times 2^4 \times 3^3}$$

$$\frac{2^3 \times 3^5 \times (7^2)^2}{7^4 \times 2^4 \times 3^3} = 2^{3-4} \times 3^{5-3} \times 7^{2 \times 2 - 4}$$

$$= 2^{-1} \times 3^2 \times 7^0 = \frac{1}{2} \times 3^2 \times 1$$

$$= \frac{9}{2} = 4\frac{1}{2}$$

Problem 10 Evaluate

$$\frac{4^{1.5} \times 8^{1/3}}{2^2 \times 32^{-2/5}}$$

$$4^{1.5} = 4^{3/2} = \sqrt{4^3} = 2^3 = 8, \quad 8^{1/3} = \sqrt[3]{8} = 2, \quad 2^2 = 4,$$

$$32^{-2/5} = \frac{1}{32^{2/5}} = \frac{1}{\sqrt[5]{32^2}} = \frac{1}{2^2} = \frac{1}{4}$$

Hence $\dfrac{4^{1.5} \times 8^{1/3}}{2^2 \times 32^{-2/5}} = \dfrac{8 \times 2}{4 \times \frac{1}{4}} = \dfrac{16}{1} = \mathbf{16}$

Alternatively,

$$\frac{4^{1.5} \times 8^{1/3}}{2^2 \times 32^{-2/5}} = \frac{[(2)^2]^{3/2} \times (2^3)^{1/3}}{2^2 \times (2^5)^{-2/5}} = \frac{2^3 \times 2^1}{2^2 \times 2^{-2}}$$
$$= 2^{3+1-2-(-2)}$$
$$= 2^4 = \mathbf{16}$$

Problem 11 Evaluate:

$$\frac{3^2 \times 5^5 + 3^3 \times 5^3}{3^4 \times 5^4}$$

Dividing each term by the HCF of the three terms, i.e. $3^2 \times 5^3$, gives:

$$\frac{3^2 \times 5^5 + 3^3 \times 5^3}{3^4 \times 5^4} = \frac{\dfrac{3^2 \times 5^5}{3^2 \times 5^3} + \dfrac{3^3 \times 5^3}{3^2 \times 5^3}}{\dfrac{3^4 \times 5^4}{3^2 \times 5^3}}$$

$$= \frac{3^{(2-2)} \times 5^{(5-3)} + 3^{(3-2)} \times 5^{(3-3)}}{3^{(4-2)} \times 5^{(4-3)}}$$

$$= \frac{3^0 \times 5^2 + 3^1 \times 5^0}{3^2 \times 5^1}$$

$$= \frac{1 \times 25 + 3 \times 1}{9 \times 5} = \frac{\mathbf{28}}{\mathbf{45}}$$

Problem 12 Find the value of

$$\frac{3^2 \times 5^5}{3^4 \times 5^4 + 3^3 \times 5^3}$$

To simplify the arithmetic, each term is divided by the HCF of all the terms, i.e. $3^2 \times 5^3$. Thus

$$\frac{3^2 \times 5^5}{3^4 \times 5^4 + 3^3 \times 5^3} = \frac{\dfrac{3^2 \times 5^5}{3^2 \times 5^3}}{\dfrac{3^4 \times 5^4}{3^2 \times 5^3} + \dfrac{3^3 \times 5^3}{3^2 \times 5^3}}$$

$$= \frac{3^{(2-2)} \times 5^{(5-3)}}{3^{(4-2)} \times 5^{(4-3)} + 3^{(3-2)} \times 5^{(3-3)}}$$

$$= \frac{3^0 \times 5^2}{3^2 \times 5^1 + 3^1 \times 5^0}$$

$$= \frac{1 \times 5^2}{3^2 \times 5 + 3 \times 1}$$

$$= \frac{25}{45 + 3} = \frac{\mathbf{25}}{\mathbf{48}}$$

Problem 13 Simplify $\dfrac{7^{-3} \times 3^4}{3^{-2} \times 7^5 \times 5^{-2}}$, expressing the answer in index form with positive indices.

Since $7^{-3} = \dfrac{1}{7^3}, \dfrac{1}{3^{-2}} = 3^2$ and $\dfrac{1}{5^{-2}} = 5^2$ then

$$\frac{7^{-3} \times 3^4}{3^{-2} \times 7^5 \times 5^{-2}} = \frac{3^4 \times 3^2 \times 5^2}{7^3 \times 7^5} = \frac{3^{(4+2)} \times 5^2}{7^{(3+5)}}$$

$$= \frac{\mathbf{3^6 \times 5^2}}{\mathbf{7^8}}$$

Problem 14 Simplify $\dfrac{16^2 \times 9^{-2}}{4 \times 3^3 - 2^{-3} \times 8^2}$ expressing the answer in index form with positive indices.

Expressing the numbers in terms of their lowest prime numbers gives:

$$\frac{16^2 \times 9^{-2}}{4 \times 3^3 - 2^{-3} \times 8^2} = \frac{(2^4)^2 \times (3^2)^{-2}}{2^2 \times 3^3 - 2^{-3} \times (2^3)^2}$$

$$= \frac{2^8 \times 3^{-4}}{2^2 \times 3^3 - 2^{-3} \times 2^6}$$

$$= \frac{2^8 \times 3^{-4}}{2^2 \times 3^3 - 2^3}$$

Dividing each term by the HCF (i.e. 2^2) gives:

$$\frac{2^6 \times 3^{-4}}{3^2 - 2} = \frac{2^6}{3^4(3^3 - 2)}$$

> **Problem 15**   Simplify $\dfrac{\left(\dfrac{4}{3}\right)^3 \times \left(\dfrac{3}{5}\right)^{-2}}{\left(\dfrac{2}{5}\right)^{-3}}$ giving the answer with positive indices.

A fraction raised to a power means that both the numerator and the denominator of the fraction are raised to that power, i.e.

$$\left(\frac{4}{3}\right)^3 = \frac{4^3}{3^3}$$

A fraction raised to a negative power has the same value as the inverse of the fraction raised to a positive power.

Thus, $\left(\dfrac{3}{5}\right)^{-2} = \dfrac{1}{\left(\dfrac{3}{5}\right)^2} = \dfrac{1}{\dfrac{3^2}{5^2}} = 1 \times \dfrac{5^2}{3^2} = \dfrac{5^2}{3^2} = \left(\dfrac{5}{3}\right)^2$

Similarly, $\left(\dfrac{2}{5}\right)^{-3} = \left(\dfrac{5}{2}\right)^3$

Thus, $\dfrac{\left(\dfrac{4}{3}\right)^3 \times \left(\dfrac{3}{5}\right)^{-2}}{\left(\dfrac{2}{5}\right)^{-3}} = \dfrac{\dfrac{4^3}{3^3} \times \dfrac{5^2}{3^2}}{\dfrac{5^3}{2^3}} = \dfrac{4^3}{3^3} \times \dfrac{5^2}{3^2} \times \dfrac{2^3}{5^3}$

$$= \frac{(2^2)^3 \times 2^3}{3^{(3+2)} \times 5^{(3-2)}} = \frac{2^9}{3^5 \times 5}$$

Further problems on indices may be found in section 2.3, Problems 1 to 22, page 17.

2.2   Standard form

A number written with one digit to the left of the decimal point and multiplied by 10 raised to some power is said to be written in **standard form**. Thus: 5837 is written as 5.837×10^3 in standard form, and 0.041 5 is written as 4.15×10^{-2} in standard form.

When a number is written in standard form, the first factor is called the **mantissa** and the second factor is called the **exponent**. Thus the number 5.8×10^3 has a mantissa of 5.8 and an exponent of 10^3.

(i) Numbers having the same exponent can be added or subtracted in standard form by adding or subtracting the mantissae and keeping the exponent the same. Thus:
$2.3 \times 10^4 + 3.7 \times 10^4 = (2.3 + 3.7) \times 10^4 = 6.0 \times 10^4$,
and
$5.9 \times 10^{-2} - 4.6 \times 10^{-2} = (5.9 - 4.6) \times 10^{-2}$
$= 1.3 \times 10^{-2}$
When numbers have different exponents, one way of adding or subtracting the numbers is to express one of the numbers in non-standard form, so that both numbers have the same exponent. Thus:

$$2.3 \times 10^4 + 3.7 \times 10^3 = 2.3 \times 10^4 + 0.37 \times 10^4$$
$$= (2.3 + 0.37) \times 10^4$$
$$= 2.67 \times 10^4$$

Alternatively, $2.3 \times 10^4 + 3.7 \times 10^3 = 23\ 000 + 3700$
$= 26\ 700 = 2.67 \times 10^4$

(ii) The laws of indices are used when multiplying or dividing numbers given in standard form. For example,

$$(2.5 \times 10^3) \times (5 \times 10^2) = (2.5 \times 5) \times (10^{3+2})$$
$$= 12.5 \times 10^5 \text{ or } 1.25 \times 10^6$$

Similarly,

$$\frac{6 \times 10^4}{1.5 \times 10^2} = \frac{6}{1.5} \times (10^{4-2}) = 4 \times 10^2$$

> **Problem 16**   Express in standard form: (a) 38.71 (b) 3746 (c) 0.012 4.

For a number to be in standard form, it is expressed with only one digit to the left of the decimal point. Thus:

(a) 38.71 must be divided by 10 to achieve one digit to the left of the decimal point and it must also be multiplied by 10 to maintain the equality, i.e.

$$38.71 = \frac{38.71}{10} \times 10 = \mathbf{3.871 \times 10} \text{ in standard form}$$

(b) $3746 = \dfrac{3746}{1000} \times 1000 = \mathbf{3.746 \times 10^3}$ in standard form

(c) $0.012\ 4 = 0.012\ 4 \times \dfrac{100}{100} = \dfrac{1.24}{100} = \mathbf{1.24 \times 10^{-2}}$ in standard form

> **Problem 17**   Express the following numbers, which are in standard form, as decimal numbers:
> (a) 1.725×10^{-2} (b) 5.491×10^4 (c) 9.84×10^0.

(a) $1.725 \times 10^{-2} = \dfrac{1.725}{100} = \mathbf{0.017\ 25}$

(b) $5.491 \times 10^4 = 5.491 \times 10\ 000 = \mathbf{54\ 910}$

(c) $9.84 \times 10^0 = 9.84 \times 1 = \mathbf{9.84}$ (since $10^0 = 1$)

Problem 18 Express in standard form, correct to 3 significant figures: (a) $\dfrac{3}{8}$ (b) $19\dfrac{2}{3}$ (c) $741\dfrac{9}{16}$.

(a) $\dfrac{3}{8} = 0.375$, and expressing it in standard form gives:

$0.375 = \mathbf{3.75 \times 10^{-1}}$

(b) $19\dfrac{2}{3} = 19.\dot{6} = \mathbf{1.97 \times 10}$ in standard form, correct to 3 significant figures

(c) $741\dfrac{9}{16} = 741.562\ 5 = \mathbf{7.42 \times 10^2}$ in standard form, correct to 3 significant figures

---

Problem 19 Express the following numbers, given in standard form, as fractions or mixed numbers:
(a) 2.5×10^{-1} (b) 6.25×10^{-2} (c) 1.354×10^2

(a) $2.5 \times 10^{-1} = \dfrac{2.5}{10} = \dfrac{25}{100} = \dfrac{\mathbf{1}}{\mathbf{4}}$

(b) $6.25 \times 10^{-2} = \dfrac{6.25}{100} = \dfrac{625}{10\ 000} = \dfrac{\mathbf{1}}{\mathbf{16}}$

(c) $1.354 \times 10^2 = 135.4 = 135\dfrac{4}{10} = \mathbf{135\dfrac{2}{5}}$

Problem 20 Find the value of
(a) $7.9 \times 10^{-2} - 5.4 \times 10^{-2}$ (b) $8.3 \times 10^3 + 5.415 \times 10^3$
and (c) $9.293 \times 10^2 + 1.3 \times 10^3$,
expressing the answers in standard form.

Numbers having the same exponent can be added or subtracted by adding or subtracting the mantissae and keeping the exponent the same. Thus:

(a) $7.9 \times 10^{-2} - 5.4 \times 10^{-2} = (7.9 - 5.4) \times 10^{-2}$
$= \mathbf{2.5 \times 10^{-2}}$

(b) $8.3 \times 10^3 + 5.415 \times 10^3$
$= (8.3 + 5.415) \times 10^3$
$= 13.715 \times 10^3$
$= \mathbf{1.371\ 5 \times 10^4}$ in standard form

(c) Since only numbers having the same exponents can be added by straight addition of the mantissae, the numbers are converted to this form before adding. Thus:

$9.293 \times 10^2 + 1.3 \times 10^3 = 9.293 \times 10^2 + 13 \times 10^2$
$= (9.293 + 13) \times 10^2$

$= 22.293 \times 10^2$
$= \mathbf{2.229\ 3 \times 10^3}$ in standard form

Alternatively, the numbers can be expressed as decimal fractions, giving

$9.293 \times 10^2 + 1.3 \times 10^3$
$= 929.3 + 1300$
$= 2229.3$
$= \mathbf{2.229\ 3 \times 10^3}$ in standard form

as obtained previously. This method is often the 'safest' way of doing this type of problem.

Problem 21 Evaluate (a) $(3.75 \times 10^3)(6 \times 10^4)$ and
(b) $\dfrac{3.5 \times 10^5}{7 \times 10^2}$ expressing answers in standard form.

(a) $(3.75 \times 10^3)(6 \times 10^4) = (3.75 \times 6)(10^{3+4}) = 22.50 \times 10^7$
$= \mathbf{2.25 \times 10^8}$

(b) $\dfrac{3.5 \times 10^5}{7 \times 10^2} = \dfrac{3.5}{7} \times 10^{5-2} = 0.5 \times 10^3 = \mathbf{5 \times 10^2}$

Further problems on standard form may be found in section 2.3, Problems 23 to 34, page 18.

2.3 Further problems on indices and standard form

Indices

In *Problems 1* to *14*, simplify the expressions given, expressing the answers in index form and with positive indices:

1   (a) $3^3 \times 3^4$ (b) $4^2 \times 4^3 \times 4^4$
    [(a) 3^7 (b) 4^9]
2   (a) $2^3 \times 2 \times 2^2$ (b) $7^2 \times 7^4 \times 7 \times 7^3$
    [(a) 2^6 (b) 7^{10}]
3   (a) $\dfrac{2^4}{2^3}$ (b) $\dfrac{3^7}{3^2}$
    [(a) 2 (b) 3^5]
4   (a) $5^6 \div 5^3$ (b) $7^{13}/7^{10}$
    [(a) 5^3 (b) 7^3]
5   (a) $(7^2)^3$ (b) $(3^3)^2$
    [(a) 7^6 (b) 3^6]
6   (a) $(15^3)^5$ (b) $(17^2)^4$
    [(a) 15^{15} (b) 17^8]
7   (a) $\dfrac{2^2 \times 2^3}{2^4}$ (b) $\dfrac{3^7 \times 3^4}{3^5}$
    [(a) 2 (b) 3^6]

8 (a) $\dfrac{5^7}{5^2 \times 5^3}$ (b) $\dfrac{13^5}{13 \times 13^2}$

   [(a) 5^2 (b) 13^2]

9 (a) $\dfrac{(9 \times 3^2)^3}{(3 \times 27)^2}$ (b) $\dfrac{(16 \times 4)^2}{(2 \times 8)^3}$

   [(a) 3^4 (b) 1]

10 (a) $\dfrac{5^{-2}}{5^{-4}}$ (b) $\dfrac{3^2 \times 3^{-4}}{3^3}$

   [(a) 5^2 (b) $\dfrac{1}{3^5}$]

11 (a) $\dfrac{7^2 \times 7^{-3}}{7 \times 7^{-4}}$ (b) $\dfrac{2^3 \times 2^{-4} \times 2^5}{2 \times 2^{-2} \times 2^6}$

   [(a) 7^2 (b) $\dfrac{1}{2}$]

12 (a) $13 \times 13^{-2} \times 13^4 \times 13^{-3}$ (b) $\dfrac{5^{-7} \times 5^2}{5^{-8} \times 5^3}$

   [(a) 1 (b) 1]

13 (a) $\dfrac{3^3 \times 5^2}{5^4 \times 3^4}$ (b) $\dfrac{7^{-2} \times 3^{-2}}{3^5 \times 7^4 \times 7^{-3}}$

   $\left[\text{(a)} \dfrac{1}{3 \times 5^2} \text{ (b)} \dfrac{1}{7^3 \times 3^7} \right]$

14 (a) $\dfrac{4^2 \times 9^3}{8^3 \times 3^4}$ (b) $\dfrac{8^{-2} \times 5^2 \times 3^{-4}}{25^2 \times 2^4 \times 9^{-2}}$

   $\left[\text{(a)} \dfrac{3^2}{2^5} \text{ (b)} \dfrac{1}{2^{10} \times 5^2} \right]$

15 Evaluate (a) $\left(\dfrac{1}{3^2}\right)^{-1}$ (b) $81^{0.25}$ (c) $16^{(-1/4)}$ (d) $\left(\dfrac{4}{9}\right)^{(1/2)}$

   $\left[\text{(a) 9 (b) } \pm 3 \text{ (c) } \dfrac{1}{2} \text{ (d) } \pm \dfrac{2}{3} \right]$

In *Problems 16* to *22*, evaluate the expressions given:

16 $\dfrac{9^2 \times 7^4}{3^4 \times 7^4 + 3^3 \times 7^2}$

   $\left[\dfrac{147}{148} \right]$

17 $\dfrac{3^3 \times 5^2}{2^3 \times 3^2 - 8^2 \times 9}$

   $\left[-1\dfrac{19}{56} \right]$

18 $\dfrac{3^3 \times 7^2 - 5^2 \times 7^3}{3^2 \times 5 \times 7^2}$

   $\left[-3\dfrac{13}{45} \right]$

19 $\dfrac{(2^4)^2 - 3^{-2} \times 4^4}{2^3 \times 16^2}$

   $\left[\dfrac{1}{9} \right]$

20 $\dfrac{\left(\dfrac{1}{2}\right)^3 - \left(\dfrac{2}{3}\right)^{-2}}{\left(\dfrac{3}{5}\right)^2}$

   $\left[-5\dfrac{65}{72} \right]$

21 $\left(\dfrac{4}{3}\right)^4 \bigg/ \left(\dfrac{2}{9}\right)^2$

   [64]

22 $\dfrac{(3^2)^{3/2} \times (8^{1/3})^2}{(3)^2 \times (4^3)^{1/2} \times (9)^{-1/2}}$

   $\left[4\dfrac{1}{2} \right]$

Standard form

In *Problems 23* to *27*, express in standard form:

23 (a) 73.9 (b) 28.4 (c) 197.62
   [(a) 7.39×10 (b) 2.84×10 (c) $1.976\,2 \times 10^2$]
24 (a) 2748 (b) 33 170 (c) 274 218
   [(a) 2.748×10^3 (b) 3.317×10^4 (c) $2.742\,18 \times 10^5$]
25 (a) 0.240 1 (b) 0.017 4 (c) 0.009 23
   [(a) 2.401×10^{-1} (b) 1.74×10^{-2} (c) 9.23×10^{-3}]
26 (a) 1702.3 (b) 10.04 (c) 0.010 9
   [(a) $1.702\,3 \times 10^3$ (b) 1.004×10 (c) 1.09×10^{-2}]
27 (a) $\dfrac{1}{2}$ (b) $11\dfrac{7}{8}$ (c) $130\dfrac{3}{5}$ (d) $\dfrac{1}{32}$
   [(a) 5×10^{-1} (b) $1.187\,5 \times 10$ (c) 1.306×10^2
   (d) 3.125×10^{-2}]

In *Problems 28* and *29*, express the numbers given as integers or decimal fractions:

28 (a) 1.01×10^3 (b) 9.327×10^2 (c) 5.41×10^4 (d) 7×10^0
   [(a) 1010 (b) 932.7 (c) 54 100 (d) 7]
29 (a) 3.89×10^{-2} (b) 6.741×10^{-1} (c) 8×10^{-3}
   [(a) 0.038 9 (b) 0.674 1 (c) 0.008]

In *Problems 30* to *33*, find values of the expressions given, stating the answers in standard form:

30 (a) $3.7 \times 10^2 + 9.81 \times 10^2$
   (b) $1.431 \times 10^{-1} + 7.3 \times 10^{-1}$

(c) $2.68 \times 10^{-2} - 8.414 \times 10^{-2}$
[(a) 1.351×10^3 (b) 8.731×10^{-1} (c) -5.734×10^{-2}]
31 (a) $4.831 \times 10^2 + 1.24 \times 10^3$
(b) $3.24 \times 10^{-3} - 1.11 \times 10^{-4}$
(c) $1.81 \times 10^2 + 3.417 \times 10^2 - 5.972 \times 10^2$
[(a) $1.723\,1 \times 10^3$ (b) 3.129×10^{-3} (c) -7.45×10]
32 (a) $(4.5 \times 10^{-2})(3 \times 10^3)$ (b) $2 \times (5.5 \times 10^4)$
[(a) 1.35×10^2 (b) 1.1×10^5]
33 (a) $\dfrac{6 \times 10^{-3}}{3 \times 10^{-5}}$ (b) $\dfrac{(2.4 \times 10^3)(3 \times 10^{-2})}{(4.8 \times 10^4)}$
[(a) 2×10^2 (b) 1.5×10^{-3}]

34 Write the following statements in standard form.
(a) The density of aluminium is 2710 kg m^{-3}.
(b) Poisson's ratio for gold is 0.44.
(c) The impedance of free space is $376.73\,\Omega$.
(d) The electron rest energy is 0.511 MeV.
(e) Proton charge–mass ratio is $95\,789\,700$ C kg^{-1}.
(f) The normal volume of a perfect gas is $0.022\,41$ m^3 mol^{-1}.
[(a) 2.71×10^3 kg m^{-3} (b) 4.4×10^{-1}
(c) $3.767\,3 \times 10^2\,\Omega$ (d) 5.11×10^{-1} MeV
(e) $9.578\,97 \times 10^7$ C kg^{-1} (f) 2.241×10^{-2} m^3 mol^{-1}]

3

Calculations

3.1 Errors

(i)   In all problems in which the measurement of distance, time, mass or other quantities occurs, an exact answer cannot be given; only an answer which is correct to a stated degree of accuracy can be given. To take account of this an **error due to measurement** is said to exist.

(ii) To take account of measurement errors it is usual to limit answers so that the result given is **not more than one significant figure greater than the least accurate number given in the data**.

(iii) **Rounding-off errors** can exist with decimal fractions. For example, to state that $\pi = 3.142$ is not strictly correct, but '$\pi = 3.142$ correct to 4 significant figures' is a true statement. (Actually, $\pi = 3.141\,592\,65.\ldots$)

(iv) It is possible, through an incorrect procedure, to obtain the wrong answer to a calculation. This type of error is known as **a blunder**.

(v)   An **order of magnitude error** is said to exist if incorrect positioning of the decimal point occurs after a calculation has been completed.

(vi) Blunders and order of magnitude errors can be reduced by determining **approximate values of calculations**. Answers which do not seem feasible must be checked and the calculation must be repeated as necessary.

> *Problem 1*   The area A of a triangle is given by $A = \dfrac{1}{2}bh$. The base b when measured is found to be 3.26 cm, and the perpendicular height h is 7.5 cm. Determine the area of the triangle.

Area of triangle $= \dfrac{1}{2}bh = \dfrac{1}{2} \times 3.26 \times 7.5 = 12.225$ cm^2 (by long multiplication or by calculator).

The approximate value is $\dfrac{1}{2} \times 3 \times 8 = 12$ cm^2, so there are no obvious blunder or magnitude errors. However, it is not usual in a measurements type problem to state the answer to an accuracy greater than 1 significant figure more than the least accurate number in the data: this is 7.5 cm, so the result should not have more than 3 significant figures. Thus area of triangle $= 12.2$ cm^2.

> *Problem 2*   State which type of error has been made in the following statements:
>
> (a) $72 \times 31.429 = 2262.9$
> (b) $16 \times 0.08 \times 7 = 89.6$
> (c) $11.714 \times 0.008\,8 = 0.324\,7$ correct to 4 decimal places
> (d) $\dfrac{29.74 \times 0.051\,2}{11.89} = 0.12$, correct to 2 significant figures.

(a) $72 \times 31.429 = 2262.888$ (by long multiplication or calculator), hence a **rounding-off error** has occurred. The answer should have stated:

$72 \times 31.429 = 2262.9$ correct to 5 significant figures.

(b) $16 \times 0.08 \times 7 = \overset{4}{\cancel{16}} \times \dfrac{8}{\underset{25}{\cancel{100}}} \times 7 = \dfrac{32 \times 7}{25} = \dfrac{224}{25} = 8\dfrac{24}{25}$

$= 8.96$

Hence an **order of magnitude** error has occurred.

(c) $11.714 \times 0.008\,8$ is approximately equal to

$12 \times 9 \times 10^{-3}$, i.e. about 108×10^{-3} or 0.108. Thus a **blunder** has been made.

(d) $\dfrac{29.74 \times 0.051\ 2}{11.89} \approx \dfrac{30 \times 5 \times 10^{-2}}{12} \approx \dfrac{150}{12 \times 10^{2}} \approx \dfrac{15}{120}$

$$\approx \frac{1}{8} \text{ or } 0.125$$

hence no order of magnitude error has occurred. However,

$\dfrac{29.74 \times 0.051\ 2}{11.89} = 0.128$ correct to 3 significant figures,

which equals 0.13 correct to 2 significant figures. Hence a **rounding-off error** has occurred.

Further problems on errors may be found in section 3.5, Problems 1 to 5, page 27.

3.2 Use of calculator

The most modern aid to calculations is the pocket-sized electronic calculator. With one of these, calculations can be quickly and accurately performed, correct to about 9 significant figures. The scientific type of calculator has made the use of tables and logarithms largely redundant.

To help you to become competent at using your calculator check that you agree with the answers to the following problems:

Problem 3 Evaluate the following, correct to 4 significant figures:

(a) $4.782\ 6 + 0.027\ 13$
(b) $17.694\ 1 - 11.876\ 2$
(c) $21.93 \times 0.012\ 981$

(a) $4.782\ 6 + 0.027\ 13 = 4.809\ 73 = \mathbf{4.810}$, correct to 4 significant figures.
(b) $17.694\ 1 - 11.\ 876\ 2 = 5.817\ 9 = \mathbf{5.818}$, correct to 4 significant figures.
(c) $21.93 \times 0.012\ 981 = 0.284\ 673\ 3 \ldots = \mathbf{0.284\ 7}$, correct to 4 significant figures.

Problem 4 Evaluate the following, correct to 4 decimal places:

(a) $46.32 \times 97.17 \times 0.012\ 58$

(b) $\dfrac{4.621}{23.76}$ (c) $\dfrac{1}{2}(62.49 \times 0.017\ 2)$

(a) $46.32 \times 97.17 \times 0.012\ 58 = 56.621\ 503\ 1 \ldots = \mathbf{56.621\ 5}$, correct to 4 decimal places.

(b) $\dfrac{4.621}{23.76} = 0.194\ 486\ 53 \ldots = \mathbf{0.194\ 5}$, correct to 4 decimal places.

(c) $\dfrac{1}{2}(62.49 \times 0.017\ 2) = 0.537\ 414 = \mathbf{0.537\ 4}$, correct to 4 decimal places.

Problem 5 Evaluate the following, correct to 3 decimal places:

(a) $\dfrac{1}{52.73}$

(b) $\dfrac{1}{0.027\ 5}$

(c) $\dfrac{1}{4.92} + \dfrac{1}{1.97}$

(a) $\dfrac{1}{52.73} = 0.018\ 964\ 53 \ldots = \mathbf{0.019}$, correct to 3 decimal places.

(b) $\dfrac{1}{0.027\ 5} = 36.363\ 636\ 3 \ldots = \mathbf{36.364}$, correct to 3 decimal places.

(c) $\dfrac{1}{4.92} + \dfrac{1}{1.97} = 0.710\ 866\ 24 \ldots = \mathbf{0.711}$, correct to 3 decimal places.

Problem 6 Evaluate the following, expressing the answers in standard form, correct to 4 significant figures.

(a) $(0.004\ 51)^2$
(b) $541.7 - (6.21 + 2.95)^2$
(c) $46.27^2 - 31.79^2$

(a) $(0.004\ 51)^2 = 2.034\ 01 \times 10^{-5} = \mathbf{2.034 \times 10^{-5}}$, correct to 4 significant figures.
(b) $541.7 - (6.21 + 2.95)^2 = 547.794\ 4 = 5.477\ 944 \times 10^2 = \mathbf{5.478 \times 10^2}$, correct to 4 significant figures.
(c) $46.27^2 - 31.79^2 = 1130.308\ 8 = \mathbf{1.130 \times 10^3}$, correct to 4 significant figures.

Problem 7 Evaluate the following, correct to 3 decimal places:

(a) $\dfrac{(2.37)^2}{0.052\ 6}$ (b) $\left(\dfrac{3.60}{1.92}\right)^2 + \left(\dfrac{5.40}{2.45}\right)^2$

(c) $\dfrac{15}{7.6^2 - 4.8^2}$

(a) $\dfrac{(2.37)^2}{0.0526} = 106.785\ 171\ldots = \mathbf{106.785}$, correct to 3 decimal places.

(b) $\left(\dfrac{3.60}{1.92}\right)^2 + \left(\dfrac{5.40}{2.45}\right)^2 = 8.373\ 600\ 84\ldots = \mathbf{8.374}$, correct to 3 decimal places.

(c) $\dfrac{15}{7.6^2 - 4.8^2} = 0.432\ 027\ 64\ldots = \mathbf{0.432}$, correct to 3 decimal places.

Problem 8 Evaluate the following, correct to 4 significant figures:

(a) $\sqrt{5.462}$ (b) $\sqrt{54.62}$ (c) $\sqrt{546.2}$

(a) $\sqrt{5.462} = 2.337\ 092\ 2\ldots = \mathbf{2.337}$, correct to 4 significant figures.

(b) $\sqrt{54.62} = 7.390\ 534\ 48\ldots = \mathbf{7.391}$, correct to 4 significant figures.

(c) $\sqrt{546.2} = 23.370\ 922\ldots = \mathbf{23.37}$, correct to 4 significant figures.

Problem 9 Evaluate the following, correct to 3 decimal places:

(a) $\sqrt{0.007\ 328}$

(b) $\sqrt{52.91} - \sqrt{31.76}$

(c) $\sqrt{(1.629\ 1 \times 10^4)}$

(a) $\sqrt{0.007\ 328} = 0.085\ 603\ 73 = \mathbf{0.086}$, correct to 3 decimal places.

(b) $\sqrt{52.91} - \sqrt{31.76} = 1.638\ 324\ 91\ldots = \mathbf{1.638}$, correct to 3 decimal places.

(c) $\sqrt{(1.629\ 1 \times 10^4)} = \sqrt{(16\ 291)} = 127.636\ 201\ldots = \mathbf{127.636}$, correct to 3 decimal places.

Problem 10 Evaluate the following, correct to 4 significant figures:

(a) 4.72^3

(b) $(0.831\ 6)^4$

(c) $\sqrt{(76.21^2 - 29.10^2)}$

(a) $4.72^3 = 105.154\ 04\ldots = \mathbf{105.2}$, correct to 4 significant figures.

(b) $(0.831\ 6)^4 = 0.478\ 253\ 24\ldots = \mathbf{0.478\ 3}$, correct to 4 significant figures.

(c) $\sqrt{(76.21^2 - 29.10^2)} = 70.435\ 460\ 5\ldots = \mathbf{70.44}$, correct to 4 significant figures.

Problem 11 Evaluate the following, correct to 3 significant figures:

(a) $\sqrt{\left[\dfrac{(6.09)^2}{25.2 \times \sqrt{7}}\right]}$

(b) $\sqrt[3]{(47.291)}$

(c) $\sqrt{(7.213^2 + 6.418^3 + 3.291^4)}$

(a) $\sqrt{\left[\dfrac{(6.09)^2}{25.2 \times \sqrt{7}}\right]} = 0.745\ 834\ 57\ldots = \mathbf{0.746}$, correct to 3 significant figures.

(b) $\sqrt[3]{(47.291)} = 3.616\ 258\ 76\ldots = \mathbf{3.62}$, correct to 3 significant figures.

(c) $\sqrt{(7.213^2 + 6.418^3 + 3.291^4)} = 20.825\ 299\ 1\ldots = \mathbf{20.8}$, correct to 3 significant figures.

Problem 12 Evaluate the following, expressing the answers in standard form, correct to 4 decimal places:

(a) $(5.176 \times 10^{-3})^2$

(b) $\left(\dfrac{1.974 \times 10^1 \times 8.61 \times 10^{-2}}{3.462}\right)^4$

(c) $\sqrt{(1.792 \times 10^{-4})}$

(a) $(5.176 \times 10^{-3})^2 = 2.678\ 097\ldots \times 10^{-5} = \mathbf{2.679\ 1 \times 10^{-5}}$, correct to 4 decimal places.

(b) $\left(\dfrac{1.974 \times 10^1 \times 8.61 \times 10^{-2}}{3.462}\right)^4 = 0.058\ 088\ 87\ldots$

$= \mathbf{5.808\ 9 \times 10^{-2}}$, correct to 4 decimal places.

(c) $\sqrt{(1.792 \times 10^{-4})} = 0.013\ 386\ 5\ldots = \mathbf{1.338\ 7 \times 10^{-2}}$, correct to 4 decimal places.

Further problems on the use of a calculator may be found in section 3.5, Problems 6 to 33, page 27.

3.3 Conversion tables and charts

It is often necessary to make calculations from various conversion tables and charts. Examples include currency exchange rates, imperial to metric unit conversions, train or bus timetables, production schedules and so on.

Problem 13   Currency exchange rates for five countries are shown in Table 3.1.

Table 3.1

France	£1 = 8.50 francs (f)
Italy	£1 = 2470 lira (l)
Spain	£1 = 203.50 pesetas (pes)
Germany	£1 = 2.50 Deutschmarks (Dm)
U.S.A.	£1 = 1.46 dollars ($)

Calculate

(a) how many French francs £27.90 will buy;
(b) the number of German Deutschmarks which can be bought for £75;
(c) the pounds sterling which can be exchanged for 62 500 lira;
(d) the number of American dollars which can be purchased for £92.50; and
(e) the pounds sterling which can be exchanged for 2705 pesetas.

(a) £1 = 8.50 francs, hence £27.90 = 27.90 × 8.50 francs
$$= \mathbf{237.15\ f}$$
(b) £1 = 2.50 Deutschmarks, hence £75 = 75 × 2.50 Dm
$$= \mathbf{187.50\ Dm}$$
(c) £1 = 2470 lira, hence 62 500 lira = £$\dfrac{62\ 500}{2470}$ = **£25.30**
(d) £1 = 1.46 dollars, hence £92.50 = 92.50 × 1.46 dollars
$$= \mathbf{\$135.05}$$
(e) £1 = 203.50 pesetas, hence 2705 pesetas = £$\dfrac{2705}{203.50}$
$$= \mathbf{£13.29}$$

Problem 14   Some approximate imperial to metric conversions are shown in Table 3.2.

Table 3.2

length	1 inch = 2.54 cm
	1 mile = 1.61 km
weight	2.2 lb = 1 kg
	(1 lb = 16 oz)
capacity	1.76 pints = 1 litre
	(8 pints = 1 gallon)

Use the table to determine:

(a) the number of millimetres in 9.5 inches;
(b) a speed of 50 miles per hour in kilometres per hour;
(c) the number of miles in 300 km;
(d) the number of kilograms in 30 pounds weight;
(e) the number of pounds and ounces in 42 kilograms (correct to the nearest ounce);
(f) the number of litres in 15 gallons; and
(g) the number of gallons in 40 litres.

(a) 9.5 inches = 9.5 × 2.54 cm = 24.13 cm
24.13 cm = 24.13. × 10 mm = **241.3 mm**
(b) 50 m.p.h. = 50 × 1.61 km/h = **80.5 km/h**
(c) 300 km = (300/1.61) miles = **186.3 miles**
(d) 30 lb = (30/2.2) kg = **13.64 kg**
(e) 42 kg = 42 × 2.2 lb = 92.4 lb
0.4 lb = 0.4 × 16 oz = 6.4 oz = 6 oz, correct to the nearest ounce.
Thus 42 kg = **92 lb 6 oz**, correct to the nearest ounce.
(f) 15 gallons = 15 × 8 pints = 120 pints
120 pints = (120/1.76) litres = **68.18 litres**
(g) 40 litres = 40 × 1.76 pints = 70.4 pints
70.4 pints = (70.4/8) gallons = **8.8 gallons**

Problem 15   Deduce the following information from the BR train timetable shown in Table 3.3.

(a) At what time should a man catch a train at London Waterloo to enable him to be in Portsmouth and Southsea by 8.0 a.m.
(b) A lady leaves Guildford at 8.11 a.m. and travels to Havant.
  (i)   How long does the journey take?
  (ii) How many miles is the journey?
  (iii) What is the average speed of the journey?
(c) A German businessman flies into Heathrow Airport at 9.45 a.m. It takes him 45 minutes to clear customs and travel by bus to Woking railway station. What is the earliest he could arrive at an appointment at Petersfield if the firm he is visiting is 10 minutes away from the railway station?

(a) After locating Portsmouth and Southsea near the bottom of the timetable move across horizontally to the nearest time before 8.0 a.m., which is 0758. Move upwards from 0758 to London Waterloo and it shows 0541. Hence the man needs to catch the **5.41 a.m.** train from Waterloo to arrive at Portsmouth and Southsea by 8.0 a.m.
(b) (i) Locating the 0811 train from Guildford and moving down to Havant shows that the train arrives at 0910. Hence the journey takes **59 minutes**.
(ii) On the far left of the top portion of the timetable it shows the mileage from London Waterloo. Guildford is $30\frac{1}{4}$ miles from Waterloo and Havant is $66\frac{1}{2}$ miles

Table 3.3   *London, Guildford → Haslemere, Portsmouth*

Reproduced with permission of British Rail

from Waterloo. Hence the distance from Guildford to Havant is $66\frac{1}{2} - 30\frac{1}{4}$, i.e. $36\frac{1}{4}$ **miles**.

(iii) Average speed of journey = $\dfrac{\text{distance travelled}}{\text{time taken}}$

$$= \dfrac{36\frac{1}{4}\ \text{miles}}{\frac{59}{60}\ \text{hour}}$$

$$= 36.86\ \text{m.p.h.}$$

(c) If the businessman arrives at Heathrow at 9.45 a.m. and takes 45 minutes to arrive at Woking station then he arrives at Woking at 10.30 a.m. The next train that leaves Woking after 10.30 a.m. is 10.45 a.m. and this arrives at Petersfield at 11.33 a.m. If it takes 10 minutes further to arrive at his appointment the earliest he will arrive will be **11.43 a.m.**

Further problems on conversion tables and charts may be found in section 3.5, Problems 34 to 36, page 28.

3.4   Logarithms

With the use of calculators firmly established, logarithmic tables are now rarely used for calculations. However, the theory of logarithms is important, for there are several scientific and engineering laws that involve the rules of logarithms.

If a number y can be written in the form a^x, then the index x is called the 'logarithm of y to the base a',

i.e. $\boxed{\textbf{if } y = a^x \textbf{ then } x = \log_a y}$

Thus, since $1000 = 10^3$, then $3 = \log_{10} 1000$.
Check this using the 'log' button on your calculator.

(a) Logarithms having a base of 10 are called **common logarithms** and $\log_{10}$ is usually abbreviated to lg. The following values may be checked by using a calculator:

$\lg 17.9 = 1.252\,8\ldots$, $\lg 462.7 = 2.665\,2\ldots$ and $\lg 0.017\,3 = -1.761\,9\ldots$

(b) Logarithms having a base of e (where 'e' is a mathematical constant approximately equal to 2.718 3) are called **hyperbolic**, **Napierian** or **natural logarithms**, and $\log_e$ is usually abbreviated to ln. The following values may be checked by using a calculator:

$\ln 3.15 = 1.147\,4\ldots$, $\ln 362.7 = 5.893\,5\ldots$ and $\ln 0.156 = -1.857\,8\ldots$

There are three rules of logarithms, which apply to any base:

(i) To multiply two numbers:

$\boxed{\log (A \times B) = \log A + \log B}$

The following may be checked by using a calculator:

$\lg 10 = 1$; also $\lg 5 + \lg 2$
$= 0.698\,97\ldots + 0.301\,029\ldots = 1$

Hence $\lg (5 \times 2) = \lg 10 = \lg 5 + \lg 2$

(ii) To divide two numbers:

$\boxed{\log \dfrac{A}{B} = \log A - \log B}$

The following may be checked using a calculator:

$\ln \dfrac{5}{2} = \ln 2.5 = 0.916\,29\ldots$

Also $\ln 5 - \ln 2 = 1.609\,43\ldots - 0.693\,14\ldots$
$= 0.916\,29\ldots$

Hence $\ln \dfrac{5}{2} = \ln 5 - \ln 2$

(iii) To raise a number to a power:

$\boxed{\log A^n = n \log A}$

The following may be checked using a calculator:

$\lg 5^2 = \lg 25 = 1.397\,94\ldots$
Also $2 \lg 5 = 2 \times 0.698\,97\ldots = 1.397\,94\ldots$
Hence $\lg 5^2 = 2 \lg 5$

The laws of logarithms may be used to solve certain equations involving powers – called **indicial equations**. For example, to solve, say, $3^x = 27$, logarithms to a base of 10 are taken of both sides,
i.e. $\log_{10} 3^x = \log_{10} 27$
and $x \log_{10} 3 = \log_{10} 27$ by the third law of logarithms.

Rearranging gives $x = \dfrac{\log_{10} 27}{\log_{10} 3} = \dfrac{1.4314\ldots}{0.4771\ldots} = 3$ which

may be readily checked.
(Note, (log 8/log 2) is **not** equal to lg (8/2).)

Problem 16 Evaluate (a) $\log_3 9$ (b) $\log_{10} 10$ (c) $\log_{16} 8$.

(a) Let $x = \log_3 9$ then $3^x = 9$ from the definition of a logarithm, i.e. $3^x = 3^2$, from which $x = 2$. Hence

$$\log_3 9 = 2$$

(b) Let $x = \log_{10} 10$ then $10^x = 10$ from the definition of a logarithm, i.e. $10^x = 10^1$, from which $x = 1$. Hence

$$\log_{10} 10 = 1$$
(which may be checked by a calculator).

(c) Let $x = \log_{16} 8$ then $16^x = 8$, from the definition of a logarithm, i.e. $(2^4)^x = 2^3$, i.e. $2^{4x} = 2^3$ from the laws of indices, from which $4x = 3$ and $x = \dfrac{3}{4}$. Hence

$$\log_{16} 8 = \dfrac{3}{4}$$

Problem 17 Evaluate (a) $\lg 0.001$ (b) $\ln e$ (c) $\log_3 \dfrac{1}{81}$.

(a) Let $x = \lg 0.001 = \log_{10} 0.001$ then $10^x = 0.001$, i.e. $10^x = 10^{-3}$, from which $x = -3$. Hence

$$\lg 0.001 = -3$$
(which may be checked by a calculator).

(b) Let $x = \ln e = \log_e e$ then $e^x = e$, i.e. $e^x = e^1$ from which $x = 1$. Hence

$$\ln e = 1$$
(which may be checked by a calculator).

(c) Let $x = \log_3 \dfrac{1}{81}$ then $3^x = \dfrac{1}{81} = \dfrac{1}{3^4} = 3^{-4}$, from which $x = -4$. Hence

$$\log_3 \frac{1}{81} = -4$$

Problem 18   Solve the following equations:

(a) $\lg x = 3$ (b) $\log_2 x = 3$ (c) $\log_5 x = -2$.

(a) If $\lg x = 3$ then $\log_{10} x = 3$ and $x = 10^3$, i.e. $x = 1000$
(b) If $\log_2 x = 3$ then $x = 2^3 = 8$
(c) If $\log_5 x = -2$ then $x = 5^{-2} = \dfrac{1}{5^2} = \dfrac{1}{25}$

Problem 19   Write (a) log 30 (b) log 450 in terms of log 2, log 3 and log 5 to any base.

(a) log 30 = log (2 × 15) = log (2 × 3 × 5)
     = **log 2 + log 3 + log 5** by the first law of logarithms.
(b) log 450 = log (2 × 225) = log (2 × 3 × 75)
     = log (2 × 3 × 3 × 25) = log (2 × 3² × 5²)
     = log 2 + log 3² + log 5² by the first law of logarithms
     i.e. log 450 = **log 2 + 2 log 3 + 2 log 5** by the third law of logarithms.

Problem 20   Write $\log \left(\dfrac{8 \times \sqrt[4]{5}}{81} \right)$ in terms of log 2, log 3 and log 5 to any base.

$\log \left(\dfrac{8 \times \sqrt[4]{5}}{81} \right) = \log 8 + \log \sqrt[4]{5} - \log 81$ by the first and second laws of logarithms
$= \log 2^3 + \log 5^{(1/4)} - \log 3^4$ by the laws of indices, i.e.
$\log \left(\dfrac{8 \times \sqrt[4]{5}}{81} \right) = \mathbf{3 \log 2 + \dfrac{1}{4} \log 5 - 4 \log 3}$ by the third law of logarithms.

Problem 21   Simplify log 64 − log 128 + log 32.

$64 = 2^6$, $128 = 2^7$ and $32 = 2^5$

Hence

    log 64 − log 128 + log 32
    = log 2⁶ − log 2⁷ + log 2⁵
    = 6 log 2 − 7 log 2 + 5 log 2 by the third law of logarithms
    = **4 log 2**

Problem 22   Evaluate

$$\frac{\log 25 - \log 125 + \dfrac{1}{2} \log 625}{3 \log 5}$$

$$\frac{\log 25 - \log 125 + \dfrac{1}{2} \log 625}{3 \log 5}$$

$$= \frac{\log 5^2 - \log 5^3 + \dfrac{1}{2} \log 5^4}{3 \log 5}$$

$$= \frac{2 \log 5 - 3 \log 5 + \dfrac{4}{2} \log 5}{3 \log 5}$$

$$= \frac{1 \log 5}{3 \log 5} = \frac{1}{3}$$

Problem 23   Solve the equation:

$\log (x - 1) + \log (x + 1) = 2 \log (x + 2)$

$\log (x - 1) + \log (x + 1)$
$= \log (x - 1)(x + 1)$ from the first law of logarithms
$= \log (x^2 - 1)$

$2 \log (x + 2) = \log (x + 2)^2 = \log (x^2 + 4x + 4)$

Hence if $\log (x^2 - 1) = \log (x^2 + 4x + 4)$
then $x^2 - 1 = x^2 + 4x + 4$
i.e. $- 1 = 4x + 4$
i.e. $- 5 = 4x$
i.e. $x = \dfrac{-5}{4}$ or $-1\dfrac{1}{4}$

Problem 24   Solve the equation $2^x = 3$, correct to 4 significant figures.

Taking logarithms to base 10 of both sides of $2^x = 3$ gives:

$$\log_{10} 2^x = \log_{10} 3$$
i.e. $x \log_{10} 2 = \log_{10} 3$

Rearranging gives:

$$x = \frac{\log_{10} 3}{\log_{10} 2} = \frac{0.477\ 121\ 25 \ldots}{0.301\ 029\ 99 \ldots} = \mathbf{1.585\ 0}$$

correct to 4 significant figures.

Problem 25 Solve the equation $2^{x+1} = 3^{2x-5}$ correct to 2 decimal places.

Taking logarithms to base 10 of both sides gives:

$$\log_{10} 2^{x+1} = \log_{10} 3^{2x-5}$$

i.e. $(x + 1) \log_{10} 2 = (2x - 5) \log_{10} 3$

$$x \log_{10} 2 + \log_{10} 2 = 2x \log_{10} 3 - 5 \log_{10} 3$$
$$x(0.301\,0) + (0.301\,0) = 2x(0.477\,1) - 5(0.477\,1)$$

i.e. $\quad 0.301\,0x + 0.301\,0 = 0.954\,2x - 2.385\,5$

Hence $\quad 2.385\,5 + 0.301\,0 = 0.954\,2x - 0.301\,0x$

$$2.686\,5 = 0.653\,2x$$

from which $x = \dfrac{2.686\,5}{0.653\,2} = \mathbf{4.11}$ correct to 2 decimal places.

Problem 26 Solve the equation $x^{3.2} = 41.15$, correct to 4 significant figures.

Taking logarithms to base 10 of both sides gives:

$$\log_{10} x^{3.2} = \log_{10} 41.15$$
$$3.2 \log_{10} x = \log_{10} 41.15$$

Hence $\log_{10} x = \dfrac{\log_{10} 41.15}{3.2} = 0.504\,49$

Thus $x = $ antilog $0.504\,49 = 10^{0.504\,49} = \mathbf{3.195}$ correct to 4 significant figures.

Further problems on logarithms may be found in section 3.5, Problems 37 to 74, page 29.

3.5 Further problems on calculations

Errors

In *Problems 1* to *5* state which type of error, or errors, have been made:

1 $25 \times 0.06 \times 1.4 = 0.21$
   [order of magnitude error]
2 $137 \times 6.842 = 937.4$
   [rounding-off error – should add 'correct to 4 significant figures']
3 $\dfrac{204 \times 0.008}{12.6} = 10.42$
   [blunder]
4 For a gas $pV = c$. When pressure $p = 103\,400$ Pa and $V = 0.54$ m^3 then $c = 55\,836$ Pa m^3
   [measured values, hence $c = 55\,800$ Pa m^3]

5 $\dfrac{4.6 \times 0.07}{52.3 \times 0.274} = 0.225$
   [order of magnitude error and rounding-off error – should be 0.022 5 correct to 3 significant figures]

Use of calculator

In *Problems 6* to *9*, use a calculator to evaluate the quantities shown correct to 4 significant figures:

6 (a) 3.249^2 (b) 73.78^2 (c) 311.4^2 (d) $0.063\,9^2$
   [(a) 10.56 (b) 5444 (c) 96 970 (d) 0.004 083]

7 (a) $\sqrt{4.735}$ (b) $\sqrt{35.46}$ (c) $\sqrt{73\,280}$ (d) $\sqrt{0.025\,6}$
   [(a) 2.176 (b) 5.955 (c) 270.7 (d) 0.160]

8 (a) $\dfrac{1}{7.768}$ (b) $\dfrac{1}{48.46}$ (c) $\dfrac{1}{0.081\,6}$ (d) $\dfrac{1}{1.118}$
   [(a) 0.128 8 (b) 0.020 63 (c) 12.25 (d) 0.892 9]

9 (a) lg 3.764 (b) lg 241.8 (c) lg 1.0 (d) lg 0.076 32
   [(a) 0.575 7 (b) 2.383 4 (c) 0 (d) −1.117 4]

10 Evaluate correct to 3 decimal places:

   (a) ln 41.62 (b) ln 0.017 9 (c) $\dfrac{\text{lg } 5.29}{\text{ln } 5.29}$
   [(a) 3.729 (b) −4.023 (c) 0.434]

In *Problems 11* to *18*, use a calculator to evaluate correct to 4 significant figures:

11 (a) 43.27×12.91 (b) $54.31 \times 0.572\,4$
   [(a) 558.6 (b) 31.09]
12 (a) $127.8 \times 0.043\,1 \times 19.8$ (b) $15.76 \div 4.329$
   [(a) 109.1 (b) 3.641]
13 (a) $\dfrac{137.6}{552.9}$ (b) $\dfrac{11.82 \times 1.736}{0.041}$
   [(a) 0.248 9 (b) 500.3]
14 (a) $\dfrac{1}{17.31}$ (b) $\dfrac{1}{0.034\,6}$ (c) $\dfrac{1}{147.9}$
   [(a) 0.057 77 (b) 28.90 (c) 0.006 763]
15 (a) 13.6^3 (b) 3.476^4 (c) 0.124^5
   [(a) 2515 (b) 146.0 (c) 0.000 029 31]
16 (a) $\sqrt{347.1}$ (b) $\sqrt{7632}$ (c) $\sqrt{0.027}$ (d) $\sqrt{0.004\,168}$
   [(a) 18.63 (b) 87.36 (c) 0.164 4 (d) 0.064 55]
17 (a) $\left(\dfrac{24.68 \times 0.053\,2}{7.412}\right)^3$ (b) $\left(\dfrac{0.268\,1 \times 41.2^2}{32.6 \times 11.89}\right)^4$
   [(a) 0.005 558 (b) 1.900]
18 (a) $\dfrac{14.32^3}{21.68^2}$ (b) $\dfrac{4.821^3}{17.33^2 - 15.86 \times 11.6}$
   [(a) 6.244 (b) 0.966 8]
19 Evaluate correct to 3 decimal places:

   (a) $\dfrac{29.12}{(5.81)^2 - (2.96)^2}$ (b) $\sqrt{53.96} - \sqrt{21.78}$
   [(a) 1.165 (b) 2.679]

20 Evaluate correct to 4 significant figures:

(a) $\sqrt{\left[\dfrac{(15.62)^2}{29.21 \times \sqrt{10.52}}\right]}$

(b) $\sqrt{(6.921^2 + 4.816^3 - 2.161^4)}$

[(a) 1.605 (b) 11.74]

21 Evaluate the following, expressing the answers in standard form, correct to 3 decimal places:

(a) $(8.291 \times 10^{-2})^2$ (b) $\sqrt{(7.623 \times 10^{-3})}$

[(a) 2.879×10^{-1} (b) 8.732×10^{-2}]

22 The area A of a rectangle is given by $A = lb$. The length l when measured is found to be 23.1 mm and the breadth b is 7.8 mm. Determine the area of the rectangle.
[180 mm^2]

23 The velocity of a body is given by $v = u + at$. The initial velocity u is measured when time t is 15 seconds and found to be 12 m/s. If the acceleration a is 9.81 m/s^2 calculate the final velocity v.
[159 m/s]

24 Calculate the current I in an electrical circuit, when $I = V/R$ amperes when the voltage V is measured and found to be 7.2 V and the resistance R is 17.7 Ω.
[0.407 A]

25 Find the distance s, given that $s = \dfrac{1}{2}gt^2$. Time $t = 0.032$ seconds and acceleration due to gravity $g = 9.81$ m/s^2.
[0.005 02 m or 5.02 mm]

26 The energy stored in a capacitor is given by $E = \dfrac{1}{2}CV^2$ joules. Determine the energy when capacitance $C = 5 \times 10^{-6}$ farads and voltage $V = 240$ V.
[0.144 0 J]

27 Find the area A of a triangle, given $A = \dfrac{1}{2}bh$, when the base length l is 23.42 m and the height h is 53.7 m.
[629.0 m^2]

28 Resistance R_2 is given by $R_2 = R_1(1 + \alpha t)$. Find R_2, correct to 4 significant figures, when $R_1 = 220$, $\alpha = 0.000\ 27$ and $t = 75.6$.
[224.5]

29 Density $= \dfrac{\text{mass}}{\text{volume}}$. Find the density when the mass is 2.462 kg and the volume is 173 cm^3. Give the answer in units of kg/m^3.
[14 230 kg/m^3]

30 Velocity = frequency × wavelength. Find the velocity when the frequency is 1825 Hz and the wavelength is 0.154 m.
[281.0 m/s]

31 Evaluate resistance R_T, given $\dfrac{1}{R_T} = \dfrac{1}{R_1} + \dfrac{1}{R_2} + \dfrac{1}{R_3}$, when $R_1 = 5.5$ Ω, $R_2 = 7.42$ Ω and $R_3 = 12.6$ Ω.
[2.526 Ω]

32 Find the total cost of 37 calculators costing £12.65 each and 19 drawing sets costing £6.38 each.
[£589.30]

33 Power $= \dfrac{\text{force} \times \text{distance}}{\text{time}}$. Find the power when a force of 3760 N raises an object a distance of 4.73 m in 35 s.
[508.2 W]

Conversion tables and charts

34 Currency exchange rates listed in a newspaper included the following:

France   £1 = 8.50 francs
Japan   £1 = 157.50 yen
Germany   £1 = 2.50 Deutschmarks
U.S.A.   £1 = $1.50
Spain   £1 = 203.0 pesetas

Calculate (a) how many French francs £32.50 will buy, (b) the number of American dollars that can be purchased for £74.80, (c) the pounds sterling which can be exchanged for 14 000 yen, (d) the pounds sterling which can be exchanged for 1750 pesetas, and (e) the German Deutschmarks which can be bought for £55.
[(a) 276.25 f (b) $112.20 (c) £88.89 (d) £8.62 (e) 137.50 Dm]

35 Below is a table of some metric to imperial conversions.

Length   2.54 cm = 1 inch
                1.61 km = 1 mile
Weight   1 kg = 2.2 lb (1 lb = 16 ounces)
Capacity 1 litre = 1.76 pints (8 pints = 1 gallon)

Use the table to determine (a) the number of millimetres in 15 inches, (b) a speed of 35 mph in km/h, (c) the number of kilometres in 235 miles, (d) the number of pounds and ounces in 24 kg (correct to the nearest ounce), (e) the number of kilograms in 15 lb, (f) the number of litres in 12 gallons, and (g) the number of gallons in 25 litres.
[(a) 381 mm (b) 56.35 km/h (c) 376 km (d) 52 lb 13 oz (e) 6.82 kg (f) 54.55 l (g) 5.5 gallons]

36 Deduce the following information from the BR train timetable shown in Table 3.4:

(a) At what time should a man catch a train at Mossley Hill to enable him to be in Manchester Piccadilly by 8.15 a.m.?

(b) A girl leaves Hunts Cross at 8.17 a.m. and travels to Manchester Oxford Road. How long does the journey take? What is the average speed of the journey?

(c) A man living at Edge Hill has to be at work at Trafford Park by 8.45 a.m. It takes him 10 minutes to walk to his work from Trafford Park station. What time train should he catch from Edge Hill?
[(a) 7.09 a.m. (b) 51 minutes, 32 m.p.h. (c) 7.04 a.m.]

Table 3.4 *Liverpool, Hunt's Cross and Warrington → Manchester*

Miles		MX	MO A ⊟	SX ◇ C ⌖	SO ◇ C ⌖	SX BHX	BHX	◇ ⌖	BHX	BHX		SX ◇ D ⌖	SX BHX	BHX	◇ E	BHX	◇ C	SX BHX		◇ ⌖		BHX
0	Liverpool Lime Street 82, 99 d			05 25	05 37		06 03	06 23		06 30		06 54	07 00	07 17		07 30	07 52			08 00	08 23		08 30
1¼	Edge Hill 82, 99 d							06 34					07 04			07 34				08 04			08 34
3¼	Mossley Hill 82 d							06 39					07 09			07 39				08 09			08 39
4¼	West Allerton 82 d							06 41					07 11			07 41				08 11			08 41
5¼	Allerton 82 d							06 43					07 13			07 43				08 13			08 43
—	Liverpool Central 101 d									06 15		06 45				07 15	07 45						08 15
—	Garston (Merseyside) 101 d									06 26		06 56				07 26	07 56						08 26
7¼	Hunt's Cross d			05u38	05u50		06 17			06 47		07u07	07 17			07 47	08u05			08 17			08 47
8¼	Halewood d							06 20		06 50			07 20			07 50				08 20			08 50
10¼	Hough Green d							06 24		06 54			07 24			07 54				08 24			08 54
12½	Widnes d							06 27		06 57			07 27	07 35		07 57	08 12			08 27			08 57
16	Sankey for Penketh d	00 02						06 32		07 02			07 32			08 02				08 32			09 02
18¼	Warrington Central a	00 07		05 50	06 02		06 37	06 46		07 07		07 19	07 37	07 43		08 07	08 20			08 37	08 46		09 07
	d			05 51	06 03	06 30		06 46	07 00		07 20	07 30	07 43	08 06		08 20		08 30		08 46		09 00
20¼	Padgate d							06 33			07 03			07 33			08 03			08 33			09 03
21¼	Birchwood d			05 56	06 08	06 36		06 51	07 06		07 25	07 36	07 48	08 06		08 25		08 36		08 51		09 06
24¼	Glazebrook d							06 41			07 11			07 41			08 11			08 41			09 11
25¼	Irlam d			06 02		06 44			07 14			07 44	07 54	08 14		08 34		08 44					09 14
28	Flixton d			06 06		06 48			07 18			07 48		08 18		08 38		08 48					09 18
28½	Chassen Road d			06 08		06 50			07 20			07 50		08 20		08 40		08 50					09 20
29	Urmston d		00 03	06 10		06 52			07 22			07 52		08 22		08 42		08 52					09 22
30¼	Humphrey Park d		00 13	06 13		06 55			07 25			07 55		08 25		08 45		08 55					09 25
31	Trafford Park d			06 15		06 57			07 27			07 57		08 27		08 47		08 57					09 27
34	Deansgate 81 d		00 23			07 03			07 33			08 03		08 33		08 52	09 03					09 33
34½	Manchester Oxford Road 81 a		00 27	06 17	07 05		07 07	08 07	35		07 40	08 05		08 37		08 40	09 05		09 08		09 35
	d		00 27	06 23	06 23		07 09				07 41			08 09	08 37		08 41	08 55		09 09		
35	Manchester Piccadilly 81 a		00 34	06 25	06 25		07 11				07 43			08 13	08 39		08 43	08 57		09 11		
—	Stockport 81, 90 a		06 34	06 34		07 32				07 54			08 32			08 54	09 19		09 32		
—	Sheffield ⊡ 90 a		07 30	07 30						08 42						09 42			

		◇ ⌖		◇ ⌖		BHX ◇ ⌖	◇ ⌖	◇ ⌖		BHX ◇ ⌖	◇ ⌖	◇ ⌖		BHX ◇ ⌖	◇ ⌖		BHX ◇ ⌖
Liverpool Lime Street 82, 99 d	08 54		09 00	09 23		09 30	09 56	10 00	10 23		10 30	10 56	11 00	11 23		11 30	11 56	12 00	12 23		12 30	12 56	13 00
Edge Hill 82, 99 d			09 04			09 34		10 04			10 34		11 04			11 34		12 04			12 34		13 04
Mossley Hill 82 d			09 09			09 39		10 09			10 39		11 09			11 39		12 09			12 39		13 09
West Allerton 82 d			09 11			09 41		10 11			10 41		11 11			11 41		12 11			12 41		13 11
Allerton 82 d			09 13			09 43		10 13			10 43		11 13			11 43		12 13			12 43		13 13
Liverpool Central 101 d	08 45					09 15	09 45				10 15	10 45				11 15	11 45				12 15	12 45
Garston (Merseyside) 101 d	08 56					09 26	09 56				10 26	10 56				11 26	11 56				12 26	12 56
Hunt's Cross d	09u09		09 17			09 47	10u09	10 17			10 47	11u09	11 17			11 47	12u09	12 17			12 47	13u09	13 17
Halewood d			09 20			09 50		10 20			10 50		11 20			11 50		12 20			12 50		13 20
Hough Green d			09 24			09 54		10 24			10 54		11 24			11 54		12 24			12 54		13 24
Widnes d			09 27			09 57		10 27			10 57		11 27			11 57		12 27			12 57		13 27
Sankey for Penketh d			09 32			10 02		10 32			11 02		11 32			12 02		12 32			13 02		13 32
Warrington Central a	09 21		09 37			10 07	10 21	10 37			11 07	11 21	11 37			12 07	12 21	12 37			13 07	13 21	13 37
d	09 22	09 30	09 46	10 00		10 22		10 46	11 00		11 22		11 46	12 00		12 22		12 46	13 00		13 22
Padgate d			09 33			10 03					11 03					12 03					13 03
Birchwood d			09 36	09 51	10 06				10 51		11 06		11 51	12 06				12 51	13 06			
Glazebrook d			09 41			10 11					11 11					12 11					13 11	
Irlam d			09 44			10 14					11 14					12 14					13 14	
Flixton d			09 48			10 18					11 18					12 18					13 18	
Chassen Road d			09 50			10 20					11 20					12 20					13 20	
Urmston d			09 52			10 22					11 22					12 22					13 22	
Humphrey Park d			09 55			10 25					11 25					12 25					13 25	
Trafford Park d			09 57			10 27					11 27					12 27					13 27	
Deansgate 81 d			10 01			10 33					11 33					12 33					13 33	
Manchester Oxford Road 81 a	09 40		10 05	10 08	10 35		10 40	11 08		11 35		12 08	12 35		12 40	13 08	13 35		13 40
d	09 41	10 06		10 09		10 41		11 09		11 41		12 09		12 41		13 09		13 41	
Manchester Piccadilly 81 a	09 43	10 08		10 11		10 43		11 11		11 43		12 11		12 43		13 11		13 43	
Stockport 81, 90 a	09 54	10 23		10 32		10 54		11 32		11 54		12 32		12 54		13 32		13 54	
Sheffield ⊡ 90 a	10 42					11 42					12 41					13 42					14 39

Reproduced with permission of British Rail

Logarithms

In *Problems 37* to *47* evaluate the given expression:

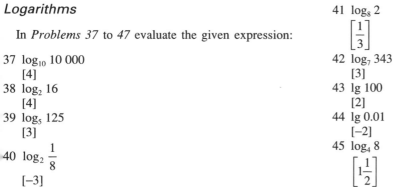

37 $\log_{10} 10\,000$
[4]

38 $\log_2 16$
[4]

39 $\log_5 125$
[3]

40 $\log_2 \dfrac{1}{8}$
[−3]

41 $\log_8 2$
$\left[\dfrac{1}{3}\right]$

42 $\log_7 343$
[3]

43 $\lg 100$
[2]

44 $\lg 0.01$
[−2]

45 $\log_4 8$
$\left[1\dfrac{1}{2}\right]$

46 $\log_{27} 3$

$$\left[\frac{1}{3}\right]$$

47 $\ln e^2$

[2]

In *Problems 48* to *54* solve the equations:

48 $\log_{10} x = 4$
[10 000]

49 $\lg x = 5$
[100 000]

50 $\log_3 x = 2$
[9]

51 $\log_4 x = -2\frac{1}{2}$

$$\left[\pm\frac{1}{32}\right]$$

52 $\lg x = -2$
[0.01]

53 $\log_8 x = -\frac{4}{3}$

$[\frac{1}{16}]$

54 $\ln x = 3$
$[e^3]$

In *Problems 55* to *58* write the given expressions in terms of log 2, log 3 and log 5 to any base:

55 $\log 60$
[2 log 2 + log 3 + log 5]

56 $\log 300$
[2 log 2 + log 3 + 2 log 5]

57 $\log\left(\frac{16 \times \sqrt[4]{5}}{27}\right)$

$$\left[4 \log 2 + \frac{1}{4}\log 5 - 3\log 3\right]$$

58 $\log\left(\frac{125 \times \sqrt[4]{16}}{\sqrt[4]{81^3}}\right)$
[log 2 − 3 log 3 + 3 log 5]

Simplify the expressions given in *Problems 59* to *61*:

59 $\log 27 - \log 9 + \log 81$
[5 log 3]

60 $\log 64 + \log 32 - \log 128$
[4 log 2]

61 $\log 8 - \log 4 + \log 32$
[6 log 2]

Evaluate the expressions given in *Problems 62* and *63*:

62 $\dfrac{\frac{1}{2}\log 16 - \frac{1}{3}\log 8}{\log 4}$

$$\left[\frac{1}{2}\right]$$

63 $\dfrac{\log 9 - \log 3 + \frac{1}{2}\log 81}{2\log 3}$

$$\left[-\frac{1}{2}\right]$$

Solve the equations given in *Problems 64* to *66*:

64 $\log x^4 - \log x^3 = \log 5x - \log 2x$

$$\left[x = 1\frac{1}{2}\right]$$

65 $\log 2t^3 - \log t = \log 16 - \log t$
$[t = 2]$

66 $2\log b^2 - 3\log b = \log 8b - \log 4b$
$[b = 2]$

In *Problems 67* to *74* solve the indicial equations for x, each correct to 4 significant figures:

67 $3^x = 6.41$
[1.691]

68 $2^x = 9$
[3.170]

69 $2^{x-1} = 3^{2x-1}$
[0.269 6]

70 $x^{1.5} = 14.91$
[6.058]

71 $25.28 = 4.2^x$
[2.251]

72 $4^{2x-1} = 5^{x+2}$
[3.959]

73 $x^{-0.25} = 0.792$
[1.499]

74 $0.027^x = 3.26$
[−0.327 2]

4

Introduction to algebra

4.1 Basic operations

Algebra is that part of mathematics in which the relations and properties of numbers are investigated by means of general symbols. For example, the area of a rectangle is found by multiplying the length by the breadth; this is expressed algebraically as $A = l \times b$, where A represents the area, l the length and b the breadth.

The basic laws introduced in arithmetic are generalized in algebra. Let a, b, c and d represent any four numbers. Then:

(i) $a + (b + c) = (a + b) + c$
(ii) $a(bc) = (ab)c$
(iii) $a + b = b + a$
(iv) $ab = ba$
(v) $a(b + c) = ab + ac$
(vi) $\dfrac{a + b}{c} = \dfrac{a}{c} + \dfrac{b}{c}$
(vii) $(a + b)(c + d) = ac + ad + bc + bd$

Problem 1   Evaluate $3ab - 2bc + abc$ when $a = 1$, $b = 3$ and $c = 5$.

Replacing a, b and c with their numerical values gives:

$3ab - 2bc + abc$
$= 3 \times 1 \times 3 - 2 \times 3 \times 5 + 1 \times 3 \times 5$
$= 9 - 30 + 15$
$= -6$

Problem 2   Find the value of $4p^2qr^3$, given that $p = 2$, $q = \dfrac{1}{2}$ and $r = 1\dfrac{1}{2}$.

Replacing p, q and r with their numerical values gives:

$$4p^2qr^3 = 4(2)^2 \left(\frac{1}{2}\right)\left(\frac{3}{2}\right)^3$$

$$= 4 \times 2 \times 2 \times \frac{1}{2} \times \frac{3}{2} \times \frac{3}{2} \times \frac{3}{2} = 27$$

Problem 3   Find the sum of $3x$, $2x$, $-x$ and $-7x$.

The sum of the positive terms is $3x + 2x = 5x$.
The sum of the negative terms is $x + 7x = 8x$.
Taking the sum of the negative terms from the sum of the positive terms gives:

$5x - 8x = -3x$

Alternatively

$3x + 2x + (-x) + (-7x) = 3x + 2x - x - 7x = -3x$

Problem 4   Find the sum of $4a$, $3b$, c, $-2a$, $-5b$ and $6c$.

Each symbol must be dealt with individually.

for the 'a' terms: $+4a - 2a = 2a$
for the 'b' terms: $+3b - 5b = -2b$
for the 'c' terms: $+c + 6c = 7c$

Thus $4a + 3b + c + (-2a) + (-5b) + 6c$
$= 4a + 3b + c - 2a - 5b + 6c$
$= 2a - 2b + 7c$

Problem 5   Find the sum of $5a - 2b$, $2a + c$, $4b - 5d$ and $b - a + 3d - 4c$.

The algebraic expressions may be tabulated as shown below, forming columns for the a's, b's, c's and d's. Thus:

$$
\begin{array}{rrrr}
+5a & - 2b & & \\
+2a & & + c & \\
& + 4b & & - 5d \\
- a & + b & - 4c & + 3d \\
\hline
\mathbf{6a} & \mathbf{+ 3b} & \mathbf{- 3c} & \mathbf{- 2d}
\end{array}
$$

Adding gives:

Problem 6   Subtract $2x + 3y - 4z$ from $x - 2y + 5z$.

$$
\begin{array}{r}
x - 2y + 5z \\
2x + 3y - 4z \\
\hline
\mathbf{-x - 5y + 9z}
\end{array}
$$

Subtracting gives:

(Note that $+5z - -4z = +5z + 4z = 9z$.)

An alternative method of subtracting algebraic expressions is to 'change the signs of the bottom line and add'. Hence:

$$
\begin{array}{r}
x - 2y + 5z \\
-2x - 3y + 4z \\
\hline
\mathbf{-x - 5y + 9z}
\end{array}
$$

Adding gives:

Problem 7   Multiply $2a + 3b$ by $a + b$.

Each term in the first expression is multiplied by a, then each term in the first expression is multiplied by b, and the two results are added. The usual layout is shown below.

$$
\begin{array}{r}
2a + 3b \\
a + b \\
\hline
2a^2 + 3ab \\
+ 2ab + 3b^2 \\
\hline
\mathbf{2a^2 + 5ab + 3b^2}
\end{array}
$$

Multiplying by $a \rightarrow$
Multiplying by $b \rightarrow$
Adding gives:

Problem 8   Multiply $3x - 2y^2 + 4xy$ by $2x - 5y$.

$$
\begin{array}{r}
3x - 2y^2 + 4xy \\
2x - 5y \\
\hline
6x^2 - 4xy^2 + 8x^2y \\
- 20xy^2 - 15xy + 10y^3 \\
\hline
\mathbf{6x^2 - 24xy^2 + 8x^2y - 15xy + 10y^3}
\end{array}
$$

Multiplying by $2x \rightarrow$
Multiplying by $-5y \rightarrow$
Adding gives:

Problem 9   Simplify $2p \div 8pq$.

$2p \div 8pq$ means $\dfrac{2p}{8pq}$. This can be reduced by cancelling as in arithmetic. Thus:

$$\frac{2p}{8pq} = \frac{\overset{1}{\cancel{2}} \times \cancel{p} \, \overset{1}{}}{\underset{4}{\cancel{8}} \times \cancel{p} \times q \, \underset{1}{}} = \frac{1}{4q}$$

Problem 10   Divide $2x^2 + x - 3$ by $x - 1$.

$2x^2 + x - 3$ is called the **dividend** and $x - 1$ the **divisor**. The usual layout is shown below with the dividend and divisor both arranged in descending powers of the symbols.

$$
\begin{array}{r}
2x + 3 \\
x - 1 \overline{) 2x^2 + x - 3} \\
\underline{2x^2 - 2x} \\
3x - 3 \\
\underline{3x - 3} \\
\cdot \quad \cdot
\end{array}
$$

Dividing the first term of the dividend by the first term of the divisor, i.e. $(2x^2/x)$ gives $2x$, which is put above the first term of the dividend as shown. The divisor is then multiplied by $2x$, i.e. $2x(x - 1) = 2x^2 - 2x$, which is placed under the dividend as shown. Subtracting gives $3x - 3$. The process is then repeated, i.e. the first term of the divisor is divided into $3x$, giving 3, which is placed above the dividend as shown. Then $3(x - 1) = 3x - 3$, which is placed under the $3x - 3$. The remainder, on subtraction, is zero, which completes the process.

Thus $(2x^2 + x - 3) \div (x - 1) = (\mathbf{2x + 3})$

(A check can be made on this answer by multiplying $(2x + 3)$ by $(x - 1)$, which should equal $2x^2 + x - 3$.)

Problem 11   Simplify $\dfrac{x^3 + y^3}{x + y}$

$$
\begin{array}{r}
\overset{①}{x^2} - \overset{④}{xy} + \overset{⑦}{y^2} \\
x + y \overline{) x^3 + 0 + 0 + y^3} \\
\underline{x^3 + x^2y} \\
- x^2y + y^3 \\
\underline{- x^2y - xy^2} \\
xy^2 + y^3 \\
\underline{xy^2 + y^3} \\
\cdot \quad \cdot
\end{array}
$$

① x into x^3 goes x^2. Put x^2 above x^3
② $x^2(x + y) = x^3 + x^2y$
③ Subtract
④ x into $-x^2y$ goes $-xy$. Put $-xy$ above dividend
⑤ $-xy(x + y) = -x^2y - xy^2$
⑥ Subtract
⑦ x into xy^2 goes y^2. Put y^2 above dividend
⑧ $y^2(x + y) = xy^2 + y^3$
⑨ Subtract

Thus $\dfrac{x^3 + y^3}{x + y} = x^2 - xy + y^2$

The zeros shown in the dividend are not normally shown, but are included to clarify the subtraction process and to keep similar terms in their respective columns.

Problem 12   Divide $4a^3 - 6a^2b + 5b^3$ by $2a - b$.

$$
\begin{array}{r}
2a^2 - 2ab\ - b^2 \\
2a - b\overline{)4a^3 - 6a^2b\qquad + 5b^3} \\
\underline{4a^3 - 2a^2b} \\
-4a^2b\qquad + 5b^3 \\
\underline{-4a^2b + 2ab^2} \\
-2ab^2 + 5b^3 \\
\underline{-2ab^2 +\ b^3} \\
4b^3
\end{array}
$$

Thus $\dfrac{4a^3 - 6a^2b + 5b^3}{2a - b} = 2a^2 - 2ab - b^2$, **remainder** $4b^3$

Alternatively, the answer may be expressed as

$2a^2 - 2ab - b^2 + \dfrac{4b^3}{2a - b}$

Further problems on basic operations may be found in section 4.6, Problems 1 to 14, page 39.

4.2   Laws of indices

The **laws of indices** are:

(i) $a^m \times a^n = a^{m+n}$

(ii) $\dfrac{a^m}{a^n} = a^{m-n}$

(iii) $(a^m)^n = a^{mn}$

(iv) $a^{m/n} = \sqrt[n]{a^m}$

(v) $a^{-n} = \dfrac{1}{a^n}$

(vi) $a^0 = 1$

Problem 13   Simplify $a^3b^2c \times ab^3c^5$.

Grouping like terms gives: $a^3 \times a \times b^2 \times b^3 \times c \times c^5$
Using the first law of indices gives: $a^{3+1} \times b^{2+3} \times c^{1+5}$
i.e. $a^4 \times b^5 \times c^6$ i.e. $\boldsymbol{a^4b^5c^6}$

Problem 14   Simplify $a^{1/2}b^2c^{-2} \times a^{1/6}b^{1/2}c$.

Using the first law of indices, $a^{1/2}b^2c^{-2} \times a^{(1/6)}b^{(1/2)}c$
$= a^{(1/2)+(1/6)} \times b^{2+1/2} \times c^{-2+1}$
$= \boldsymbol{a^{2/3}b^{5/2}c^{-1}}$

Problem 15   Simplify $\dfrac{a^3b^2c^4}{abc^{-2}}$ and evaluate when

$a = 3$, $b = \dfrac{1}{8}$ and $c = 2$.

Using the second law of indices,

$\dfrac{a^3}{a} = a^{3-1} = a^2$, $\dfrac{b^2}{b} = b^{2-1} = b$

and $\dfrac{c^4}{c^{-2}} = c^{4--2} = c^6$ Thus $\dfrac{a^3b^2c^4}{abc^{-2}} = a^2bc^6$

When $a = 3$, $b = \dfrac{1}{8}$ and $c = 2$,

$a^2bc^6 = (3)^2\,\dfrac{1}{8}\,(2)^6 = (9)\,\dfrac{1}{8}\,(64) = \boldsymbol{72}$

Problem 16   Simplify $\dfrac{p^{1/2}q^2r^{2/3}}{p^{1/4}q^{1/2}r^{1/6}}$ and evaluate when $p = 16$, $q = 9$ and $r = 4$, taking positive roots only.

Using the second law of indices gives:

$p^{(1/2)-(1/4)}q^{2-1/2}r^{2/3-1/6} = p^{1/4}q^{3/2}r^{1/2}$

When $p = 16$, $q = 9$ and $r = 4$,

$p^{1/4}q^{3/2}r^{1/2} = (16)^{1/4}(9)^{3/2}(4)^{1/2}$
$= (\sqrt[4]{16})(\sqrt{9^3})(\sqrt{4}) = (2)(3^3)(2) = \boldsymbol{108}$

Problem 17   Simplify $\dfrac{x^2y^3 + xy^2}{xy}$

Algebraic expressions of the form $\dfrac{a+b}{c}$ can be split into $\dfrac{a}{c} + \dfrac{b}{c}$. Thus:

$\dfrac{x^2y^3 + xy^2}{xy} = \dfrac{x^2y^3}{xy} + \dfrac{xy^2}{xy} = x^{2-1}y^{3-1} + x^{1-1}y^{2-1}$
$= \boldsymbol{xy^2 + y}$

(since $x^0 = 1$, from the sixth law of indices).

Problem 18   Simplify $\dfrac{x^2y}{xy^2 - xy}$

The highest common factor (HCF) of each of the three terms comprising the numerator and denominator is xy. Dividing each term by xy gives:

$$\frac{x^2y}{xy^2 - xy} = \frac{\dfrac{x^2y}{xy}}{\dfrac{xy^2}{xy} - \dfrac{xy}{xy}} = \frac{x}{y - 1}$$

Problem 19   Simplify $\dfrac{a^2b}{ab^2 - a^{1/2}b^3}$

The HCF of each of the three terms is $a^{1/2}b$. Dividing each term by $a^{1/2}b$ gives:

$$\frac{a^2b}{ab^2 - a^{1/2}b^3} = \frac{\dfrac{a^2b}{a^{1/2}b}}{\dfrac{ab^2}{a^{1/2}b} - \dfrac{a^{1/2}b^3}{a^{1/2}b}} = \frac{a^{3/2}}{a^{1/2}b - b^2}$$

Problem 20   Simplify $(p^3)^{1/2}(q^2)^4$.

Using the third law of indices gives:

$$p^{3\times(1/2)}q^{2\times4}$$

i.e. $p^{(3/2)}q^8$

Problem 21   Simplify $\dfrac{(mn^2)^3}{(m^{(1/2)}n^{(1/4)})^4}$

The brackets indicate that each letter in the bracket must be raised to the power outside.

Using the third law of indices gives:

$$\frac{(mn^2)^3}{(m^{1/2}n^{1/4})^4} = \frac{m^{1\times3}n^{2\times3}}{m^{(1/2)\times4}n^{(1/4)\times4}} = \frac{m^3n^6}{m^2n^1}$$

Using the second law of indices gives: $\dfrac{m^3n^6}{m^2n^1} = m^{3-2}n^{6-1}$

$$= mn^5$$

Problem 22   Simplify $(a^3\sqrt{b}\sqrt{c^5})(\sqrt{a}\sqrt[3]{b^2}\,c^3)$ and evaluate when $a = \dfrac{1}{4}$, $b = 64$ and $c = 1$.

Using the fourth law of indices, the expression can be written as:

$$(a^3b^{1/2}c^{5/2})(a^{1/2}b^{2/3}c^3)$$

Using the first law of indices gives:

$$a^{3+(1/2)}b^{(1/2)+(2/3)}c^{(5/2)+3} = a^{7/2}b^{7/6}c^{11/2}$$

It is usual to express the answer in the same form as the question. Hence

$$a^{7/2}b^{7/6}c^{11/2} = \sqrt{a^7}\sqrt[6]{b^7}\sqrt{c^{11}}$$

When $a = \dfrac{1}{4}$, $b = 64$ and $c = 1$,

$$\sqrt{a^7}\sqrt[6]{b^7}\sqrt{c^{11}} = \sqrt{\left(\frac{1}{4}\right)^7}\left(\sqrt[6]{64^7}\right)\sqrt{1^{11}}$$

$$= \left(\frac{1}{2}\right)^7 (2)^7(1) = 1$$

Problem 23   Simplify $(a^3b)(a^{-4}b^{-2})$, expressing the answer with positive indices only.

Using the first law of indices gives: $a^{3+-4}b^{1+-2} = a^{-1}b^{-1}$

Using the fifth law of indices gives: $a^{-1}b^{-1} = \dfrac{1}{a^{+1}b^{+1}} = \dfrac{1}{ab}$

Problem 24   Simplify $\dfrac{d^2e^2f^{(1/2)}}{\left(d^{(3/2)}ef^{(5/2)}\right)^2}$ expressing the answer with positive indices only.

Using the third law of indices gives:

$$\frac{d^2e^2f^{1/2}}{d^{(3/2)\times2}e^{1\times2}f^{(5/2)\times2}} = \frac{d^2e^2f^{1/2}}{d^3e^2f^5}$$

Using the second law of indices gives:

$$d^{2-3}e^{2-2}f^{(1/2)-5}$$
$$= d^{-1}e^0f^{-9/2}$$
$$= d^{-1}f^{(-9/2)} \text{ since } e^0 = 1 \text{ from the sixth law of indices,}$$
$$= \frac{1}{df^{(9/2)}} \text{ from the fifth law of indices.}$$

Problem 25   Simplify $\dfrac{(x^2y^{1/2})(\sqrt{x}\sqrt[3]{y^2})}{(x^5y^3)^{1/2}}$

Using the third and fourth laws of indices gives:

$$\frac{(x^2y^{1/2})(x^{1/2}y^{2/3})}{x^{5/2}y^{2/3}}$$

Using the first and second laws of indices gives:

$$x^{2+(1/2)-(5/2)}y^{(1/2)+(2/3)-(3/2)} = x^0y^{-1/3} = y^{-1/3} \text{ or } \frac{1}{y^{1/3}}$$

from the fifth and sixth laws of indices.

Further problems on laws of indices may be found in section 4.6, Problems 15 to 25, page 39.

4.3 Brackets and factorization

When two or more terms in an algebraic expression contain a common factor, then this factor can be shown outside of a bracket. For example

$$ab + ac = a(b + c)$$

which is simply the reverse of law (v) on page 31, and

$$6px + 2py - 4pz = 2p(3x + y - 2z)$$

This process is called **factorization**.

> *Problem 26*   Remove the brackets and simplify the expression $(3a + b) + 2(b + c) - 4(c + d)$.

Both b and c in the second bracket have to be multiplied by 2, and c and d in the third bracket by −4 when the brackets are removed (see law (v) on page 31). Thus:

$$(3a + b) + 2(b + c) - 4(c + d)$$
$$= 3a + b + 2b + 2c - 4c - 4d$$

Collecting similar terms together gives: $\mathbf{3a + 3b - 2c - 4d}$

> *Problem 27*   Simplify $a^2 - (2a - ab) - a(3b + a)$.

When the brackets are removed, both $2a$ and $-ab$ in the first bracket must be multiplied by −1 and both $3b$ and a in the second bracket by $-a$. Thus:

$$a^2 - (2a - ab) - a(3b + a)$$
$$= a^2 - 2a + ab - 3ab - a^2$$

Collecting similar terms together gives: $-2a - 2ab$

Since $-2a$ is a common factor the answer can be expressed as $\mathbf{-2a(1 + b)}$

> *Problem 28*   Simplify $(a + b)(a - b)$.

Each term in the second bracket has to be multiplied by each term in the first bracket. Thus:

$$(a + b)(a - b) = a(a - b) + b(a - b)$$
$$= a^2 - ab + ab - b^2 = \mathbf{a^2 - b^2}$$

Alternatively

$$
\begin{array}{r}
a + b \\
a - b \\
\hline
\end{array}
$$

Multiplying by $a \rightarrow$ $a^2 + ab$
Multiplying by $-b \rightarrow$ $\underline{\quad - ab - b^2}$
Adding gives: $a^2 \qquad - b^2$

> *Problem 29*   Remove the brackets from the expression $(x - 2y)(3x + y^2)$.

$$(x - 2y)(3x + y^2) = x(3x + y^2) - 2y(3x + y^2)$$
$$= \mathbf{3x^2 + xy^2 - 6xy - 2y^3}$$

> *Problem 30*   Simplify $(2x - 3y)^2$.

$$(2x - 3y)^2 = (2x - 3y)(2x - 3y)$$
$$= 2x(2x - 3y) - 3y(2x - 3y)$$
$$= 4x^2 - 6xy - 6xy + 9y^2$$
$$= \mathbf{4x^2 - 12xy + 9y^2}$$

Alternatively

$$
\begin{array}{r}
2x - 3y \\
2x - 3y \\
\hline
\end{array}
$$

Multiplying by $2x \rightarrow$ $4x^2 - 6xy$
Multiplying by $-3y \rightarrow$ $\underline{\quad - 6xy + 9y^2}$
Adding gives: $4x^2 - 12xy + 9y^2$

> *Problem 31*   Remove the brackets from the expression $2[p^2 - 3(q + r) + q^2]$.

In this problem there are two brackets and the 'inner' one is removed first.

Hence $2[p^2 - 3(q + r) + q^2] = 2[p^2 - 3q - 3r + q^2]$
$$= \mathbf{2p^2 - 6q - 6r + 2q^2}$$

> *Problem 32*   Remove the brackets and simplify the expression $2a - [3\{2(4a - b) - 5(a + 2b)\} + 4a]$.

Removing the innermost brackets gives:

$$2a - [3\{8a - 2b - 5a - 10b\} + 4a]$$

Collecting together similar terms gives:

$$2a - [3\{3a - 12b\} + 4a]$$

Removing the 'curly' brackets gives:

$$2a - [9a - 36b + 4a]$$

Collecting together similar terms gives:

$$2a - [13a - 36b]$$

Removing the outer brackets gives:

$2a - 13a + 36b$

i.e.

$-11a + 36b$ or $36b - 11a$ (see law (v), page 31)

Problem 33 Simplify $x(2x - 4y) - 2x(4x + y)$.

Removing brackets gives:

$2x^2 - 4xy - 8x^2 - 2xy$

Collecting together similar terms gives:

$-6x^2 - 6xy$

Factorizing gives:

$-6x(x + y)$

since $-6x$ is common to both terms.

Problem 34 Factorize (a) $xy - 3xz$, (b) $4a^2 + 16ab^3$, (c) $3a^2b - 6ab^2 + 15ab$.

For each part of this problem, the HCF of the terms will become one of the factors. Thus:

(a) $xy - 3xz$ $\qquad = x(y - 3z)$
(b) $4a^2 + 16ab^3$ $\qquad = 4a(a + 4b^3)$
(c) $3a^2b - 6ab^2 + 15ab = 3ab(a - 2b + 5)$

Problem 35 Factorize $ax - ay + bx - by$.

The first two terms have a common factor of a and the last two terms a common factor of b. Thus:

$ax - ay + bx - by = a(x - y) + b(x - y)$

The two newly formed terms have a common factor of $(x - y)$. Thus:

$a(x - y) + b(x - y) = (x - y)(a + b)$

Problem 36 Factorize $2ax - 3ay + 2bx - 3by$.

---

a is a common factor of the first two terms and b a common factor of the last two terms. Thus:

$2ax - 3ay + 2bx - 3by = a(2x - 3y) + b(2x - 3y)$

$(2x - 3y)$ is now a common factor thus:

$a(2x - 3y) + b(2x - 3y) = \mathbf{(2x - 3y)(a + b)}$

Alternatively, $2x$ is a common factor of the original first and third terms and $-3y$ is a common factor of the second and fourth terms. Thus:

$2ax - 3ay + 2bx - 3by = 2x(a + b) - 3y(a + b)$

$(a + b)$ is now a common factor thus:

$2x(a + b) - 3y(a + b) = \mathbf{(a + b)(2x - 3y)}$

as before.

Problem 37 Factorize $x^3 + 3x^2 - x - 3$.

---

x^2 is a common factor of the first two terms, thus:

$x^3 + 3x^2 - x - 3 = x^2(x + 3) - x - 3$

-1 is a common factor of the last two terms, thus:

$x^2(x + 3) - x - 3 = x^2(x + 3) - 1(x + 3)$

$(x + 3)$ is now a common factor, thus:

$x^2(x + 3) - 1(x + 3) = \mathbf{(x + 3)(x^2 - 1)}$

Further problems on brackets and factorization may be found in section 4.6, Problems 26 to 42, page 40.

4.4 Fundamental laws and precedence

The **laws of precedence** which apply to arithmetic also apply to algebraic expressions. The order is *B*rackets, *O*f, *D*ivision, *M*ultiplication, *A*ddition and *S*ubtraction (i.e. **BODMAS**).

Problem 38 Simplify $2a + 5a \times 3a - a$.

Multiplication is performed before addition and subtraction thus:

$2a + 5a \times 3a - a = 2a + 15a^2 - a$
$\qquad\qquad\qquad\quad = a + 15a^2 = \mathbf{a(1 + 15a)}$

Problem 39 Simplify $(a + 5a) \times 2a - 3a$.

The order of precedence is brackets, multiplication, then subtraction. Hence:

$(a + 5a) \times 2a - 3a = 6a \times 2a - 3a = 12a^2 - 3a$
$\qquad\qquad\qquad\qquad\quad = \mathbf{3a(4a - 1)}$

Problem 40 Simplify $a + 5a \times (2a - 3a)$.

The order of precedence is brackets, multiplication, then subtraction. Hence:

$$a + 5a \times (2a - 3a) = a + 5a \times -a = a + -5a^2$$
$$= a - 5a^2 = a(1 - 5a)$$

Problem 41 Simplify $a \div 5a + 2a - 3a$.

The order of precedence is division, then addition and subtraction. Hence:

$$a \div 5a + 2a - 3a = \frac{a}{5a} + 2a - 3a = \frac{1}{5} + 2a - 3a$$

$$= \frac{1}{5} - a$$

Problem 42 Simplify $a \div (5a + 2a) - 3a$.

The order of precedence is brackets, division and subtraction. Hence:

$$a \div (5a + 2a) - 3a = a \div 7a - 3a = \frac{a}{7a} - 3a$$

$$= \frac{1}{7} - 3a$$

Problem 43 Simplify $a \div (5a + 2a - 3a)$.

The order of precedence is brackets, then division. Hence:

$$a \div (5a + 2a - 3a) = a \div 4a = \frac{a}{4a} = \frac{1}{4}$$

Problem 44 Simplify $3c + 2c \times 4c + c \div 5c - 8c$.

The order of precedence is division, multiplication, addition and subtraction. Hence:

$$3c + 2c \times 4c + c \div 5c - 8c$$

$$= 3c + 2c \times 4c + \left(\frac{c}{5c}\right) - 8c$$

$$= 3c + 8c^2 + \frac{1}{5} - 8c$$

$$= 8c^2 - 5c + \frac{1}{5} \text{ or } c(8c - 5) + \frac{1}{5}$$

Problem 45 Simplify $(3c + 2c)4c + c \div 5c - 8c$.

The order of precedence is brackets, division, multiplication, addition and subtraction. Hence:

$$(3c + 2c)4c + c \div 5c - 8c$$
$$= 5c \times 4c + c \div 5c - 8c$$

$$= 5c \times 4c + \frac{c}{5c} - 8c$$

$$= 20c^2 + \frac{1}{5} - 8c \text{ or } 4c(5c - 2) + \frac{1}{5}$$

Problem 46 Simplify $3c + 2c \times 4c + c \div (5c - 8c)$.

The order of precedence is brackets, division, multiplication and addition. Hence:

$$3c + 2c \times 4c + c \div (5c - 8c)$$
$$= 3c + 2c \times 4c + c \div -3c$$

$$= 3c + 2c \times 4c + \frac{c}{-3c}$$

Now $\dfrac{c}{-3c} = \dfrac{1}{-3}$ Multiplying numerator and denominator

by -1 gives $\dfrac{1 \times -1}{-3 \times -1}$ i.e. $-\dfrac{1}{3}$. Hence:

$$3c + 2c \times 4c + \frac{c}{-3c}$$

$$= 3c + 2c \times 4c - \frac{1}{3}$$

$$= 3c + 8c^2 - \frac{1}{3} \text{ or } c(3 + 8c) - \frac{1}{3}$$

Problem 47 Simplify $(3c + 2c)(4c + c) \div (5c - 8c)$.

The order of precedence is brackets, division and multiplication. Hence:

$$(3c + 2c)(4c + c) \div (5c - 8c) = 5c \times 5c \div -3c$$

$$= 5c \times \frac{5c}{-3c}$$

$$= 5c \times -\frac{5}{3} = -\frac{25}{3}c$$

Problem 48 Simplify $(2a - 3) \div 4a + 5 \times 6 - 3a$.

The bracket around the $(2a - 3)$ shows that both $2a$ and -3 have to be divided by $4a$, and to remove the bracket the expression is written in fraction form. Hence:

$$(2a - 3) \div 4a + 5 \times 6 - 3a = \frac{2a - 3}{4a} + 5 \times 6 - 3a$$

$$= \frac{2a - 3}{4a} + 30 - 3a = \frac{2a}{4a} - \frac{3}{4a} + 30 - 3a$$

$$= \frac{1}{2} - \frac{3}{4a} + 30 - 3a = \mathbf{30}\frac{1}{2} - \frac{3}{4a} - \mathbf{3a}$$

Problem 49   Simplify $\frac{1}{3}$ of $3p + 4p(3p - p)$.

Applying BODMAS, the expression becomes $\frac{1}{3}$ of $3p + 4p$

$\times 2p$, and changing 'of' to '×', gives: $\frac{1}{3} \times 3p + 4p \times 2p$

i.e. $\boldsymbol{p + 8p^2}$ or $\boldsymbol{p(1 + 8p)}$

Further problems on fundamental laws and precedence may be found in section 4.6, Problems 43 to 54, page 40.

4.5   Direct and inverse proportionality

An expression such as $y = 3x$ contains two variables. For every value of x there is a corresponding value of y. The variable x is called the **independent variable** and y is called the **dependent variable**.

When an increase or decrease in an independent variable leads to an increase or decrease of the same proportion in the dependent variable this is termed **direct proportion**. If $y = 3x$ then y is directly proportional to x, which may be written as $y \propto x$ or $y = kx$, where k is called the **coefficient of proportionality** (in this case, k being equal to 3).

When an increase in an independent variable leads to a decrease of the same proportion in the dependent variable (or vice versa) this is termed **inverse proportion**. If y is inversely proportional to x then $y \propto (1/x)$ or $y = k/x$. Alternatively, $k = xy$, that is, for inverse proportionality the product of the variables is constant.

Examples of laws involving direct and inverse proportion in science include:

(i)   **Hooke's law**, which states that within the elastic limit of a material, the strain ϵ produced is directly proportional to the stress, σ, producing it, i.e. $\epsilon \propto \sigma$ or $\epsilon = k\sigma$.

(ii)   **Charles's law**, which states that for a given mass of gas at constant pressure the volume V is directly proportional to its thermodynamic temperature T, i.e. $V \propto T$ or $V = kT$.

(iii)   **Ohm's law**, which states that the current I flowing through a fixed resistor is directly proportional to the applied voltage V, i.e. $I \propto V$ or $I = kV$.

(iv)   **Boyle's law**, which states that for a gas at constant temperature, the volume V of a fixed mass of gas is inversely proportional to its absolute pressure p, i.e.

$$p \propto (1/V) \text{ or } p = k/V, \text{ i.e. } pV = k.$$

Problem 50   If y is directly proportional to x and $y = 2.48$ when $x = 0.4$, determine (a) the coefficient of proportionality and (b) the value of y when $x = 0.65$.

(a) $y \propto x$, i.e. $y = kx$. When $y = 2.48$ and $x = 0.4$, $2.48 = k(0.4)$. Hence the coefficient of proportionality

$$k = \frac{2.48}{0.4} = \mathbf{6.2}$$

(b) $y = kx$. Hence, when $x = 0.65$, $y = (6.2)(0.65) = \mathbf{4.03}$

Problem 51   Hooke's law states that stress σ is directly proportional to strain ϵ within the elastic limit of a material. When, for mild steel, the stress is 25×10^6 pascals, the strain is 0.000 125. Determine (a) the coefficient of proportionality and (b) the value of strain when the stress is 18×10^6 pascals.

(a) $\sigma \propto \epsilon$, i.e. $\sigma = k\epsilon$, from which $k = \sigma/\epsilon$. Hence the coefficient of proportionality

$$k = \frac{25 \times 10^6}{0.000 \ 125} = \mathbf{200 \times 10^9 \ pascals}$$

(The coefficient of proportionality k in this case is called Young's Modulus of Elasticity.)

(b) Since $\sigma = k\epsilon$, $\epsilon = \sigma/k$. Hence

when $\sigma = 18 \times 10^6$, strain $\epsilon = \dfrac{18 \times 10^6}{200 \times 10^9} = \mathbf{0.000 \ 09}$

Problem 52   The electrical resistance R of a piece of wire is inversely proportional to the cross-sectional area A. When $A = 5$ mm^2, $R = 7.2$ ohms. Determine (a) the coefficient of proportionality and (b) the cross-sectional area when the resistance is 4 ohms.

(a) $R \propto 1/A$, i.e. $R = k/A$ or $k = RA$. Hence, when $R = 7.2$ and $A = 5$, the coefficient of proportionality, $k = (7.2)(5)$ = **36**

(b) Since $k = RA$ then $A = k/R$. When $R = 4$, the cross sectional area $A = \dfrac{36}{4} = $ **9 mm²**

Problem 53 Boyle's law states that at constant temperature, the volume V of a fixed mass of gas is inversely proportional to its absolute pressure p. If a gas occupies a volume of 0.08 m³ at a pressure of 1.5×10^6 pascals determine (a) the coefficient of proportionality and (b) the volume if the pressure is changed to 4×10^6 pascals.

(a) $V \propto 1/p$, i.e. $V = k/p$ or $k = pV$. Hence the coefficient of proportionality
$k = (1.5 \times 10^6)(0.08) = $ **0.12 × 10⁶**

(b) Volume $V = \dfrac{k}{p} = \dfrac{0.12 \times 10^6}{4 \times 10^6} = $ **0.03 m³**

Further problems on direct and inverse proportionality may be found in section 4.6, Problems 55 to 59, page 40.

4.6 Further problems on algebra

Basic operations

1 Find the value of $2xy + 3yz - xyz$, when $x = 2$, $y = -2$ and $z = 4$.
[−16]

2 Evaluate $3pq^2r^3$ when $p = \dfrac{2}{3}$, $q = -2$ and $r = -1$.

[−8]

3 Find the sum of $3a$, $-2a$, $-6a$, $5a$ and $4a$.
[$4a$]

4 Simplify $\dfrac{4}{3}c + 2c - \dfrac{1}{6}c - c$.

$\left[2\dfrac{1}{6}c \right]$

5 Find the sum of $3x$, $2y$, $-5x$, $2z$, $-\dfrac{1}{2}y$, $-\dfrac{1}{4}x$.

$\left[-2\dfrac{1}{4}x + 1\dfrac{1}{2}y + 2z \right]$

6 Add together $2a + 3b + 4c$, $-5a - 2b + c$, $4a - 5b - 6c$.
[$a - 4b - c$]

7 Add together $3d + 4e$, $-2e + f$, $2d - 3f$, $4d - e + 2f - 3e$.
[$9d - 2e$]

8 From $4x - 3y + 2z$ subtract $x + 2y - 3z$.
[$3x - 5y + 5z$]

9 Subtract $\dfrac{3}{2}a - \dfrac{b}{3} + c$ from $\dfrac{b}{2} - 4a - 3c$.

$\left[-5\dfrac{1}{2}a + \dfrac{5}{6}b - 4c \right]$

10 Multiply $3x + 2y$ by $x - y$.
[$3x^2 - xy - 2y^2$]

11 Multiply $2a - 5b + c$ by $3a + b$.
[$6a^2 - 13ab + 3ac - 5b^2 + bc$]

12 Simplify (i) $3a \div 9ab$ (ii) $4a^2b \div 2a$.

$\left[\text{(i) } \dfrac{1}{3b} \text{ (ii) } 2ab \right]$

13 Divide $2x^2 + xy - y^2$ by $x + y$.
[$2x - y$]

14 Divide $p^3 + q^3$ by $p + q$.
[$p^2 - pq + q^2$]

Laws of indices

15 Simplify $(x^2y^3z)(x^3yz^3)$ and evaluate when $x = \dfrac{1}{2}$, $y = 2$ and $z = 3$.

$\left[x^5y^4z^3, 13\dfrac{1}{2} \right]$

16 Simplify $(a^{3/2}bc^{-3})(a^{1/2}b^{-1/2}c)$ and evaluate when $a = 3$, $b = 4$ and $c = 2$.

$\left[a^2b^{1/2}c^{-2}, 4\dfrac{1}{2} \right]$

17 Simplify $\dfrac{a^5bc^3}{a^2b^3c^2}$ and evaluate when $a = \dfrac{3}{2}$, $b = \dfrac{1}{2}$ and $c = \dfrac{2}{3}$.
[$a^3b^{-2}c$, 9]

In *Problems 18* to *25*, simplify the given expressions:

18 $\dfrac{x^{1/5}y^{1/2}z^{1/3}}{x^{-1/2}y^{1/3}z^{-1/6}}$
[$x^{7/10}y^{1/6}z^{1/2}$]

19 $\dfrac{a^2b + a^3b}{a^2b^2}$

$\left[\dfrac{1 + a}{b} \right]$

20 $\dfrac{p^3q^2}{pq^2 - p^2q}$

$\left[\dfrac{p^2q}{q - p} \right]$

21 $(a^2)^{1/2}(b^2)^3(c^{1/2})^3$
[$ab^6c^{3/2}$]

22 $\dfrac{(abc)^2}{(a^2b^{-1}c^{-3})^3}$
[$a^{-4}b^5c^{11}$]

23 $\left(\sqrt{x}\sqrt{y^3}\sqrt[3]{z^2}\right)\left(\sqrt{x}\sqrt{y^3}\sqrt{z^3}\right)$
[$xy^3\sqrt[6]{z^{13}}$]

24 $(e^2f^3)(e^{-3}f^{-5})$, expressing the answer with positive indices only.
$\left[\dfrac{1}{ef^2}\right]$

25 $\dfrac{(a^3b^{1/2}c^{-1/2})(ab)^{1/3}}{(\sqrt{a^3}\sqrt{bc})}$
$\left[a^{11/6}b^{1/3}c^{-3/2}\text{ or }\dfrac{\sqrt[6]{a^{11}}\sqrt[3]{b}}{\sqrt{c^3}}\right]$

Brackets and factorization

In *Problems 26 to 38*, remove the brackets and simplify where possible:

26 $(x + 2y) + (2x - y)$
[$3x + y$]

27 $(4a + 3y) - (a - 2y)$
[$3a + 5y$]

28 $2(x - y) - 3(y - x)$
[$5(x - y)$]

29 $2x^2 - 3(x - xy) - x(2y - x)$
[$x(3x - 3 + y)$]

30 $2(p + 3q - r) - 4(r - q + 2p) + p$
[$-5p + 10q - 6r$]

31 $(a + b)(a + 2b)$
[$a^2 + 3ab + 2b^2$]

32 $(p + q)(3p - 2q)$
[$3p^2 + pq - 2q^2$]

33 (i) $(x - 2y)^2$ (ii) $(3a - b)^2$
$\left[\begin{array}{l}\text{(i) } x^2 - 4xy + 4y^2 \\ \text{(ii) } 9a^2 - 6ab + b^2\end{array}\right]$

34 $3a(b + c) + 4c(a - b)$
[$3ab + 7ac - 4bc$]

35 $2x + \{y - (2x + y)\}$
[0]

36 $3a + 2\{a - (3a - 2)\}$
[$4 - a$]

37 $2 - 5\{a(a - 2b) - (a - b)^2\}$
[$2 + 5b^2$]

38 $24p - [2\{3(5p - q) - 2(p + 2q)\} + 3q]$
[$11q - 2p$]

In *Problems 39 to 42*, factorize:

39 (i) $pb + 2pc$ (ii) $2l^2 + 8ln$
$\left[\begin{array}{l}\text{(i) } p(b + 2c) \\ \text{(ii) } 2l(l + 4n)\end{array}\right]$

40 (i) $21a^2b^2 - 28ab$ (ii) $2xy^2 + 6x^2y + 8x^3y$
$\left[\begin{array}{l}\text{(i) } 7ab(3ab - 4) \\ \text{(ii) } 2xy(y + 3x + 4x^2)\end{array}\right]$

41 (i) $ay + by + a + b$ (ii) $px + qx + py + qy$
$\left[\begin{array}{l}\text{(i) } (a + b)(y + 1) \\ \text{(ii) } (p + q)(x + y)\end{array}\right]$

42 (i) $ax - ay + bx - by$ (ii) $2ax + 3ay - 4bx - 6by$.
$\left[\begin{array}{l}\text{(i) } (x - y)(a + b) \\ \text{(ii) } (a - 2b)(2x + 3y)\end{array}\right]$

Fundamental laws and precedence

In *Problems 43 to 54*, simplify:

43 $2x \div 4x + 6x$
$\left[\dfrac{1}{2} + 6x\right]$

44 $2x \div (4x + 6x)$
$\left[\dfrac{1}{5}\right]$

45 $3a - 2a \times 4a + a$
[$4a(1 - 2a)$]

46 $(3a - 2a)4a + a$
[$a(4a + 1)$]

47 $3a - 2a(4a + a)$
[$a(3 - 10a)$]

48 $2y + 4 \div 6y + 3 \times 4 - 6y$
$\left[\dfrac{2}{3y} - 3y + 12\right]$

49 $(2y + 4) \div 6y + 3 \times 4 - 5y$
$\left[\dfrac{2}{3y} + 12\dfrac{1}{3} - 5y\right]$

50 $2y + 4 \div 6y + 3(4 - 5y)$
$\left[\dfrac{2}{3y} + 12 - 13y\right]$

51 $3 \div y + 2 \div y + 1$
$\left[\dfrac{5}{y} + 1\right]$

52 $p^2 - 3pq \times 2p \div 6q + pq$
[pq]

53 $(x + 1)(x - 4) \div (2x + 2)$
$\left[\dfrac{1}{2}(x - 4)\right]$

54 $\dfrac{1}{4}$ of $2y + 3y(2y - y)$.
$\left[y(\dfrac{1}{2} + 3y)\right]$

Direct and inverse proportionality

55 If p is directly proportional to q and $p = 37.5$ when $q = 2.5$, determine (a) the constant of proportionality and (b) the value of p when q is 5.2.
[(a) 15 (b) 78]

56 Charles's law states that for a given mass of gas at constant pressure the volume is directly proportional to its thermodynamic temperature. A gas occupies a volume of 2.25 litres at 300 K. Determine (a) the constant of proportionality, (b) the volume at 420 K and (c) the temperature when the volume is 2.625 litres.
[(a) 0.007 5 (b) 3.15 l (c) 350 K]

57 Ohm's law states that the current flowing in a fixed resistor is directly proportional to the applied voltage. When 30 volts is applied across a resistor the current flowing through the resistor is 2.4×10^{-3} amperes. Determine (a) the constant of proportionality, (b) the current when the voltage is 52 volts, and (c) the voltage when the current is 3.6×10^{-3} amperes.
[(a) 0.000 08 (b) 4.16×10^{-3} A (c) 45 V]

58 If y is inversely proportional to x and $y = 15.3$ when $x = 0.6$, determine (a) the coefficient of proportionality, (b) the value of y when x is 1.5, and (c) the value of x when y is 27.2.
[(a) 9.18 (b) 6.12 (c) 0.337 5]

59 Boyle's law states that for a gas at constant temperature, the volume of a fixed mass of gas is inversely proportional to its absolute pressure. If a gas occupies a volume of $1.5 \, m^3$ at a pressure of 200×10^3 pascals, determine (a) the constant of proportionality, (b) the volume when the pressure is 800×10^3 pascals and (c) the pressure when the volume is $1.25 \, m^3$.
[(a) 300×10^3 (b) $0.375 \, m^3$ (c) 240×10^3 Pa]

5

Simple equations

5.1 Expressions, equations and identities

$(3x - 5)$ is an example of an **algebraic expression**, whereas $3x - 5 = 1$ is an example of an **equation** (i.e. it contains an 'equals' sign).

An equation is simply a statement that two quantities are equal. For example, $1\ m = 1000\ mm$ or $y = mx + c$.

An **identity** is a relationship which is true for all values of the unknown, whereas an equation is only true for particular values of the unknown. For example, $3x - 5 = 1$ is an equation since it is only true when $x = 2$, whereas $3x \equiv 8x - 5x$ is an identity since it is true for all values of x. (Note '$\equiv$' means 'is identical to'.)

Simple linear equations (or equations of the first degree) are those in which an unknown quantity is raised only to the power 1.

To '**solve an equation**' means 'to find the value of the unknown'.

Any arithmetic operation may be applied to an equation **as long as the equality of the equation is maintained**.

5.2 Worked problems on simple equations

Problem 1   Solve the equation $4x = 20$.

Dividing each side of the equation by 4 gives: $\dfrac{4x}{4} = \dfrac{20}{4}$

(Note that the same operation has been applied to both the left-hand side (LHS) and the right-hand side (RHS) of the equation so the equality has been maintained.)

Cancelling gives $x = \mathbf{5}$, which is the solution to the equation. Solutions to simple equations should always be checked and this is accomplished by substituting the solution into the original equation. In this case, LHS = 4(5) = 20 = RHS.

Problem 2   Solve $\dfrac{2x}{5} = 6$

The LHS is a fraction and this can be removed by multiplying both sides of the equation by 5. Hence $5\left(\dfrac{2x}{5}\right) = 5(6)$

Cancelling gives: $2x = 30$

Dividing both sides of the equation by 2 gives: $\dfrac{2x}{2} = \dfrac{30}{2}$

i.e. $x = \mathbf{15}$

Problem 3   Solve $a - 5 = 8$.

Adding 5 to both sides of the equation gives:

$$a - 5 + 5 = 8 + 5$$

i.e. $a = \mathbf{13}$

The result of the above procedure is to move the '−5' from the LHS of the original equation, across the equals sign, to the RHS, but the sign is changed to +.

> *Problem 4*   Solve $x + 3 = 7$.

Subtracting 3 from both sides of the equation gives:

$$x + 3 - 3 = 7 - 3$$

i.e. $x = 4$

The result of the above procedure is to move the '+3' from the LHS of the original equation, across the equals sign, to the RHS, but the sign is changed to $-$. Thus a term can be moved from one side of an equation to the other as long as a change in sign is made.

> *Problem 5*   Solve $6x + 1 = 2x + 9$.

In such equations the terms containing x are grouped on one side of the equation and the remaining terms grouped on the other side of the equation. As in *Problems 3* and *4*, changing from one side of an equation to the other must be accompanied by a change of sign.

Thus since $6x + 1 = 2x + 9$

then
$$6x - 2x = 9 - 1$$
$$4x = 8$$
$$\frac{4x}{4} = \frac{8}{4}$$

i.e. $x = 2$

Check: LHS of original equation $= 6(2) + 1 = 13$
RHS of original equation $= 2(2) + 9 = 13$

Hence the solution $x = 2$ is correct.

> *Problem 6*   Solve $4 - 3p = 2p - 11$.

In order to keep the p term positive the terms in p are moved to the RHS and the constant terms to the LHS.

Hence $4 + 11 = 2p + 3p$
$$15 = 5p$$
$$\frac{15}{5} = \frac{5p}{5}$$
Hence $p = 3$

Check: LHS $= 4 - 3(3) = 4 - 9 = -5$
RHS $= 2(3) - 11 = 6 - 11 = -5$

Hence the solution $p = 3$ is correct.
If, in this example, the unknown quantities had been grouped initially on the LHS instead of the RHS then:

$$-3p - 2p = -11 - 4$$
i.e. $-5p = -15$

$$\frac{-5p}{-5} = \frac{-15}{-5}$$
and $p = 3$, as before

It is often easier, however, to work with positive values where possible.

> *Problem 7*   Solve $3(x - 2) = 9$.

Removing the bracket gives: $3x - 6 = 9$
Rearranging gives: $3x = 9 + 6$
$$3x = 15$$
$$\frac{3x}{3} = \frac{15}{3}$$
i.e. $x = 5$

Check: LHS $= 3(5 - 2) = 3(3) = 9 =$ RHS

Hence the solution $x = 5$ is correct.

> *Problem 8*   Solve $4(2r - 3) - 2(r - 4) = 3(r - 3) - 1$.

Removing brackets gives:

$$8r - 12 - 2r + 8 = 3r - 9 - 1$$

Rearranging gives: $8r - 2r - 3r = -9 - 1 + 12 - 8$
i.e. $3r = -6$
$$r = \frac{-6}{3} = -2$$

Check: LHS $= 4(-4 - 3) - 2(-2 - 4) = -28 + 12 = -16$
RHS $= 3(-2 - 3) - 1 = -15 - 1 = -16$

Hence the solution $r = -2$ is correct.

> *Problem 9*   Solve $\dfrac{3}{x} = \dfrac{4}{5}$.

The lowest common multiple (LCM) of the denominators, i.e. the lowest algebraic expression that both x and 5 will divide into, is $5x$. Multiplying both sides by $5x$ gives:

$$5x\left(\frac{3}{x}\right) = 5x\left(\frac{4}{5}\right)$$

Cancelling gives:

$$15 = 4x \qquad\qquad (1)$$
$$\frac{15}{4} = \frac{4x}{4}$$

i.e. $x = 3\dfrac{3}{4}$

Check: LHS $= \dfrac{3}{3\frac{3}{4}} = \dfrac{3}{\frac{15}{4}} = 3\left(\dfrac{4}{15}\right) = \dfrac{12}{15} = \dfrac{4}{5} =$ RHS

(Note that when there is only one fraction on each side of an equation, 'cross-multiplication' can be applied. In this example if $3/x = 4/5$ then $(3)(5) = 4x$, which is a quicker way of arriving at equation (1) above.)

Problem 10 Solve $\dfrac{2y}{5} + \dfrac{3}{4} + 5 = \dfrac{1}{20} - \dfrac{3y}{2}$

The LCM of the denominators is 20. Multiplying each term by 20 gives:

$$20\left(\frac{2y}{5}\right) + 20\left(\frac{3}{4}\right) + 20(5) = 20\left(\frac{1}{20}\right) - 20\left(\frac{3y}{2}\right)$$

Cancelling gives: $4(2y) + 5(3) + 100 = 1 - 10(3y)$
i.e. $8y + 15 + 100 = 1 - 30y$
Rearranging gives: $8y + 30y = 1 - 15 - 100$
                               $38y = -114$
                               $y = \dfrac{-114}{38} = -3$

Check: LHS $= \dfrac{2(-3)}{5} + \dfrac{3}{4} + 5 = \dfrac{-6}{5} + \dfrac{3}{4} + 5 = \dfrac{-9}{20} + 5$

$= 4\dfrac{11}{20}$

RHS $= \dfrac{1}{20} - \dfrac{3(-3)}{2} = \dfrac{1}{20} + \dfrac{9}{2} = 4\dfrac{11}{20}$

Hence the solution $y = -3$ is correct.

Problem 11 Solve $\dfrac{3}{t-2} = \dfrac{4}{3t+4}$

By 'cross-multiplication': $3(3t + 4) = 4(t - 2)$
Removing brackets gives: $9t + 12 = 4t - 8$
Rearranging gives: $9t - 4t = -8 - 12$
i.e. $5t = -20$
                               $t = \dfrac{-20}{5} = -4$

Check: LHS $= \dfrac{3}{-4-2} = \dfrac{3}{-6} = -\dfrac{1}{2}$

RHS $= \dfrac{4}{3(-4)+4} = \dfrac{4}{-12+4} = \dfrac{4}{-8} = -\dfrac{1}{2}$

Hence the solution $t = -4$ is correct.

Problem 12 Solve $\sqrt{x} = 2$.

Wherever square root signs are involved with the unknown quantity, both sides of the equation must be squared. Hence

$$(\sqrt{x})^2 = (2)^2$$
i.e. $x = 4$

Problem 13 Solve $2\sqrt{d} = 8$.

To avoid possible errors it is usually best to arrange the term containing the square root on its own. Thus

$$\frac{2\sqrt{d}}{2} = \frac{8}{2}$$

i.e. $\sqrt{d} = 4$

Squaring both sides gives: $d = 16$, which may be checked in the original equation.

Problem 14 Solve $\left(\dfrac{\sqrt{b}+3}{\sqrt{b}}\right) = 2$

To remove the fraction each term is multiplied by $\sqrt{b}$. Hence

$$\sqrt{b}\left(\frac{\sqrt{b}+3}{\sqrt{b}}\right) = \sqrt{b}(2)$$

Cancelling gives: $\sqrt{b} + 3 = 2\sqrt{b}$
Rearranging gives: $3 = 2\sqrt{b} - \sqrt{b} = \sqrt{b}$
Squaring both sides gives: $9 = b$

Check: LHS $= \dfrac{\sqrt{9}+3}{\sqrt{9}} = \dfrac{3+3}{3} = \dfrac{6}{3} = 2 =$ RHS

Problem 15 Solve $x^2 = 25$.

This problem (and *Problem 16*) involves a square term and thus are not simple equations (they are, in fact, quadratic equations). However the solution of such equations is often required and is therefore included for completeness.

Whenever a square of the unknown is involved, the square root of both sides of the equation is taken. Hence

$$\sqrt{x^2} = \sqrt{25}$$
i.e. $x = 5$

However, $x = -5$ is also a solution of the equation because $(-5) \times (-5) = +25$.

Therefore, whenever the square root of a number is required there are always two answers, one positive, the other negative.

The solution of $x^2 = 25$ is thus written as $x = \pm 5$

Problem 16 Solve $\dfrac{15}{4t^2} = \dfrac{2}{3}$

'Cross-multiplying' gives: $15(3) = 2(4t^2)$

$$45 = 8t^2$$

$$\frac{45}{8} = t^2$$

i.e. $\qquad t^2 = 5\frac{5}{8} = 5.625$

Hence $t = \sqrt{(5.625)} = \pm 2.372$ correct to 4 significant figures.

Further problems involving simple equations may be found in section 5.4, Problems 1 to 44.

5.3 Practical problems involving simple equations

Problem 17 A copper wire has a length l of 1.5 km, a resistance R of 5 Ω and a resistivity ρ of 17.2×10^{-6} Ω mm. Find the cross-sectional areas, a, of the wire, given that $R = \rho l/a$.

Since $R = \rho l/a$ then

$$5 \ \Omega = \frac{(1.72 \times 10^{-6} \ \Omega \ \text{mm})(1\,500 \times 10^3 \ \text{mm})}{a}$$

From the units given, a is measured in mm². Thus

$$5a = 17.2 \times 10^{-6} \times 1500 \times 10^3$$

$$a = \frac{17.2 \times 10^{-6} \times 1500 \times 10^3}{5}$$

$$= \frac{17.2 \times 1500 \times 10^3}{10^6 \times 5} = \frac{17.2 \times \cancel{15}^{3}}{10 \times \cancel{5}_1} = 5.16$$

Hence the cross-sectional area of the wire is 5.16 mm².

Problem 18 A rectangular box with square ends has its length 15 cm greater than its breadth and the total length of its edges is 2.04 m. Find the width of the box and its volume.

Let x cm = width = height of box. Then the length of the box is $(x + 15)$ cm. The length of the edges of the box is

$$2(4x) + 4(x + 15) \ \text{cm}$$

Hence $\qquad 204 = 2(4x) + 4(x + 15)$

$$204 = 8x + 4x + 60$$

$$204 - 60 = 12x$$

i.e. $\qquad 144 = 12x$

and $\qquad x = 12$ cm

Hence the width of the box is 12 cm.

Volume of box = length × width × height

$$= (x + 15)(x)(x) = (27)(12)(12)$$

$$= \textbf{3888 cm}^3$$

Problem 19 The temperature coefficient of resistance α may be calculated from the formula $R_t = R_0(1 + \alpha t)$. Find α given $R_t = 0.928$, $R_0 = 0.8$ and $t = 40$.

Since $R_t = R_0(1 + \alpha t)$ then $0.928 = 0.8[1 + \alpha(40)]$

$$0.928 = 0.8 + (0.8)(\alpha)(40)$$

$$0.928 - 0.8 = 32\alpha$$

$$0.128 = 32\alpha$$

Hence $\qquad \alpha = \dfrac{0.128}{32} = \textbf{0.004}$

Problem 20 The distance s metres travelled in time t seconds is given by the formula $s = ut + \dfrac{1}{2}at^2$, where u is the initial velocity in m/s and a is the acceleration in m/s². Find the acceleration of the body if it travels 168 m in 6 s, with an initial velocity of 10 m/s.

$s = ut + \dfrac{1}{2}at^2$, and $s = 168$, $u = 10$ and $t = 6$

Hence $\qquad 168 = (10)(6) + \dfrac{1}{2}a(6)^2$

$$168 = 60 + 18a$$

$$168 - 60 = 18a$$

$$108 = 18a$$

$$a = \frac{108}{18} = 6$$

Hence the acceleration of the body is 6 m/s².

Problem 21 When three resistors in an electrical circuit are connected in parallel the total resistance R_T is given by:

$$\frac{1}{R_T} = \frac{1}{R_1} + \frac{1}{R_2} + \frac{1}{R_3}$$

Find the total resistance when $R_1 = 5 \ \Omega$, $R_2 = 10 \ \Omega$ and $R_3 = 30 \ \Omega$.

$$\frac{1}{R_T} = \frac{1}{5} + \frac{1}{10} + \frac{1}{30} = \frac{6 + 3 + 1}{30} = \frac{10}{30} = \frac{1}{3}$$

Taking the reciprocal of both sides gives: $\boldsymbol{R_T = 3 \ \Omega}$

Alternatively, if $\dfrac{1}{R_T} = \dfrac{1}{5} + \dfrac{1}{10} + \dfrac{1}{30}$ the LCM of the denominators is $30R_T$.

Hence $30R_T\left(\dfrac{1}{R_T}\right) = 30R_T\left(\dfrac{1}{5}\right) + 30R_T\left(\dfrac{1}{10}\right) + 30R_T\left(\dfrac{1}{30}\right)$

Cancelling gives: $\quad 30 = 6R_T + 3R_T + R_T$
i.e. $\qquad\qquad\quad 30 = 10R_T$

$$R_T = \frac{30}{10} = 3\,\Omega,\ \text{as above}$$

Problem 22 The extension x m of an aluminium tie bar of length l m and cross-sectional area A m^2 when carrying a load of F newtons is given by the modulus of elasticity $E = Fl/Ax$. Find the extension of the tie bar (in mm) if $E = 70 \times 10^9$ N/m^2, $F = 20 \times 10^6$ N, $A = 0.1$ m^2 and $l = 1.4$ m.

$E = Fl/Ax$. Hence $70 \times 10^9\ \dfrac{\text{N}}{\text{m}^2} = \dfrac{(20 \times 10^6\ \text{N})(1.4\ \text{m})}{(0.1\ \text{m}^2)(x)}$

(the unit of x is thus metres).

$$70 \times 10^9 \times 0.1 \times x = 20 \times 10^6 \times 1.4$$
$$x = \frac{20 \times 10^6 \times 1.4}{70 \times 10^9 \times 0.1}$$

Cancelling gives: $x = \dfrac{2 \times 1.4}{7 \times 100}\text{m} = \dfrac{2 \times 1.4}{7 \times 100} \times 1000\ \text{mm}$

Hence the extension of the tie bar, x = 4 mm.

Problem 23 Power in a d.c. circuit is given by $P = V^2/R$, where V is the supply voltage and R is the circuit resistance. Find the supply voltage if the circuit resistance is 1.25 Ω and the power measured is 320 W.

Since $P = \dfrac{V^2}{R}$ then $320 = \dfrac{V^2}{1.25}$
$$(320)(1.25) = V^2$$
i.e. $\qquad\qquad V^2 = 400$

Supply voltage, $\quad V = \sqrt{400} = \pm\mathbf{20V}$

Problem 24 A painter is paid £4.20 per hour for a basic 36 hour week, and overtime is paid at one and a third times this rate. Determine how many hours the painter has to work in a week to earn £212.80.

Basic rate per hour = £4.20; overtime rate per hour

$$= 1\frac{1}{3} \times £4.20 = £5.60$$

Let the number of overtime hours worked = x. Then

$$(36)(4.20) + (x)(5.60) = 212.80$$
$$151.2 + 5.6x = 212.80$$
$$5.6x = 212.80 - 151.2 = 61.6$$
$$x = \frac{61.6}{5.60} = 11$$

Thus 11 hours overtime would have to be worked to earn £212.80 per week. Hence the total number of hours worked is 36 + 11, i.e. **47 hours**.

Problem 25 A formula relating initial and final states of pressures, P_1 and P_2, volumes, V_1 and V_2, and absolute temperatures, T_1 and T_2, of an ideal gas is

$$\frac{P_1 V_1}{T_1} = \frac{P_2 V_2}{T_2}$$

Find the value of P_2 given $P_1 = 100 \times 10^3$, $V_1 = 1.0$, $V_2 = 0.266$, $T_1 = 423$ and $T_2 = 293$.

Since $\dfrac{P_1 V_1}{T_1} = \dfrac{P_2 V_2}{T_2}$ then $\dfrac{(100 \times 10^3)(1.0)}{423} = \dfrac{P_2(0.266)}{293}$

'Cross-multiplying' gives:

$$(100 \times 10^3)(1.0)(293) = P_2\,(0.266)(423)$$
$$P_2 = \frac{(100 \times 10^3)(1.0)(293)}{(0.266)(423)}$$

Hence $\qquad\qquad P_2 = \mathbf{260 \times 10^3}$ or $\mathbf{2.6 \times 10^5}$

Problem 26 The stress f in a material of a thick cylinder can be obtained from $\dfrac{D}{d} = \sqrt{\left(\dfrac{f+p}{f-p}\right)}$ Calculate the stress given that $D = 21.5$, $d = 10.75$ and $p = 1800$.

Since $\dfrac{D}{d} = \sqrt{\left(\dfrac{f+p}{f-p}\right)}$ then $\dfrac{21.5}{10.75} = \sqrt{\left(\dfrac{f+1800}{f-1800}\right)}$

i.e. $\qquad\qquad 2 = \sqrt{\left(\dfrac{f+1800}{f-1800}\right)}$

Squaring both sides gives: $\qquad 4 = \dfrac{f+1800}{f-1800}$

$$4(f - 1800) = f + 1800$$
$$4f - 7200 = f + 1800$$
$$4f - f = 1800 + 7200$$
$$3f = 9000$$
$$f = \frac{9000}{3} = 3000$$

Hence the stress, f, is 3000.

Problem 27 12 workmen employed on a building site earn between them a total of £2015 per week. Labourers are paid £158 per week and craftsmen are paid £175 per week. How many craftsmen and how many labourers are employed?

Let the number of craftsmen be c. The number of labourers are thus $(12 - c)$. The wage bill equation is:

$$175c + 158(12 - c) = 2015$$
$$175c + 1896 - 158c = 2015$$
$$175c - 158c = 2015 - 1896$$
$$17c = 119$$
$$c = \frac{119}{17} = 7$$

Hence there are 7 craftsmen and (12 − 7), i.e. 5 labourers on the site.

Further examples on practical problems involving simple equations may be found in section 5.4, Problems 45 to 58, page 48.

5.4 Further problems on simple equations

Simple equations

Solve the following equations:

1 $2x + 5 = 7$
[1]

2 $8 - 3t = 2$
[2]

3 $\frac{2}{3}c - 1 = 3$
[6]

4 $2x - 1 = 5x + 11$
[−4]

5 $7 - 4p = 2p - 3$
$[1\frac{2}{3}]$

6 $2.6x - 1.3 = 0.9x + 0.4$
[1]

7 $2a + 6 - 5a = 0$
[2]

8 $3x - 2 - 5x = 2x - 4$
$\left[\frac{1}{2}\right]$

9 $20d - 3 + 3d = 11d + 5 - 8$
[0]

10 $2(x - 1) = 4$
[3]

11 $16 = 4 (t + 2)$
[2]

12 $5(f - 2) - 3(2f + 5) + 15 = 0$
[−10]

13 $2x = 4(x - 3)$
[6]

14 $6(2 - 3y) - 42 = -2(y - 1)$
[−2]

15 $2(3g - 5) - 5 = 0$
$\left[2\frac{1}{2}\right]$

16 $4(3x + 1) = 7(x + 4) - 2(x + 5)$
[2]

17 $10 + 3(r - 7) = 16 - (r + 2)$
$\left[6\frac{1}{4}\right]$

18 $8 + 4(x - 1) - 5(x - 3) = 2(5 - 2x)$
[−3]

19 $\frac{1}{5}d + 3 = 4$
[5]

20 $2 + \frac{3}{4}y = 1 + \frac{2}{3}y + \frac{5}{6}$
[−2]

21 $\frac{1}{4}(2x - 1) + 3 = \frac{1}{2}$
$\left[-4\frac{1}{2}\right]$

22 $\frac{1}{5}(2f - 3) + \frac{1}{6}(f - 4) + \frac{2}{15} = 0$
[2]

23 $\frac{1}{3}(3m - 6) - \frac{1}{4}(5m + 4) + \frac{1}{5}(2m - 9) = -3$
[12]

24 $\frac{x}{3} - \frac{x}{5} = 2$
[15]

25 $1 - \frac{y}{3} = 3 + \frac{y}{3} - \frac{y}{6}$
[−4]

26 $\frac{2}{a} = \frac{3}{8}$
$\left[5\frac{1}{3}\right]$

27 $\frac{1}{3n} + \frac{1}{4n} = \frac{7}{24}$
[2]

28 $\frac{x + 3}{4} = \frac{x - 3}{5} + 2$
[13]

29 $\dfrac{3t}{20} = \dfrac{6-t}{12} + \dfrac{2t}{15} - \dfrac{3}{2}$

[−10]

30 $\dfrac{y}{5} + \dfrac{7}{20} = \dfrac{5-y}{4}$

[2]

31 $\dfrac{v-2}{2v-3} = \dfrac{1}{3}$

[3]

32 $\dfrac{2}{a-3} = \dfrac{3}{2a+1}$

[−11]

33 $\dfrac{1}{3m-2} + \dfrac{1}{5m+3} = 0$

$[-\dfrac{1}{8}]$

34 $\dfrac{x}{4} - \dfrac{x+6}{5} = \dfrac{x+3}{2}$

[−6]

35 $\dfrac{2c-3}{4} - \dfrac{1-c}{5} - 1 = \dfrac{2c+3}{3} + \dfrac{43}{60}$

[110]

36 $3\sqrt{t} = 9$

[9]

37 $2\sqrt{y} = 5$

$[6\dfrac{1}{4}]$

38 $4 = \sqrt{\left(\dfrac{3}{a}\right)} + 3$

[3]

39 $\dfrac{3\sqrt{x}}{1-\sqrt{x}} = -6$

[4]

40 $10 = 5\sqrt{\left(\dfrac{x}{2} - 1\right)}$

[10]

41 $16 = \dfrac{t^2}{9}$

[±12]

42 $\sqrt{\left(\dfrac{y+2}{y-2}\right)} = \dfrac{1}{2}$

$[-3\dfrac{1}{3}]$

43 $\dfrac{6}{a} = \dfrac{2a}{3}$

[±3]

44 $\dfrac{11}{2} = 5 + \dfrac{8}{x^2}$

[±4]

Practical problems involving simple equations

45 A formula used for calculating resistance of a cable is $R = (\rho l)/a$.
Given $R = 1.25$, $l = 2500$ and $a = 2 \times 10^{-4}$ find the value of ρ.
[10^{-7}]

46 Force F newtons is given by $F = ma$, where m is the mass in kilograms and a is the acceleration in metres per second squared. Find the acceleration when a force of 4 kN is applied to a mass of 500 kg.
[8 m/s^2]

47 $PV = mRT$ is the characteristic gas equation. Find the value of m when $P = 100 \times 10^3$, $V = 3.00$, $R = 288$ and $T = 300$.
[3.472]

48 When three resistors R_1, R_2 and R_3 are connected in parallel the total resistance R_T is determined from

$$\dfrac{1}{R_T} = \dfrac{1}{R_1} + \dfrac{1}{R_2} + \dfrac{1}{R_3}$$

(a) Find the total resistance when $R_1 = 3\,\Omega$, $R_2 = 6\,\Omega$ and $R_3 = 18\,\Omega$.
(b) Find the value of R_3 given that $R_T = 3\,\Omega$, $R_1 = 5\,\Omega$ and $R_2 = 10\,\Omega$.
[(a) 1.8 Ω (b) 30 Ω]

49 Five pens and two rulers cost 94p. If a ruler costs 5p more than a pen, find the cost of each.
[12p, 17p]

50 Ohm's law may be represented by $I = V/R$, where I is the current in amperes, V is the voltage in volts and R is the resistance in ohms. A soldering iron takes a current of 0.30 A from a 240 V supply. Find the resistance of the element.
[800 Ω]

51 A rectangle has a length of 20 cm and a width b cm. When its width is reduced by 4 cm its area becomes 160 cm^2. Find the original width and area of the rectangle.
[12 cm, 240 cm^2]

52 Given $R_2 = R_1(1 + \alpha t)$, find α given $R_1 = 5.0$, $R_2 = 6.03$ and $t = 51.5$.
[0.004]

53 If $v^2 = u^2 + 2as$, find u given $v = 24$, $a = -40$ and $s = 4.05$.
[30]

54 The relationship between the temperature on a Fahrenheit scale and that on a Celsius scale is given by $F = \dfrac{9}{5}C + 32$. Express 113°F in degrees Celsius.
[45 °C]

55 If $t = 2\pi\sqrt{(w/Sg)}$, find the value of S given $w = 1.219$, $g = 9.81$ and $t = 0.313\,2$.
[50]

56 Two joiners and five mates earn £1134 between them for a particular job. If a joiner earns £28 more than a mate, calculate the earnings for a joiner and for a mate.
[£182, £154]

57 An alloy contains 60% by weight of copper, the remainder being zinc. How much copper must be mixed with 50 kg of this alloy to give an alloy containing 75% copper?
[30 kg]

58 A rectangular laboratory has a length equal to one and a half times its width and a perimeter of 40 m. Find its length and width.
[12 m, 8 m]

6

Simultaneous equations

6.1 Introduction to simultaneous equations

Only one equation is necessary when finding the value of a **single unknown quantity** (as with simple equations in chapter 5).

When an equation contains **two unknown quantities** it has an infinite number of solutions. When two equations are available connecting the same two unknown values then a unique solution is possible. Similarly, for three unknown quantities it is necessary to have three equations in order to solve for a particular value of each of the unknown quantities, and so on.

Equations which have to be solved together to find the unique values of the unknown quantities, which are true for each of the equations, are called **simultaneous equations**.

There are two methods of solving simultaneous equations analytically: (a) by **substitution**, and (b) by **elimination**.

(A **graphical solution** of simultaneous equations is shown in Chapter 13, p. 114).

6.2 Worked problems on simultaneous equations in two unknowns

Problem 1 Solve the following equations for x and y, (a) by substitution, and (b) by elimination:

$$x + 2y = -1 \tag{1}$$
$$4x - 3y = 18 \tag{2}$$

(a) *By substitution*:
From equation (1): $x = -1 - 2y$
Substituting this expression for x into equation (2) gives:

$$4(-1 - 2y) - 3y = 18$$

This is now a simple equation in y.
Removing the bracket gives:

$$-4 - 8y - 3y = 18$$
$$-11y = 18 + 4 = 22$$
$$y = \frac{22}{-11} = -2$$

Substituting $y = -2$ into equation (1) gives:

$$x + 2(-2) = -1$$
$$x - 4 = -1$$
$$x = -1 + 4 = 3$$

Thus $x = 3$ and $y = -2$ is the solution to the simultaneous equations.

(Check: In equation (2), since $x = 3$ and $y = -2$,
LHS = $4(3) - 3(-2)$
= $12 + 6 = 18$ = RHS.)

(b) *By elimination*

$$x + 2y = -1 \tag{1}$$
$$4x - 3y = 18 \tag{2}$$

If equation (1) is multiplied throughout by 4 the coefficient of x will be the same as in equation (2), giving:

$$4x + 8y = -4 \tag{3}$$

Subtracting equation (3) from equation (2) gives:

$$4x - 3y = 18 \qquad (2)$$
$$\underline{4x + 8y = -4} \qquad (3)$$
$$0 - 11y = 22$$

Hence $y = \dfrac{22}{-11} = -2$

(Note, in the above subtraction, $18 - -4 = 18 + 4 = 22$.)
Substituting $y = -2$ into either equation (1) or equation (2) will give $x = 3$ as in method (a). The solution $x = 3$, $y = -2$ is the only pair of values that satisfies both of the original equations.

Problem 2 Solve, by a substitution method, the simultaneous equations

$$3x - 2y = 12 \qquad (1)$$
$$x + 3y = -7 \qquad (2)$$

From equation (2) $x = -7 - 3y$
Substituting for x in equation (1) gives:

$$3(-7 - 3y) - 2y = 12$$
i.e. $\qquad -21 - 9y - 2y = 12$
$$-11y = 12 + 21 = 33$$

Hence $\qquad y = \dfrac{33}{-11} = -3$

Substituting $y = -3$ in equation (2) gives:

$$x + 3(-3) = -7$$
i.e. $\qquad x - 9 = -7$
Hence $\qquad x = -7 + 9 = 2$

Thus $x = 2$, $y = -3$ is the solution of the simultaneous equations. (Such solutions should always be checked by substituting values into each of the original two equations.)

Problem 3 Use an elimination method to solve the simultaneous equations

$$3x + 4y = 5 \qquad (1)$$
$$2x - 5y = -12 \qquad (2)$$

If equation (1) is multiplied throughout by 2 and equation (2) by 3, then the coefficient of x will be the same in the newly formed equations. Thus

2 × equation (1) gives: $\qquad 6x + 8y = 10 \qquad (3)$
3 × equation (2) gives: $\qquad 6x - 15y = -36 \qquad (4)$
Equation (3) − equation (4) gives: $0 + 23y = 46$
i.e. $\qquad y = \dfrac{46}{23} = 2$

(Note $+8y - -15y = 8y + 15y = 23y$ and $10 - -36 = 10 + 36 = 46$. Alternatively, 'change the signs of the bottom line and add'.)

Substituting $y = 2$ in equation (1) gives: $3x + 4(2) = 5$, from which $\qquad 3x = 5 - 8 = -3$
and $\qquad x = -1$

Checking in equation (2), left-hand side $= 2(-1) - 5(2) = -2 - 10 = -12 = $ right-hand side.

Hence $x = -1$ and $y = 2$ is the solution of the simultaneous equations.

The elimination method is the most common method of solving simultaneous equations.

Problem 4 Solve $7x - 2y = 26 \qquad (1)$
$\qquad \qquad \qquad 6x + 5y = 29 \qquad (2)$

When equation (1) is multiplied by 5 and equation (2) by 2 the coefficients of y in each equation are numerically the same, i.e. 10, but are of opposite sign.

5 × equation (1) gives: $\qquad 35x - 10y = 130 \quad (3)$
2 × equation (2) gives: $\qquad \underline{12x + 10y = 58} \quad (4)$
Adding equations (3) and (4) gives: $47x + 0 = 188$

Hence $x = \dfrac{188}{47} = 4$

(Note that when the signs of common coefficients are **different** the two equations are **added**, and when the signs of common coefficients are the **same** the two equations are **subtracted** (as in *Problems 1* and *3*).)
Substituting $x = 4$ into equation (1) gives:

$$7(4) - 2y = 26$$
$$28 - 2y = 26$$
$$28 - 26 = 2y$$
$$2 = 2y$$
Hence $\qquad y = 1$

Checking, by substituting $x = 4$, $y = 1$ into equation (2), gives:

$$\text{LHS} = 6(4) + 5(1) = 24 + 5 = 29 = \text{RHS}$$

Thus the solution is $x = 4$, $y = 1$, since these values maintain the equality when substituted in both equations.

Problem 5 Solve $3p = 2q \qquad (1)$
$\qquad \qquad \qquad 4p + q + 11 = 0 \qquad (2)$

Rearranging gives:

$$3p - 2q = 0 \qquad (3)$$
$$4p + q = -11 \qquad (4)$$

Multiplying equation (4) by 2 gives:

$$8p + 2q = -22 \qquad (5)$$

Adding equations (3) and (5) gives:

$$11p + 0 = -22$$
$$p = \frac{-22}{11} = -2$$

Substituting $p = -2$ into equation (1) gives:

$$3(-2) = 2q$$
$$-6 = 2q$$
$$q = \frac{-6}{2} = -3$$

Checking, by substituting $p = -2$ and $q = -3$ into equation (2), gives:

$$\text{LHS} = 4(-2) + (-3) + 11 = -8 - 3 + 11 = 0$$
$$= \text{RHS}$$

Hence the solution is $p = -2$, $q = -3$.

Problem 6 Solve $\dfrac{x}{8} + \dfrac{5}{2} = y$ $\qquad (1)$

$$13 - \frac{y}{3} = 3x \qquad (2)$$

Whenever fractions are involved in simultaneous equations it is usual first to remove them. Thus, multiplying equation (1) by 8 gives:

$$8\left(\frac{x}{8}\right) + 8\left(\frac{5}{2}\right) = 8y$$

i.e. $x + 20 = 8y$ $\qquad (3)$

Multiplying equation (2) by 3 gives:

$$39 - y = 9x \qquad (4)$$

Rearranging equations (3) and (4) gives:

$$x - 8y = -20 \qquad (5)$$
$$9x + y = 39 \qquad (6)$$

Multiplying equation (6) by 8 gives:

$$72x + 8y = 312 \qquad (7)$$

Adding equations (5) and (7) gives:

$$73x + 0 = 292$$
$$x = \frac{292}{73} = 4$$

Substituting $x = 4$ into equation (5) gives:

$$4 - 8y = -20$$
$$4 + 20 = 8y$$
$$24 = 8y$$
$$y = \frac{24}{8} = 3$$

Checking, substituting $x = 4$, $y = 3$ in the original equations, gives:

Equation (1):

$$\text{LHS} = \frac{4}{8} + \frac{5}{2} = \frac{1}{2} + 2\frac{1}{2} = 3 = y = \text{RHS}$$

Equation (2):

$$\text{LHS} = 13 - \frac{3}{3} = 13 - 1 = 12$$
$$\text{RHS} = 3x = 3(4) = 12$$

Hence the solution is $x = 4$, $y = 3$.

Problem 7 Solve $2.5x + 0.75 - 3y = 0$
$$1.6x = 1.08 - 1.2y$$

It is often easier initially to remove decimal fractions. Thus multiplying equations (1) and (2) by 100 gives:

$$250x + 75 - 300y = 0 \qquad (1)$$
$$160x = 108 - 120y \qquad (2)$$

Rearranging gives:

$$250x - 300y = -75 \qquad (3)$$
$$160x + 120y = 108 \qquad (4)$$

Multiplying equation (3) by 2 gives:

$$500x - 600y = -150 \qquad (5)$$

Multiplying equation (4) by 5 gives:

$$800x + 600y = 540 \qquad (6)$$

Adding equations (5) and (6) gives:

$$1300x + 0 = 390$$
$$x = \frac{390}{1300} = \frac{39}{130} = \frac{3}{10} = 0.3$$

Substituting $x = 0.3$ into equation (1) gives:

$$250(0.3) + 75 - 300y = 0$$
$$75 + 75 = 300y$$
$$150 = 300y$$
$$y = \frac{150}{300} = 0.5$$

Checking $x = 0.3$, $y = 0.5$ in equation (2) gives:

$$\text{LHS} = 160(0.3) = 48 \qquad \text{RHS} = 108 - 120(0.5)$$
$$= 108 - 60 = 48$$

Hence the solution is $x = 0.3$, $y = 0.5$.

Problem 8   Solve $\dfrac{2}{x} + \dfrac{3}{y} = 7$ (1)

$$\frac{1}{x} - \frac{4}{y} = -2 \qquad (2)$$

In this type of equation a substitution can initially be made.

Let $\dfrac{1}{x} = a$ and $\dfrac{1}{y} = b$

Thus equation (1) becomes:

$$2a + 3b = 7 \qquad (3)$$

and equation (2) becomes:

$$a - 4b = -2 \qquad (4)$$

Multiplying equation (4) by 2 gives:

$$2a - 8b = -4 \qquad (5)$$

Subtracting equation (5) from equation (3) gives:

$$0 + 11b = 11$$
i.e. $\qquad b = 1$

Substituting $b = 1$ into equation (3) gives:

$$2a + 3 = 7$$
$$2a = 7 - 3 = 4$$
i.e. $\qquad a = 2$

Checking, substituting $a = 2$, $b = 1$ in equation (4), gives:

$$\text{LHS} = 2 - 4(1) = 2 - 4 = -2 = \text{RHS}$$

Hence $a = 2$, $b = 1$

However, since $\dfrac{1}{x} = a$ then $x = \dfrac{1}{a} = \dfrac{1}{2}$

and since $\qquad \dfrac{1}{y} = b$ then $y = \dfrac{1}{b} = \dfrac{1}{1} = 1$

Hence the solution is $x = \dfrac{1}{2}$, $y = 1$, which may be checked in the original equations.

Problem 9   Solve $\dfrac{1}{2a} + \dfrac{3}{5b} = 4$ (1)

$$\frac{4}{a} + \frac{1}{2b} = 10.5 \qquad (2)$$

Let $\dfrac{1}{a} = x$ and $\dfrac{1}{b} = y$

Then $\dfrac{x}{2} + \dfrac{3}{5}y = 4$ (3)

$$4x + \frac{1}{2}y = 10.5 \qquad (4)$$

To remove fractions, equation (3) is multiplied by 10 giving:

$$10\left(\frac{x}{2}\right) + 10\left(\frac{3}{5}y\right) = 10(4)$$

i.e. $\quad 5x + 6y = 40$ (5)

Multiplying equation (4) by 2 gives:

$$8x + y = 21 \qquad (6)$$

Multiplying equation (6) by 6 gives:

$$48x + 6y = 126 \qquad (7)$$

Subtracting equation (5) from equation (7) gives:

$$43x + 0 = 86$$
$$x = \frac{86}{43} = 2$$

Substituting $x = 2$ into equation (3) gives:

$$\frac{2}{2} + \frac{3}{5}y = 4$$

$$\frac{3}{5}y = 4 - 1 = 3$$

$$y = \frac{5}{3}(3) = 5$$

Since $\dfrac{1}{a} = x$ then $a = \dfrac{1}{x} = \dfrac{1}{2}$

and since $\dfrac{1}{b} = y$ then $b = \dfrac{1}{y} = \dfrac{1}{5}$

Hence the solution is $a = \dfrac{1}{2}, b = \dfrac{1}{5}$, which may be checked in the original equations.

Problem 10   Solve $\dfrac{1}{x+y} = \dfrac{4}{27}$ (1)

$\dfrac{1}{2x-y} = \dfrac{4}{33}$ (2)

To eliminate fractions both sides of equation (1) are multiplied by $27(x + y)$ giving

$$27(x + y)\dfrac{1}{x+y} = 27(x + y)\dfrac{4}{27}$$

i.e. $27(1) = 4(x + y)$

$\quad 27 \quad = 4x + 4y$ (3)

Similarly, in equation (2) $33 = 4(2x - y)$
i.e. $33 = 8x - 4y$ (4)

Equation (3) + equation (4) gives:

$$60 = 12x, \text{ i.e. } x = \dfrac{60}{12} = 5$$

Substituting $x = 5$ in equation (3) gives:

$\qquad\qquad 27 = 4(5) + 4y$
from which $4y = 27 - 20 = 7$
and $\qquad\qquad y = \dfrac{7}{4} = 1\dfrac{3}{4}$

Hence $x = 5, y = 1\dfrac{3}{4}$ is the required solution, which may be checked in the original equations.

Problem 11   Solve $\dfrac{x-1}{3} + \dfrac{y+2}{5} = \dfrac{2}{15}$ (1)

$\dfrac{1-x}{6} + \dfrac{5+y}{2} = \dfrac{5}{6}$ (2)

Before equations (1) and (2) can be simultaneously solved, the fractions need to be removed and the equations rearranged.

Multiplying equation (1) by 15 gives:

$$15\left(\dfrac{x-1}{3}\right) + 15\left(\dfrac{y+2}{5}\right) = 15\left(\dfrac{2}{15}\right)$$

i.e. $5(x - 1) + 3(y + 2) = 2$
$\qquad\qquad 5x - 5 + 3y + 6 = 2$
$\qquad\qquad\qquad 5x + 3y = 2 + 5 - 6$
Hence $\qquad\qquad\quad 5x + 3y = 1$ (3)

Multiplying equation (2) by 6 gives:

$$6\left(\dfrac{1-x}{6}\right) + 6\left(\dfrac{5+y}{2}\right) = 6\left(\dfrac{5}{6}\right)$$

i.e. $(1 - x) + 3(5 + y) = 5$
$\qquad\qquad 1 - x + 15 + 3y = 5$
$\qquad\qquad\qquad -x + 3y = 5 - 1 - 15$
Hence $\qquad\qquad\quad -x + 3y = -11$ (4)

Thus the initial problem containing fractions can be expressed as:

$5x + 3y = 1$ (3)
$-x + 3y = -11$ (4)

Subtracting equation (4) from equation (3) gives:

$\qquad\qquad 6x + 0 = 12$
$\qquad\qquad\quad x = \dfrac{12}{6} = 2$

Substituting $x = 2$ into equation (3) gives:

$\qquad\qquad 5(2) + 3y = 1$
$\qquad\qquad\quad 10 + 3y = 1$
$\qquad\qquad\qquad 3y = 1 - 10 = -9$
$\qquad\qquad\qquad\quad y = \dfrac{-9}{3} = -3$

Checking, substituting $x = 2, y = -3$ in equation (4), gives:

$$\text{LHS} = -2 + 3(-3) = -2 - 9 = -11 = \text{RHS}$$

Hence the solution is $x = 2, y = -3$, which may be checked in the original equations.

Further problems on simultaneous equations may be found in section 6.4, Problems 1 to 27, page 57.

6.3 Practical problems involving simultaneous equations in two unknowns

---

Problem 12 The law connecting friction F and load L for an experiment is of the form $F = aL + b$, where a and b are constants. When $F = 5.6$, $L = 8.0$ and when $F = 4.4$, $L = 2.0$. Find the values of a and b and the value of F when $L = 6.5$.

Substituting $F = 5.6$, $L = 8.0$ into $F = aL + b$ gives:

$$5.6 = 8.0a + b \tag{1}$$

Substituting $F = 4.4$, $L = 2.0$ into $F = aL + b$ gives:

$$4.4 = 2.0a + b \tag{2}$$

Subtracting equation (2) from equation (1) gives:

$$1.2 = 6.0a$$
$$a = \frac{1.2}{6.0} = \frac{1}{5}$$

Substituting $a = \dfrac{1}{5}$ into equation (1) gives:

$$5.6 = 8.0\left(\frac{1}{5}\right) + b$$
$$5.6 = 1.6 + b$$
$$5.6 - 1.6 = b$$

i.e. $\qquad b = 4$

Checking, substituting $a = \dfrac{1}{5}$, $b = 4$ in equation (2), gives:

$$\text{RHS} = 2.0\left(\frac{1}{5}\right) + 4 = 0.4 + 4 = 4.4 = \text{LHS}$$

Hence $a = \dfrac{1}{5}$ and $b = 4$.

When $L = 6.5$, $F = aL + b = \dfrac{1}{5}(6.5) + 4 = 1.3 + 4$, i.e. $F = 5.30$.

Problem 13 The equation of a straight line, of slope m and intercept on the y-axis c, is $y = mx + c$. If a straight line passes through the point where $x = 1$ and $y = -2$, and also through the point where $x = 3\frac{1}{2}$ and $y = 10\frac{1}{2}$, find the values of the slope and the y-axis intercept.

Substituting $x = 1$ and $y = -2$ into $y = mx + c$ gives:

$$-2 = m + c \tag{1}$$

Substituting $x = 3\frac{1}{2}$ and $y = 10\frac{1}{2}$ into $y = mx + c$ gives:

$$10\frac{1}{2} = 3\frac{1}{2}m + c \tag{2}$$

Subtracting equation (1) from equation (2) gives:

$$12\frac{1}{2} = 2\frac{1}{2}m$$

$$m = \frac{12\frac{1}{2}}{2\frac{1}{2}} = 5$$

Substituting $m = 5$ into equation (1) gives:

$$-2 = 5 + c$$
$$c = -2 - 5 = -7$$

Checking, substituting $m = 5$, $c = -7$ in equation (2), gives:

$$\text{RHS} = \left(3\frac{1}{2}\right)(5) + (-7) = 17\frac{1}{2} - 7 = 10\frac{1}{2} = \text{LHS}$$

Hence the slope, $m = 5$ and the y-axis intercept, $c = -7$.

Problem 14 When Kirchhoff's laws are applied to a particular electrical circuit the currents I_1 and I_2 are connected by the equations:

$$27 = 1.5I_1 + 8(I_1 - I_2) \tag{1}$$
$$-26 = 2I_2 - 8(I_1 - I_2) \tag{2}$$

Solve the equations to find the values of currents I_1 and I_2.

Removing brackets from equation (1) gives:

$$27 = 1.5I_1 + 8I_1 - 8I_2$$

Rearranging gives:

$$9.5I_1 - 8I_2 = 27 \tag{3}$$

Removing brackets from equation (2) gives:

$$-26 = 2I_2 - 8I_1 + 8I_2$$

Rearranging gives:

$$-8I_1 + 10I_2 = -26 \qquad (4)$$

Multiplying equation (3) by 5 gives:

$$47.5I_1 - 40I_2 = 135 \qquad (5)$$

Multiplying equation (4) by 4 gives:

$$-32I_1 + 40I_2 = -104 \qquad (6)$$

Adding equations (5) and (6) gives:

$$15.5I_1 + 0 = 31$$
$$I_2 = \frac{31}{15.5} = 2$$

Substituting $I_1 = 2$ into equation (3) gives:

$$9.5(2) - 8I_2 = 27$$
$$19 - 8I_2 = 27$$
$$19 - 27 = 8I_2$$
$$-8 = 8I_2$$
$$I_2 = -1$$

Hence the solution is $I_1 = 2$ and $I_2 = -1$ (which may be checked in the original equations).

Problem 15 The distance s metres from a fixed point of a vehicle travelling in a straight line with constant acceleration, a m/s², is given by $s = ut + \frac{1}{2}at^2$, where u is the initial velocity in m/s and t the time in seconds. Determine the initial velocity and the acceleration given that $s = 42$ m when $t = 2$ s and $s = 144$ m when $t = 4$ s. Find also the distance travelled after 3 s.

Substituting $s = 42$, $t = 2$ into $s = ut + \frac{1}{2}at^2$ gives:

$$42 = 2u + \frac{1}{2}a(2)^2$$

i.e. $42 = 2u + 2a \qquad (1)$

Substituting $s = 144$, $t = 4$ into $s = ut + \frac{1}{2}at^2$ gives:

$$144 = 4u + \frac{1}{2}a(4)^2$$

i.e. $144 = 4u + 8a \qquad (2)$

Multiplying equation (1) by 2 gives:

$$84 = 4u + 4a \qquad (3)$$

Subtracting equation (3) from equation (2) gives:

$$60 = 0 + 4a$$
$$a = \frac{60}{4} = 15$$

Substituting $a = 15$ into equation (1) gives:

$$42 = 2u + 2(15)$$
$$42 - 30 = 2u$$
$$u = \frac{12}{2} = 6$$

Substituting $a = 15$, $u = 6$ in equation (2) gives:

$$\text{RHS} = 4(6) + 8(15) = 24 + 120 = 144 = \text{LHS}$$

Hence the initial velocity, $u = 6$ m/s and the acceleration, $a = 15$ m/s².

Distance travelled after 3 s is given by $s = ut + \frac{1}{2}at^2$, where $t = 3$, $u = 6$ and $a = 15$

Hence $s = 6(3) + \frac{1}{2}(15)(3)^2$

$$= 18 + 67\frac{1}{2}$$

i.e. **distance travelled after 3 s = $85\frac{1}{2}$ m.**

Problem 16 A craftsman and 4 labourers together earn £737 basic per week, whilst 4 craftsmen and 9 labourers together earn £1982 basic per week. Determine the basic weekly wage of a craftsman and a labourer.

Let C represent the wage of a craftsman and L that of a labourer. Thus

$$C + 4L = 737 \qquad (1)$$
$$4C + 9L = 1982 \qquad (2)$$

Multiplying equation (1) by 4 gives:

$$4C + 16L = 2948 \qquad (3)$$

Subtracting equation (2) from equation (3) gives:

$$7L = 966$$
$$L = \frac{966}{7} = 138$$

Substituting $L = 138$ into equation (1) gives:

$$C + 4(138) = 737$$
$$C + 552 = 737$$
$$C = 737 - 552 = 185$$

Checking, substituting $C = 185$, $L = 138$ in equation (2), gives:

$$\text{LHS} = 4(185) + 9(138) = 740 + 1242 = 1982$$
$$= \text{RHS}$$

Thus the solution is that the basic weekly wage of a craftsman is £185 and that of a labourer is £138.

Problem 17 The resistance R Ω of a length of wire at $t°C$ is given by $R = R_0(1 + \alpha t)$, where R_0 is the resistance at $0°C$ and α is the temperature coefficient of resistance in $/°C$. Find the values of α and R_0 if $R = 30$ ohms at $50°C$ and $R = 35$ Ω at $100°C$.

Substituting $R = 30$, $t = 50$ into $R = R_0(1 + \alpha t)$ gives:

$$30 = R_0(1 + 50\alpha) \tag{1}$$

Substituting $R = 35$, $t = 100$ into $R = R_0(1 + \alpha t)$ gives:

$$35 = R_0(1 + 100\alpha) \tag{2}$$

Although these equations may be solved by the conventional substitution method, an easier way is to eliminate R_0 by division. Thus, dividing equation (1) by equation (2) gives:

$$\frac{30}{35} = \frac{R_0(1 + 50\alpha)}{R_0(1 + 100\alpha)} = \frac{1 + 50\alpha}{1 + 100\alpha}$$

'Cross-multiplying' gives:

$$30(1 + 100\alpha) = 35(1 + 50\alpha)$$
$$30 + 3000\alpha = 35 + 1750\alpha$$
$$3000\alpha - 1750\alpha = 35 - 30$$
$$1250\alpha = 5$$

i.e.
$$\alpha = \frac{5}{1250} = \frac{1}{250} \text{ or } 0.004$$

Substituting $\alpha = \frac{1}{250}$ into equation (1) gives:

$$30 = R_0\left[1 + \frac{1}{250}(50)\right]$$
$$30 = R_0(1.2)$$
$$R_0 = \frac{3.0}{1.2} = 2.5$$

Checking, substituting $\alpha = \frac{1}{250}$, $R_0 = 25$ in equation (2), gives:

$$\text{RHS} = 25\left[1 + \frac{1}{250}(100)\right] = 25(1.4) = 35 = \text{LHS}$$

Thus the solution is $\alpha = 0.004/°C$ and $R_0 = 25$ Ω.

Problem 18 The molar heat capacity of a solid compound is given by the equation $c = a + bT$, where a and b are constants. When $c = 52$, $T = 100$ and when $c = 172$, $T = 400$. Determine the values of a and b.

When $c = 52$, $T = 100$. Hence

$$52 = a + 100b \tag{1}$$

When $c = 172$, $T = 400$. Hence

$$172 = a + 400b \tag{2}$$

Equation (2) − equation (1) gives:

$$120 = 300b$$

from which

$$b = \frac{120}{300} = 0.4$$

Substituting $b = 0.4$ in equation (1) gives:

$$52 = a + 100(0.4)$$
$$a = 52 - 40 = 12$$

Hence **$a = 12$ and $b = 0.4$**.

Further examples on practical problems involving simultaneous equations in two unknowns may be found in section 6.4, Problems 28 to 35, page 58.

6.4 Further problems involving simultaneous equations in two unknowns

In *Problems 1* to 27, solve the simultaneous equations and verify the results.

1   $a + b = 7$
    $a - b = 3$
    $[a = 5, b = 2]$

2   $2x + 5y = 7$
   $x + 3y = 4$
   $[x = 1, y = 1]$

3   $3s + 2t = 12$
   $4s - t = 5$
   $[s = 2, t = 3]$

4   $3x - 2y = 13$
   $2x + 5y = -4$
   $[x = 3, y = -2]$

5   $5m - 3n = 11$
   $3m + n = 8$
   $[m = 2\frac{1}{2}, n = \frac{1}{2}]$

6   $8a - 3b = 51$
   $3a + 4b = 14$
   $[a = 6, b = -1]$

7   $5x = 2y$
   $3x + 7y = 41$
   $[x = 2, y = 5]$

8   $5c = 1 - 3d$
   $2d + c + 4 = 0$
   $[c = 2, d = -3]$

9   $7p + 11 + 2q = 0$
   $-1 = 3q - 5p$
   $[p = -1, q = -2]$

10   $\dfrac{x}{2} + \dfrac{y}{3} = 4; \quad \dfrac{x}{6} - \dfrac{y}{9} = 0$
   $[x = 4, y = 6]$

11   $\dfrac{a}{2} - 7 = -2b; \quad 12 = 5a + \dfrac{2}{3}b$
   $[a = 2, b = 3]$

12   $\dfrac{3}{2}s - 2t = 8; \quad \dfrac{s}{4} + 3t = -2$
   $[s = 4, t = -1]$

13   $\dfrac{x}{5} + \dfrac{2y}{3} = \dfrac{49}{15}; \quad \dfrac{3x}{7} - \dfrac{y}{2} + \dfrac{5}{7} = 0$
   $[x = 3, y = 4]$

14   $v - 1 = \dfrac{u}{12}; \quad u + \dfrac{v}{4} - \dfrac{25}{2} = 0$
   $[u = 12, v = 2]$

15   $1.5x - 2.2y = -18$
   $2.4x + 0.6y = 33$
   $[x = 10, y = 15]$

16   $3b - 2.5a = 0.45$
   $1.6a + 0.8b = 0.8$
   $[a = 0.30, b = 0.40]$

17   $10.1 + 1.7y = 0.8x$
   $2.5x + 1.5 + 1.3y = 0$
   $[x = 2, y = -5]$

18   $0.4b - 0.7 = 0.5a$
   $1.2a - 3.6 = 0.3b$
   $[a = 5, b = 8]$

19   $2.30c - 1.70d = 9.11$
   $3.68 + 8.80c + 4.20d = 0$
   $[c = 1.3, d = -3.6]$

20   $\dfrac{3}{x} + \dfrac{2}{y} = 14; \quad \dfrac{5}{x} - \dfrac{3}{y} = -2$
   $\left[x = \dfrac{1}{2}, y = \dfrac{1}{4}\right]$

21   $\dfrac{4}{a} - \dfrac{3}{b} = 18; \quad \dfrac{2}{a} + \dfrac{5}{b} = -4$
   $\left[a = \dfrac{1}{3}, b = -\dfrac{1}{2}\right]$

22   $\dfrac{1}{2p} + \dfrac{3}{5q} = 5; \quad \dfrac{5}{p} - \dfrac{1}{2q} = \dfrac{35}{2}$
   $\left[p = \dfrac{1}{4}, q = \dfrac{1}{5}\right]$

23   $\dfrac{5}{x} + \dfrac{3}{y} = 1.1; \quad \dfrac{3}{x} - \dfrac{7}{y} = -1.1$
   $[x = 10, y = 5]$

24   $\dfrac{c+1}{4} - \dfrac{d+2}{3} + 1 = 0; \quad \dfrac{1-c}{5} + \dfrac{3-d}{4} + \dfrac{13}{20} = 0$
   $[c = 3, d = 4]$

25   $\dfrac{3r+2}{5} - \dfrac{2s-1}{4} = \dfrac{11}{5}; \quad \dfrac{3+2r}{4} + \dfrac{5-s}{3} = \dfrac{15}{4}$
   $\left[r = 3, s = \dfrac{1}{2}\right]$

26   $\dfrac{5}{x+y} = \dfrac{20}{27}; \quad \dfrac{4}{2x-y} = \dfrac{16}{33}$
   $\left[x = 5, y = 1\dfrac{3}{4}\right]$

27   If $5x - \dfrac{3}{y} = 1$ and $x + \dfrac{4}{y} = \dfrac{5}{2}$ find the value of $\dfrac{xy+1}{y}$
   [1]

Practical problems involving simultaneous equations in two unknowns

28   In a system of pulleys, the effort P required to raise a load W is given by $P = aW + b$, where a and b are constants. If $W = 40$ when $P = 12$ and $W = 90$ when $P = 22$, find the values of a and b.
   $\left[a = \dfrac{1}{5}, b = 4\right]$

29   Applying Kirchhoff's laws to an electrical circuit produces the following equations:

$$5 = 0.2I_1 + 2(I_1 - I_2)$$
$$12 = 3I_2 + 0.4I_2 - 2(I_1 - I_2)$$

Determine the values of currents I_1 and I_2.

$[I_1 = 6.47, I_2 = 4.62]$

30 Velocity v is given by the formula $v = u + at$. If $v = 20$ when $t = 2$ and $v = 40$ when $t = 7$ find the values of u and a. Hence find the velocity when $t = 3.5$.

$[u = 12; a = 4, v = 26]$

31 Three new cars and 4 new vans supplied to a dealer together cost £83 700 and 5 new cars and 2 new vans of the same models cost £89 100. Find the cost of a car and a van.

[£13 500, £10 800]

32 $y = mx + c$ is the equation of a straight line of slope m and y-axis intercept c. If the line passes through the point where $x = 2$ and $y = 2$, and also through the point where $x = 5$ and $y = \dfrac{1}{2}$, find the slope and y-axis intercept of the straight line.

$$\left[m = -\frac{1}{2}, c = 3 \right]$$

33 The resistance R ohms of copper wire at $t°C$ is given by $R = R_0(1 + \alpha t)$, where R_0 is the resistance at $0°C$ and α is the temperature coefficient of resistance. If $R = 25.44 \ \Omega$ at $30°C$ and $R = 32.17 \ \Omega$ at $100°C$, find α and R_0.

$[\alpha = 0.004 \ 26, R_0 = 22.56 \ \Omega]$

34 The molar heat capacity of a solid compound is given by the equation $c = a + bT$. When $c = 52$, $T = 100$ and when $c = 172$, $T = 400$. Find the values of a and b.

$[a = 12, b = 0.40]$

35 In an engineering process two variables p and q are related by: $q = ap + b/p$, where a and b are constants. Evaluate a and b if $q = 13$ when $p = 2$ and $q = 22$ when $p = 5$.

$[a = 4, b = 10]$

7

Evaluation and transposition of formulae

7.1 Evaluation of formulae

The statement $v = u + at$ is said to be a **formula** for v in terms of u, a and t.

v, u, a and t are called **symbols**.

The single term on the left-hand side of the equation, v, is called the **subject of the formula**.

Provided values are given for all the symbols in a formula except one, the remaining symbol can be made the subject of the formula and may be evaluated by using a calculator.

Problem 1 In an electrical circuit the voltage V is given by Ohm's law, i.e. $V = IR$. Find, correct to 4 significant figures, the voltage when $I = 5.36$ A and $R = 14.76 \, \Omega$.

$$V = IR = (5.36)(14.76)$$

Hence **voltage V = 79.11 V, correct to 4 significant figures**.

Problem 2 The surface area A of a hollow cone is given by $A = \pi rl$. Determine the surface area when $r = 3.0$ cm, $l = 8.5$ cm and $\pi = 3.14$.

$$A = \pi rl = (3.14)(3.0)(8.5) \text{ cm}^2$$

Hence **surface area $A = 80.07$ cm^2.**

Problem 3 Velocity v is given by $v = u + at$. If $u = 9.86$ m/s, $a = 4.25$ m/s^2 and $t = 6.84$ s, find v, correct to 3 significant figures.

$$\begin{aligned} v = u + at &= 9.86 + (4.25)(6.84) \\ &= 9.86 + 29.07 \\ &= 38.93 \end{aligned}$$

Hence **velocity v = 38.9 m/s, correct to 3 significant figures.**

Problem 4 The area, A, of a circle is given by $A = \pi r^2$. Determine the area correct to 2 decimal places, given $\pi = 3.142$ and $r = 5.23$ m.

$$\begin{aligned} A = \pi r^2 &= (3.142)(5.23)^2 \\ &= (3.142)(27.352\,9) \end{aligned}$$

Hence **area, A = 85.94 m^2, correct to 2 decimal places.**

Problem 5 The power P watts dissipated in an electrical circuit may be expressed by the formula $P = V^2/R$. Evaluate the power, correct to 3 significant figures, given that $V = 17.48$ V and $R = 36.12 \, \Omega$.

$$P = \frac{V^2}{R} = \frac{(17.48)^2}{36.12} = \frac{305.6}{36.12}$$

Hence **power, P = 8.46 W, correct to 3 significant figures.**

Problem 6 The volume V cm^3 of a right circular cone is given by $V = \frac{1}{3}\pi r^2 h$. Given that $r = 4.321$ cm, $h = 18.35$ cm and $\pi = 3.142$, find the volume correct to 4 significant figures.

$$V = \frac{1}{3}\pi r^2 h = \frac{1}{3}(3.142)(4.321)^2(18.35)$$

$$= \frac{1}{3}(3.142)(18.671)(18.35)$$

Hence volume, $V = 358.8$ cm^3, correct to 4 significant figures.

Problem 7 Force F newtons is given by the formula $F = (Gm_1 m_2)/d^2$, where m_1 and m_2 are masses, d their distance apart and G is a constant. Find the value of the force given that $G = 6.67 \times 10^{-11}$, $m_1 = 7.36$, $m_2 = 15.5$ and $d = 22.6$. Express the answer in standard form, correct to 3 significant figures.

$$F = \frac{Gm_1 m_2}{d^2} = \frac{(6.67 \times 10^{-11})(7.36)(15.5)}{(22.6)^2}$$

$$= \frac{(6.67)(7.36)(15.5)}{(10^{11})(510.8)} = \frac{1.490}{10^{11}}$$

Hence force $F = 1.49 \times 10^{-11}$ newtons, correct to 3 significant figures.

Problem 8 The time of swing, t seconds, of a simple pendulum is given by $t = 2\pi\sqrt{(l/g)}$. Determine the time, correct to 3 decimal places, given that $\pi = 3.142$, $l = 12.0$ and $g = 9.81$.

$$t = 2\pi\sqrt{\left(\frac{l}{g}\right)} = (2)(3.142)\sqrt{\left(\frac{12.0}{9.81}\right)}$$

$$= (2)(3.142)\sqrt{(1.223)}$$
$$= (2)(3.142)(1.106)$$

Hence time $t = 6.950$ seconds, correct to 3 decimal places.

Problem 9 Resistance, R Ω, varies with temperature according to the formula $R = R_0(1 + \alpha t)$. Evaluate R, correct to 3 significant figures, given $R_0 = 14.59$, $\alpha = 0.004\,3$ and $t = 80$.

$$R = R_0(1 + \alpha t) = 14.59[1 + (0.004\,3)(80)]$$
$$= 14.59(1 + 0.344)$$
$$= 14.59(1.344)$$

Hence resistance, $R = 19.6$ Ω, correct to 3 significant figures.

Further problems on evaluating formulae may be found in section 7.3, Problems 1 to 12, page 67.

7.2 Transposition of formulae

When a symbol other than the subject is required to be calculated it is usual to rearrange the formula to make a new subject. This rearranging process is called **transposing the formula** or **transposition**.

The rules used for transposition of formulae are the same as those used for the solution of simple equations (see Chapter 5) – basically, **that the equality of an equation must be maintained**.

Problem 10 Transpose $p = q + r + s$ to make r the subject.

The aim is to obtain r on its own on the left-hand side (LHS) of the equation. Changing the equation around so that r is on the LHS gives:

$$q + r + s = p \tag{1}$$

Subtracting $(q + s)$ from both sides of the equation gives:

$$q + r + s - (q + s) = p - (q + s)$$
Thus $q + r + s - q - s = p - q - s$
i.e. $\quad\boldsymbol{r = p - q - s} \tag{2}$

It is shown with simple equations (Chapter 5), that a quantity can be moved from one side of an equation to the other with an appropriate change of sign. Thus equation (2) follows immediately from equation (1) above.

Problem 11 If $a + b = w - x + y$, express x as the subject.

Rearranging gives:

$$w - x + y = a + b \quad \text{and} \quad -x = a + b - w - y$$

Multiplying both sides by -1 gives:

$$(-1)(-x) = (-1)(a + b - w - y)$$
i.e. $x = -a - b + w + y$

The result of multiplying each side of the equation by -1 is to change all the signs in the equation.

It is conventional to express answers with positive quantities first. Hence rather than $x = -a - b + w + y$, $\boldsymbol{x} = \boldsymbol{w} + \boldsymbol{y} - \boldsymbol{a} - \boldsymbol{b}$, since the order of terms connected by $+$ and $-$ signs is immaterial.

Problem 12   Transpose $v = f\lambda$ to make λ the subject.

Rearranging gives:

$$f\lambda = v$$

Dividing both sides by f gives:

$$\frac{f\lambda}{f} = \frac{v}{f}$$

i.e. $\boldsymbol{\lambda = \dfrac{v}{f}}$

Problem 13   When a body falls freely through a height h, the velocity v is given by $v^2 = 2gh$. Express this formula with h as the subject.

Rearranging gives:

$$2gh = v^2$$

Dividing both sides by $2g$ gives:

$$\frac{2gh}{2g} = \frac{v^2}{2g}$$

i.e. $\boldsymbol{h = \dfrac{v^2}{2g}}$

Problem 14   If $I = V/R$, rearrange to make V the subject.

Rearranging gives:

$$\frac{V}{R} = I$$

Multiplying both sides by R gives:

$$R\left(\frac{V}{R}\right) = R(I)$$

Hence $\boldsymbol{V = IR}$

Problem 15   Transpose $a = \dfrac{F}{m}$ for m

Rearranging gives:

$$\frac{F}{m} = a$$

Multiplying both sides by m gives:

$$m\left(\frac{F}{m}\right) = m(a) \text{ i.e. } F = ma$$

Rearranging gives:

$$ma = F$$

Dividing both sides by a gives:

$$\frac{ma}{a} = \frac{F}{a}$$

i.e. $\boldsymbol{m = \dfrac{F}{a}}$

Problem 16   Rearrange the formula $R = (\rho l)/a$ to make (i) a the subject, and (ii) l the subject.

(i)   Rearranging gives:

$$\frac{\rho l}{a} = R$$

Multiplying both sides by a gives:

$$a\left(\frac{\rho l}{a}\right) = a(R) \text{ i.e. } \rho l = aR$$

Rearranging gives:

$$aR = \rho l$$

Dividing both sides by R gives:

$$\frac{aR}{R} = \frac{\rho l}{R}$$

i.e. $\boldsymbol{a = \dfrac{\rho l}{R}}$

(ii) $\rho l/a = R$

Multiplying both sides by a gives:

$$\rho l = aR$$

Dividing both sides by ρ gives:

$$\frac{\rho l}{\rho} = \frac{aR}{\rho}$$

i.e. $l = \dfrac{aR}{\rho}$

Problem 17 Transpose the formula $v = u + (ft)/m$, to make f the subject.

Rearranging gives:

$$u + \frac{ft}{m} = v$$

and

$$\frac{ft}{m} = v - u$$

Multiplying each side by m gives:

$$m\left(\frac{ft}{m}\right) = m(v - u)$$

i.e. $ft = m(v - u)$

Dividing both sides by t gives:

$$\frac{ft}{t} = \frac{m}{t}(v - u)$$

i.e. $f = \dfrac{m}{t}(v - u)$

Problem 18 The final length, l_2, of a piece of wire heated through $\theta°C$ is given by the formula $l_2 = l_1(1 + \alpha\theta)$. Make the coefficient of expansion, α, the subject.

Rearranging gives:

$$l_1(1 + \alpha\theta) = l_2$$

Removing the bracket gives:

$$l_1 + l_1\alpha\theta = l_2$$

Rearranging gives:

$$l_1\alpha\theta = l_2 - l_1$$

Dividing both sides by $l_1\theta$ gives:

$$\frac{l_1\alpha\theta}{l_1\theta} = \frac{l_2 - l_1}{l_1\theta}$$

i.e. $\alpha = \dfrac{l_2 - l_1}{l_1\theta}$

Problem 19 A formula for the distance moved by a body is given by $s = \dfrac{1}{2}(v + u)t$. Rearrange the formula to make u the subject.

Rearranging gives:

$$\frac{1}{2}(v + u)t = s$$

Multiplying both sides by 2 gives:

$$(v + u)t = 2s$$

Dividing both sides by t gives:

$$\frac{(v + u)t}{t} = \frac{2s}{t}$$

i.e. $v + u = \dfrac{2s}{t}$

Hence $\quad u = \dfrac{2s}{t} - v \quad$ or $\quad \dfrac{2s - vt}{t}$

Problem 20 A formula for kinetic energy is $k = \dfrac{1}{2}mv^2$. Transpose the formula to make v the subject.

Rearranging gives:

$$\frac{1}{2}mv^2 = k$$

Whenever the prospective new subject is a squared term, that term is isolated on the LHS, and then the square root of both sides of the equation is taken.

Multiplying both sides by 2 gives:

$$mv^2 = 2k$$

Dividing both sides by m gives:

$$\frac{mv^2}{m} = \frac{2k}{m}$$

i.e. $\quad v^2 = \dfrac{2k}{m}$

Taking the square root of both sides gives:

$$\sqrt{v^2} = \sqrt{\left(\frac{2k}{m}\right)}$$

i.e. $v = \sqrt{\left(\frac{2k}{m}\right)}$

Problem 21   In a right angled triangle having sides x, y and hypotenuse z, Pythagoras' theorem states $z^2 = x^2 + y^2$. Transpose the formula to find x.

Rearranging gives:

$$x^2 + y^2 = z^2$$

and

$$x^2 = z^2 - y^2$$

Taking the square root of both sides gives:

$$x = \sqrt{(z^2 - y^2)}$$

Problem 22   Given $t = 2\pi\sqrt{(l/g)}$, find g in terms of t, l and π.

Whenever the prospective new subject is within a square root sign, it is best to isolate that term on the LHS and then to square both sides of the equation.
Rearranging gives:

$$2\pi\sqrt{\left(\frac{l}{g}\right)} = t$$

Dividing both sides by 2π gives:

$$\sqrt{\left(\frac{l}{g}\right)} = \frac{t}{2\pi}$$

Squaring both sides gives:

$$\frac{l}{g} = \left(\frac{t}{2\pi}\right)^2 = \frac{t^2}{4\pi^2}$$

Cross-multiplying, i.e. multiplying each term by $4\pi^2 g$, gives:

$$4\pi^2 l = gt^2$$
or $gt^2 = 4\pi^2 l$

Dividing both sides by t^2 gives:

$$\frac{gt^2}{t^2} = \frac{4\pi^2 l}{t^2}$$

i.e. $g = \dfrac{4\pi^2 l}{t^2}$

Problem 23   The impedance of an a.c. circuit is given by $Z = \sqrt{(R^2 + X^2)}$. Make the reactance, X, the subject.

Rearranging gives:

$$\sqrt{(R^2 + X^2)} = Z$$

Squaring both sides gives:

$$R^2 + X^2 = Z^2$$

Rearranging gives:

$$X^2 = Z^2 - R^2$$

Taking the square root of both sides gives:

$$X = \sqrt{(Z^2 - R^2)}$$

Problem 24   The volume V of a hemisphere is given by $V = \dfrac{2}{3}\pi r^3$. Find r in terms of V.

Rearranging gives:

$$\frac{2}{3}\pi r^3 = V$$

Multiplying both sides by 3 gives:

$$2\pi r^3 = 3V$$

Dividing both sides by 2π gives:

$$\frac{2\pi r^3}{2\pi} = \frac{3V}{2\pi}$$

i.e. $r^3 = \dfrac{3V}{2\pi}$

Taking the cube root of both sides gives:

$$\sqrt[3]{r^3} = \sqrt[3]{\left(\frac{3V}{2\pi}\right)}$$

i.e. $r = \sqrt[3]{\left(\frac{3V}{2\pi}\right)}$

Problem 25 Transpose the formula $p = \dfrac{a^2x + a^2y}{r}$, to make a the subject.

Rearranging gives:

$$\frac{a^2x + a^2y}{r} = p$$

Multiplying both sides by r gives:

$$a^2x + a^2y = rp$$

Factorizing the LHS gives:

$$a^2(x + y) = rp$$

Dividing both sides by $(x + y)$ gives:

$$\frac{a^2(x + y)}{(x + y)} = \frac{rp}{(x + y)}$$

i.e. $a^2 = \dfrac{rp}{(x + y)}$

Taking the square root of both sides gives:

$$Z = a = \sqrt{\left(\frac{rp}{x + y}\right)}$$

Problem 26 Make b the subject of the formula $a = \dfrac{x - y}{\sqrt{(bd + be)}}$

Rearranging gives:

$$\frac{x - y}{\sqrt{(bd + be)}} = a$$

Multiplying both sides by $\sqrt{(bd + be)}$ gives:

$$x - y = a\sqrt{(bd + be)}$$

or $a\sqrt{(bd + be)} = x - y$

Dividing both sides by a gives:

$$\sqrt{(bd + be)} = \frac{x - y}{a}$$

Squaring both sides gives:

$$bd + be = \left(\frac{x - y}{a}\right)^2$$

Factorizing the LHS gives:

$$b(d + e) = \left(\frac{x - y}{a}\right)^2$$

Dividing both sides by $(d + e)$ gives:

$$b = \frac{\left(\dfrac{x - y}{a}\right)^2}{(d + e)}$$

i.e. $b = \dfrac{(x - y)^2}{a^2(d + e)}$

Problem 27 If $cd = 3d + e - ad$, express d in terms of a, c and e.

Rearranging to obtain the terms in d on the LHS gives:

$$cd - 3d + ad = e$$

Factorizing the LHS gives:

$$d(c - 3 + a) = e$$

Dividing both sides by $(c - 3 + a)$ gives:

$$d = \frac{e}{c - 3 + a}$$

Problem 28 If $a = b/(1 + b)$, make b the subject of the formula.

Rearranging gives:

$$\frac{b}{1 + b} = a$$

Multiplying both sides by $(1 + b)$ gives:

$$b = a(1 + b)$$

Removing the bracket gives:

$$b = a + ab$$

Rearranging to obtain terms in b on the LHS gives:

$$b - ab = a$$

Factorizing the LHS gives:

$$b(1 - a) = a$$

Dividing both sides by $(1 - a)$ gives:

$$b = \frac{a}{1 - a}$$

Problem 29 Transpose the formula $V = \dfrac{Er}{R + r}$ to make r the subject.

Rearranging gives:

$$\frac{Er}{R + r} = V$$

Multiplying both sides by $(R + r)$ gives:

$$Er = V(R + r)$$

Removing the bracket gives:

$$Er = VR + Vr$$

Rearranging to obtain terms in r on the LHS gives:

$$Er - Vr = VR$$

Factorizing gives:

$$r(E - V) = VR$$

Dividing both sides by $(E - V)$ gives:

$$r = \frac{VR}{E - V}$$

Problem 30 Transpose the formula $y = pq^2/(r + q^2)$ − t, to make q the subject.

Rearranging gives:

$$\frac{pq^2}{r + q^2} - t = y$$

and

$$\frac{pq^2}{r + q^2} = y + t$$

Multiplying both sides by $(r + q^2)$ gives:

$$pq^2 = (r + q^2)(y + t)$$

Removing brackets gives:

$$pq^2 = ry + rt + q^2y + q^2t$$

Rearranging to obtain terms in q on the LHS gives:

$$pq^2 - q^2y - q^2t = ry + rt$$

Factorizing gives:

$$q^2(p - y - t) = r(y + t)$$

Dividing both sides by $(p - y - t)$ gives:

$$q^2 = \frac{r(y + t)}{(p - y - t)}$$

Taking the square root of both sides gives:

$$q = \sqrt{\frac{r(y + t)}{(p - y - t)}}$$

Problem 31 Given that $\dfrac{D}{d} = \sqrt{\left(\dfrac{f + p}{f - p}\right)}$, express p in terms of D, d and f.

Rearranging gives:

$$\sqrt{\left(\frac{f + p}{f - p}\right)} = \frac{D}{d}$$

Squaring both sides gives:

$$\frac{f + p}{f - p} = \frac{D^2}{d^2}$$

Cross-multiplying, i.e. multiplying each term by $d^2(f - p)$, gives:

$$d^2(f + p) = D^2(f - p)$$

Removing brackets gives:

$$d^2f + d^2p = D^2f - D^2p$$

Rearranging, to obtain terms in p on the LHS, gives:

$$d^2p + D^2p = D^2f - d^2f$$

Factorizing gives:

$$p(d^2 + D^2) = f(D^2 - d^2)$$

Dividing both sides by $(d^2 + D^2)$ gives:

$$p = \frac{f(D^2 - d^2)}{(d^2 + D^2)}$$

Further problems on transposing formulae may be found in section 7.3, Problems 13 to 45.

7.3 Further problems on evaluation and transposition of formulae

Evaluation of formulae

1   The area A of a rectangle is given by the formula $A = lb$. Evaluate the area when $l = 12.4$ cm and $b = 5.37$ cm.
    $[A = 66.59$ cm$^2]$

2   The circumference C of a circle is given by the formula $C = 2\pi r$. Determine the circumference given $\pi = 3.14$ and $r = 8.40$ mm.
    $[C = 52.75$ mm$]$

3   A formula used in connection with gases is $R = (PV)/T$. Evaluate R when $P = 1500$, $V = 5$ and $T = 200$.
    $[R = 37.5]$

4   The potential difference, V volts, available at battery terminals is given by $V = E - Ir$. Evaluate V when $E = 5.62$, $I = 0.70$ and $R = 4.30$.
    $[V = 2.61$ V$]$

5   Given force $F = \frac{1}{2}m (v^2 - u^2)$, find F when $m = 18.3$, $v = 12.7$ and $u = 8.24$.
    $[F = 854.5]$

6   The current I amperes flowing in a number of cells is given by $I = (nE)/(R + nr)$. Evaluate the current when $n = 36$, $E = 2.20$, $R = 2.80$ and $r = 0.50$.
    $[I = 3.81$ A$]$

7   The time, t seconds, of oscillation for a simple pendulum is given by $t = 2\pi\sqrt{(l/g)}$. Determine the time when $\pi = 3.142$, $l = 54.32$ and $g = 9.81$.
    $[t = 14.79$ s$]$

8   Energy, E joules, is given by the formula $E = \frac{1}{2}LI^2$. Evaluate the energy when $L = 5.5$ and $I = 1.2$.
    $[E = 3.96$ J$]$

9   The current I amperes in an a.c. circuit is given by $I = \dfrac{V}{\sqrt{(R^2 + X^2)}}$. Evaluate the current when $V = 250$, $R = 11.0$ and $X = 16.2$.
    $[I = 12.77$ A$]$

10 Distance s metres is given by the formula $s = ut + \frac{1}{2}at^2$. If $u = 9.50$, $t = 4.60$ and $a = -2.50$, evaluate the distance.
    $[s = 17.25$ m$]$

11 The area, A, of any triangle is given by $A = \sqrt{[s(s - a)(s - b)(s - c)]}$, where $s = \dfrac{a + b + c}{2}$.
    Evaluate the area given $a = 3.60$ cm, $b = 4.00$ cm and $c = 5.20$ cm.
    $[A = 7.184$ cm$^2]$

12 Given that $a = 0.290$, $b = 14.86$, $c = 0.042$, $d = 31.8$ and $e = 0.650$ evaluate v given that $v = \sqrt{\left(\dfrac{ab}{c} - \dfrac{d}{e}\right)}$.
    $[v = 7.327]$

Transposition of formulae

Make the symbol indicated the subject of each of the formulae shown in *Problems 13* to *40*, and express each in its simplest form.

13 $a + b = c - d - e$ (d)
    $[d = c - e - a - b]$

14 $x + 3y = t$ (y)
    $\left[y = \frac{1}{3}(t - x)\right]$

15 $c = 2\pi r$ (r)
    $\left[r = \dfrac{c}{2\pi}\right]$

16 $y = mx + c$ (x)
    $\left[x = \dfrac{y - c}{m}\right]$

17 $I = PRT$ (T)
    $\left[T = \dfrac{I}{PR}\right]$

18 $I = \dfrac{E}{R}$ (R)
    $\left[R = \dfrac{E}{I}\right]$

19 $S = \dfrac{a}{1 - r}$ (r)
    $\left[r = \dfrac{S - a}{S}\right]$

20 $F = \dfrac{9}{5}C + 32$ (C)

$\left[C = \dfrac{5}{9}(F - 32) \right]$

21 $y = \dfrac{\lambda(x - d)}{d}$ (x)

$\left[x = \dfrac{d}{\lambda}(y + \lambda) \right]$

22 $A = \dfrac{3(F - f)}{L}$ (f)

$\left[f = \dfrac{3F - AL}{3} \right]$

23 $y = \dfrac{Ml^2}{8EI}$ (E)

$\left[E = \dfrac{Ml^2}{8yI} \right]$

24 $R = R_0(1 + \alpha t)$ (t)

$\left[t = \dfrac{R - R_0}{R_0\alpha} \right]$

25 $\dfrac{1}{R} = \dfrac{1}{R_1} + \dfrac{1}{R_2}$ (R_2)

$\left[R_2 = \dfrac{RR_1}{R_1 - R} \right]$

26 $I = \dfrac{E - e}{R + r}$ (R)

$\left[R = \dfrac{E - e - Ir}{I} \right]$

27 $y = 4ab^2c^2$ (b)

$\left[b = \sqrt{\left(\dfrac{y}{4ac^2} \right)} \right]$

28 $\dfrac{a^2}{x^2} + \dfrac{b^2}{y^2} = 1$ (x)

$\left[x = \dfrac{ay}{\sqrt{(y^2 - b^2)}} \right]$

29 $t = 2\pi\sqrt{\left(\dfrac{l}{g} \right)}$ (l)

$\left[l = \dfrac{t^2g}{4\pi^2} \right]$

30 $v^2 = u^2 + 2as$ (u)

$\left[u = \sqrt{(v^2 - 2as)} \right]$

31 $A = \dfrac{\pi R^2\theta}{360}$ (R)

$\left[R = \sqrt{\left(\dfrac{360A}{\pi\theta} \right)} \right]$

32 $N = \sqrt{\left(\dfrac{a + x}{y} \right)}$ (a)

$[a = N^2y - x]$

33 $Z = \sqrt{[R^2 + (2\pi fL)^2]}$ (L)

$\left[L = \dfrac{\sqrt{(Z^2 - R^2)}}{2\pi f} \right]$

34 $y = \dfrac{a^2m - a^2n}{x}$ (a)

$\left[a = \sqrt{\left(\dfrac{xy}{m - n} \right)} \right]$

35 $M = \pi(R^4 - r^4)$ (R)

$\left[R = \sqrt[4]{\left(\dfrac{M}{\pi} + r^4 \right)} \right]$

36 $x + y = \dfrac{r}{3 + r}$ (r)

$\left[r = \dfrac{3(x + y)}{(1 - x - y)} \right]$

37 $m = \dfrac{\mu L}{L + rCR}$ (L)

$\left[L = \dfrac{mrCR}{\mu - m} \right]$

38 $a^2 = \dfrac{b^2 - c^2}{b^2}$ (b)

$\left[b = \dfrac{c}{\sqrt{(1 - a^2)}} \right]$

39 $\dfrac{x}{y} = \dfrac{1 + r^2}{1 - r^2}$ (r)

$\left[r = \sqrt{\left(\dfrac{x - y}{x + y} \right)} \right]$

40 $\dfrac{p}{q} = \sqrt{\left(\dfrac{a + 2b}{a - 2b} \right)}$ (b)

$\left[b = \dfrac{a(p^2 - q^2)}{2(p^2 + q^2)} \right]$

41 A formula for the focal length, f, of a convex lens is $\dfrac{1}{f} = \dfrac{1}{u} + \dfrac{1}{v}$ Transpose the formula to make v the subject and evaluate v when $f = 5$ and $u = 6$.

$$\left[v = \frac{uf}{u-f}, \; 30 \right]$$

42 The quantity of heat, Q, is given by the formula $Q = mc(t_2 - t_1)$. Make t_2 the subject of the formula and evaluate t_2 when $m = 10$, $t_1 = 15$, $c = 4$ and $Q = 1600$.

$$\left[t_2 = t_1 + \frac{Q}{mc}, \; 55 \right]$$

43 The velocity, v, of water in a pipe appears in the formula

$$h = \frac{0.03Lv^2}{2dg}$$

Express v as the subject of the formula and evaluate v when $h = 0.712$, $L = 150$, $d = 0.30$ and $g = 9.81$.

$$\left[v = \sqrt{\left(\frac{2dgh}{0.03L} \right)}, \; 0.965 \right]$$

44 The sag S at the centre of a wire is given by the formula

$$S = \sqrt{\left\{ \frac{3d(l-d)}{8} \right\}}$$

Make l the subject of the formula and evaluate l when $d = 1.75$ and $S = 0.80$.

$$\left[l = \frac{8S^2}{3d} + d, \; 2.725 \right]$$

45 In an electrical alternating current circuit the impedance Z is given by

$$Z = \sqrt{\left\{ R^2 + \left(\omega L - \frac{1}{\omega C} \right)^2 \right\}}$$

Transpose the formula to make C the subject and hence evaluate C when $Z = 130$, $R = 120$, $\omega = 314$ and $L = 0.32$.

$$\left[C = \frac{1}{\omega \left\{ \omega L - \sqrt{(Z^2 - R^2)} \right\}}, \; 63.1 \times 10^{-6} \right]$$

8

Quadratic equations

8.1 Introduction to quadratic equations

An **equation** is a statement that two quantities are equal. To '**solve an equation**' means 'to find the value of the unknown'. The value of the unknown is called the **root** of the equation.

A **quadratic equation** is one in which the highest power of the unknown quantity is 2. For example, $x^2 - 3x + 1 = 0$ is a quadratic equation.

There are four methods of **solving quadratic equations**. These are: (i) by factorization (where possible), (ii) by 'completing the square', (iii) by using the 'quadratic formula', or (iv) graphically (see Chapter 13).

8.2 Solution of quadratic equations by factorization

Multiplying out $(2x + 1)(x - 3)$ gives $2x^2 - 6x + x - 3$, i.e. $2x^2 - 5x - 3$.

The reverse process of moving from $2x^2 - 5x - 3$ to $(2x + 1)(x - 3)$ is called **factorizing**.

If the quadratic expression can be factorized this provides the simplest method of solving a quadratic equation. For example, if $2x^2 - 5x - 3 = 0$, then, by factorizing:

$$(2x + 1)(x - 3) = 0$$

Hence either $(2x + 1) = 0$, i.e. $x = -\dfrac{1}{2}$

or $(x - 3) = 0$, i.e. $x = 3$

The technique of factorizing is often one of 'trial and error'.

Problem 1 Solve the equations (a) $x^2 + 2x - 8 = 0$, and (b) $3x^2 - 11x - 4 = 0$ by factorization.

(a) $x^2 + 2x - 8 = 0$. The factors of x^2 are x and x. These are placed in brackets thus: $(x \quad)(x \quad)$
The factors of -8 are $+8$ and -1, or -8 and $+1$, or $+4$ and -2, or -4 and $+2$.
The only combination to give a middle term of $+2x$ is $+4$ and -2, i.e.

$$x^2 + 2x - 8 = (x - 2)(x + 4)$$

(Note that the product of the two inner terms added to the product of the two outer terms must equal the middle term, $+2x$ in this case.)

The quadratic equation $x^2 + 2x - 8 = 0$ thus becomes $(x + 4)(x - 2) = 0$.
Since the only way that this can be true is for either the first or the second, or both factors to be zero, then

either $(x + 4) = 0$ i.e. $x = -4$
or $(x - 2) = 0$ i.e. $x = 2$

Hence the roots of $x^2 + 2x - 8 = 0$ are $x = -4$ and 2.
(b) $3x^2 - 11x - 4 = 0$
The factors of $3x^2$ are $3x$ and x. These are placed in brackets thus: $(x \quad)(3x \quad)$.

The factors of -4 are -4 and $+1$, or $+4$ and -1, or -2 and 2. Remembering that the product of the two inner terms added to the product of the two outer terms must equal $-11x$, the only combination to give this is -4 and $+1$, i.e.

$$3x^2 - 11x - 4 = (3x + 1)(x - 4)$$

The quadratic equation $3x^2 - 11x - 4 = 0$ thus becomes $(3x + 1)(x - 4) = 0$. Hence,

either $(3x + 1) = 0$ i.e. $x = -\dfrac{1}{3}$

or $(x - 4) = 0$ i.e. $x = 4$

and both solutions may be checked in the original equation.

Problem 2   Determine the roots of (a) $x^2 - 6x + 9 = 0$, and (b) $4x^2 - 25 = 0$, by factorization.

(a) $x^2 - 6x + 9 = 0$. Hence $(x - 3)(x - 3) = 0$, i.e. $(x - 3)^2 = 0$ (the left-hand side is known as **a perfect square**). Hence $x = 3$ is the only root of the equation $x^2 - 6x + 9 = 0$.

(b) $4x^2 - 25 = 0$ (the left-hand side is **the difference of two squares**, $(2x)^2$ and $(5)^2$). Hence $(2x + 5)(2x - 5) = 0$

Hence either $(2x + 5) = 0$ i.e. $x = -\dfrac{5}{2}$

or $(2x - 5) = 0$ i.e. $x = \dfrac{5}{2}$

Problem 3   Solve the following quadratic equations by factorizing:

(a) $4x^2 + 8x + 3 = 0$
(b) $15x^2 + 2x - 8 = 0$

(a) $4x^2 + 8x + 3 = 0$. The factors of $4x^2$ are $4x$ and x or $2x$ and $2x$. The factors of 3 are 3 and 1, or -3 and -1. Remembering that the product of the inner terms added to the product of the two outer terms must equal $+8x$, the only combination that is true (by trial and error) is

$$(4x^2 + 8x + 3) = (2x + 3)(2x + 1)$$

Hence $(2x + 3)(2x + 1) = 0$ from which, either $(2x + 3) = 0$
or $(2x + 1) = 0$

Thus $2x = -3$, from which, $x = -\dfrac{3}{2}$

or $2x = -1$, from which, $x = -\dfrac{1}{2}$

which may be checked in the original equation.

(b) $15x^2 + 2x - 8 = 0$. The factors of $15x^2$ are $15x$ and x or $5x$ and $3x$. The factors of -8 are -4 and $+2$, or 4 and -2, or -8 and $+1$, or 8 and -1. By trial and error the only combination that works is

$$15x^2 + 2x - 8 = (5x + 4)(3x - 2)$$

Hence $(5x + 4)(3x - 2) = 0$ from which either

       $5x + 4 = 0$
or $3x - 2 = 0$

Hence $x = -\dfrac{4}{5}$ or $x = \dfrac{2}{3}$

which may be checked in the original equation.

Problem 4   The roots of a quadratic equation are $\dfrac{1}{3}$ and -2. Determine the equation.

If the roots of a quadratic equation are α and β then $(x - \alpha)(x - \beta) = 0$. Hence if $\alpha = \dfrac{1}{3}$ and $\beta = -2$, then

$$\left(x - \dfrac{1}{3}\right)(x - (-2)) = 0$$

$$\left(x - \dfrac{1}{3}\right)(x + 2) = 0$$

$$x^2 - \dfrac{1}{3}x + 2x - \dfrac{2}{3} = 0$$

$$x^2 + \dfrac{5}{3}x - \dfrac{2}{3} = 0$$

Hence $3x^2 + 5x - 2 = 0$

Problem 5   Find the equations whose roots are (a) 5 and -5 (b) 1.2 and -0.4.

(a) If 5 and -5 are the roots of a quadratic equation then

$$(x - 5)(x + 5) = 0$$
i.e. $x^2 - 5x + 5x - 25 = 0$
i.e. $x^2 - 25 = 0$

(b) If 1.2 and -0.4 are the roots of a quadratic equation then

$$(x - 1.2)(x + 0.4) = 0$$
i.e. $x^2 - 1.2x + 0.4x - 0.48 = 0$
i.e. $x^2 - 0.8x - 0.48 = 0$

Further problems on solving quadratic equations by factorization may be found in section 8.7, Problems 1 to 6, page 77.

8.3 Solution of quadratic equations by 'completing the square'

An expression such as x^2 or $(x + 2)^2$ or $(x - 3)^2$ is called a perfect square.

If $x^2 = 3$ then $x = \pm\sqrt{3}$
If $(x + 2)^2 = 5$ then $x + 2 = \pm\sqrt{5}$ and $x = -2 \pm \sqrt{5}$
If $(x - 3)^2 = 8$ then $x - 3 = \pm\sqrt{8}$ and $x = 3 \pm \sqrt{8}$

Hence if a quadratic equation can be rearranged so that one side of the equation is a perfect square and the other side of the equation is a number, then the solution of the equation is readily obtained by taking the square root of each side as in the above examples. The process of rearranging one side of a quadratic equation into a perfect square before solving is called **'completing the square'**.

$$(x + a)^2 = x^2 + 2ax + a^2$$

Thus in order to make the quadratic expression $x^2 + 2a$ into a perfect square it is necessary to add (half the coefficient of $x)^2$, i.e.

$$\left(\frac{2a}{2}\right)^2 \text{ or } a^2$$

For example, $x^2 + 3x$ becomes a perfect square by adding $(3/2)^2$, i.e.

$$x^2 + 3x + \left(\frac{3}{2}\right)^2 = \left(x + \frac{3}{2}\right)^2$$

The method is demonstrated in the following worked problems.

Problem 6   Solve $2x^2 + 5x = 3$ by 'completing the square'.

The procedure is as follows:

1. Rearrange the equation so that all terms are on the same side of the equals sign (and the coefficient of the x^2 term is positive). Hence $2x^2 + 5x - 3 = 0$
2. Make the coefficient of the x^2 term unity. In this case this is achieved by dividing throughout by 2. Hence

$$\frac{2x^2}{2} + \frac{5x}{2} - \frac{3}{2} = 0$$

i.e. $x^2 + \dfrac{5}{2}x - \dfrac{3}{2} = 0$

3. Rearrange the equations so that the x^2 and x terms are on one side of the equals sign and the constant is on the other side. Hence

$$x^2 + \frac{5}{2}x = \frac{3}{2}$$

4. Add to both sides of the equation (half the coefficient of $x)^2$. In this case the coefficient of x is $\dfrac{5}{2}$. Half the coefficient squared is therefore $(5/4)^2$. Thus

$$x^2 + \frac{5}{2}x + \left(\frac{5}{4}\right)^2 = \frac{3}{2} + \left(\frac{5}{4}\right)^2$$

The LHS is now a perfect square, i.e.

$$\left(x + \frac{5}{4}\right)^2 = \frac{3}{2} + \left(\frac{5}{4}\right)^2$$

5. Evaluate the RHS. Thus

$$\left(x + \frac{5}{4}\right)^2 = \frac{3}{2} + \frac{25}{16} = \frac{24 + 25}{16} = \frac{49}{16}$$

6. Take the square root of both sides of the equation (remembering that the square root of a number gives a $\pm$ answer). Thus

$$\sqrt{\left(x + \frac{5}{4}\right)^2} = \sqrt{\frac{49}{16}}$$

i.e. $x + \dfrac{5}{4} = \pm\dfrac{7}{4}$

7. Solve the simple equation. Thus

$$x = -\frac{5}{4} \pm \frac{7}{4}$$

i.e. $x = -\dfrac{5}{4} + \dfrac{7}{4} = \dfrac{2}{4} = \dfrac{1}{2}$

and $x = -\dfrac{5}{4} - \dfrac{7}{4} = -\dfrac{12}{4} = -3$

Hence $x = \dfrac{1}{2}$ or -3 are the roots of the equation $2x^2 + 5x = 3$.

Problem 7   Solve $2x^2 + 9x + 8 = 0$, correct to 3 significant figures, by 'completing the square'.

Making the coefficient of x^2 unity gives:

$$x^2 + \frac{9}{2}x + 4 = 0$$

and rearranging gives:

$$x^2 + \frac{9}{2}x = -4$$

Adding to both sides (half the coefficient of x)^2 gives:

$$x^2 + \frac{9}{2}x + \left(\frac{9}{4}\right)^2 = \left(\frac{9}{4}\right)^2 - 4$$

The LHS is now a perfect square, thus

$$\left(x + \frac{9}{4}\right)^2 = \frac{81}{16} - 4 = \frac{17}{16}$$

Taking the square root of both sides gives:

$$x + \frac{9}{4} = \sqrt{\left(\frac{17}{16}\right)} = \pm 1.031$$

Hence $x = -\dfrac{9}{4} \pm 1.031$

i.e. $x = -3.28$ or -1.22, correct to 3 significant figures.

> *Problem 8* By 'completing the square', solve the quadratic equation $4.6y^2 + 3.5y - 1.75 = 0$, correct to 3 decimal places.

$$4.6y^2 + 3.5y - 1.75 = 0$$

Making the coefficient of y^2 unity gives:

$$y^2 + \frac{3.5}{4.6}y - \frac{1.75}{4.6} = 0$$

and rearranging gives:

$$y^2 + \frac{3.5}{4.6}y = \frac{1.75}{4.6}$$

Adding to both sides (half the coefficient of y)^2 gives

$$y^2 + \frac{3.5}{4.6}y + \left(\frac{3.5}{9.2}\right)^2 = \frac{1.75}{4.6} + \left(\frac{3.5}{9.2}\right)^2$$

The LHS is now a perfect square, thus

$$\left(y + \frac{3.5}{9.2}\right)^2 = 0.525\ 165\ 4$$

Taking the square root of both sides gives:

$$y + \frac{3.5}{9.2} = \sqrt{0.525\ 165\ 4}$$

$$= \pm 0.724\ 683\ 0$$

Hence $\quad y = -\dfrac{3.5}{9.2} \pm 0.724\ 683\ 0$

i.e. $\quad y = 0.344$ or -1.105

Further problems on solving quadratic equations by 'completing the square' may be found in section 8.7, Problems 7 and 8, page 77.

8.4 Solution of quadratic equations by formula

Let the general form of a quadratic equation be given by:

$$ax^2 + bx + c = 0$$

where a, b and c are constants.
Dividing $ax^2 + bx + c = 0$ by a gives:

$$x^2 + \frac{b}{a}x + \frac{c}{a} = 0$$

Rearranging gives:

$$x^2 + \frac{b}{a}x = -\frac{c}{a}$$

Adding to each side of the equation the square of half the coefficient of the term in x to make the LHS a perfect square gives:

$$x^2 + \frac{b}{a}x + \left(\frac{b}{2a}\right)^2 = \left(\frac{b}{2a}\right)^2 - \frac{c}{a}$$

Rearranging gives:

$$\left(x + \frac{b}{2a}\right)^2 = \frac{b^2}{4a^2} - \frac{c}{a} = \frac{b^2 - 4ac}{4a^2}$$

Taking the square root of both sides gives:

$$x + \frac{b}{2a} = \sqrt{\left(\frac{b^2 - 4ac}{4a^2}\right)} = \frac{\pm\sqrt{(b^2 - 4ac)}}{2a}$$

Hence
$$x = -\frac{b}{2a} \pm \frac{\sqrt{(b^2 - 4ac)}}{2a}$$

i.e. the quadratic formula is $x = \dfrac{-b \pm \sqrt{(b^2 - 4ac)}}{2a}$

(This method of solution is 'completing the square' – as shown in section 8.3.)

Summarizing:

if $ax^2 + bx + c = 0$ then $\boxed{x = \dfrac{-b \pm \sqrt{(b^2 - 4ac)}}{2a}}$

This is known as the **quadratic formula.**

Problem 9   Solve (a) $x^2 + 2x - 8 = 0$, and
(b) $3x^2 - 11x - 4 = 0$ by using the quadratic formula.

(a) Comparing $x^2 + 2x - 8 = 0$ with $ax^2 + bx + c = 0$ gives
$a = 1$, $b = 2$ and $c = -8$.
Substituting these values into the quadratic formula

$$x = \frac{-b \pm \sqrt{(b^2 - 4ac)}}{2a}$$

gives:

$$x = \frac{-2 \pm \sqrt{[(2)^2 - 4(1)(-8)]}}{2(1)}$$

$$= \frac{-2 \pm \sqrt{(4 + 32)}}{2} = \frac{-2 \pm \sqrt{36}}{2}$$

$$= \frac{-2 \pm 6}{2} = \frac{-2 + 6}{2} \text{ or } \frac{-2 - 6}{2}$$

Hence $x = \dfrac{4}{2} = \mathbf{2}$ or $\dfrac{-8}{2} = -\mathbf{4}$ (as in *Problem 1(a)*).

(b) Comparing $3x^2 - 11x - 4 = 0$ with $ax^2 + bx + c = 0$ gives $a = 3$, $b = -11$ and $c = -4$. Hence

$$x = \frac{-(-11) \pm \sqrt{[(-11)^2 - 4(3)(-4)]}}{2(3)}$$

$$= \frac{+11 \pm \sqrt{(121 + 48)}}{6} = \frac{11 \pm \sqrt{169}}{6}$$

$$= \frac{11 \pm 13}{6} = \frac{11 + 13}{6} \text{ or } \frac{11 - 13}{6}$$

Hence $x = \dfrac{24}{6} = \mathbf{4}$ or $\dfrac{-2}{6} = -\dfrac{\mathbf{1}}{\mathbf{3}}$ (as in *Problem 1(b)*).

Problem 10   Solve $4x^2 + 7x + 2 = 0$ giving the roots correct to 2 decimal places.

Comparing $4x^2 + 7x + 2 = 0$ with $ax^2 + bx + c = 0$ gives $a = 4$, $b = 7$ and $c = 2$. Hence

$$x = \frac{-7 \pm \sqrt{[(7)^2 - 4(4)(2)]}}{2(4)} = \frac{-7 \pm \sqrt{17}}{8}$$

$$= \frac{-7 \pm 4.123}{8} = \frac{-7 + 4.123}{8}$$

or
$$\frac{-7 - 4.123}{8}$$

Hence $x = \mathbf{-0.36}$ or $\mathbf{-1.39}$, **correct to 2 decimal places.**

Problem 11   Use the quadratic formula to solve
$\dfrac{x + 2}{4} + \dfrac{3}{x - 1} = 7$ correct to 4 significant figures.

Multiplying throughout by $4(x - 1)$ gives:

$$4(x - 1)\frac{(x + 2)}{4} + 4(x-1)\frac{3}{(x - 1)} = 4(x - 1)(7)$$

i.e. $(x - 1)(x + 2) + (4)(3) = 28(x - 1)$

$$x^2 + x - 2 + 12 = 28x - 28$$

Hence $x^2 - 27x + 38 = 0$

Using the quadratic formula:

$$x = \frac{-(-27) \pm \sqrt{[(-27)^2 - 4(1)(38)]}}{2}$$

$$= \frac{27 \pm \sqrt{577}}{2} = \frac{27 \pm 24.020\ 8}{2}$$

Hence $x = \dfrac{27 + 24.020\ 8}{2} = 25.510\ 4$

or $x = \dfrac{27 - 24.020\ 8}{2} = 1.489\ 6$

Hence $x = \mathbf{25.51}$ or $\mathbf{1.490}$, correct to 4 significant figures.

Further problems on solving quadratic equations by formula may be found in section 8.7, Problems 9 and 10, page 77.

8.5 Practical problems involving quadratic equations

There are many **practical problems** where a quadratic equation has first to be obtained, from given information, before it is solved.

Problem 12 The area of a rectangle is 23.6 cm^2 and its width is 3.10 cm shorter than its length. Determine the dimensions of the rectangle, correct to 3 significant figures.

Let the length of the rectangle be x cm. Then the width is $(x - 3.10)$ cm. Area = length × width = $x(x - 3.10)$ = 23.6, i.e.

$$x^2 - 3.10x - 23.6 = 0$$

Using the quadratic formula,

$$x = \frac{-(-3.10) \pm \sqrt{[(-3.10)^2 - 4(1)(-23.6)]}}{2(1)}$$

$$= \frac{3.10 \pm \sqrt{(9.61 + 94.4)}}{2} = \frac{3.10 \pm 10.20}{2}$$

$$= \frac{13.30}{2} \text{ or } \frac{-7.10}{2}$$

Hence $x = 6.65$ cm or -3.55 cm. The latter solution is neglected since length cannot be negative. Thus length $x = 6.65$ cm and width = $x - 3.10 = 6.65 - 3.10 = 3.55$ cm. **Hence the dimensions of the rectangle are 6.61 cm by 3.55 cm.**
(Check: Area = $6.65 \times 3.55 = 23.6$ cm^2, correct to 3 significant figures.)

Problem 13 Calculate the diameter of a solid cylinder which has a height of 82.0 cm and a total surface area of 2.0 m^2.

Total surface area of a cylinder

= curved surface area + 2 circular ends
= $2\pi rh + 2\pi r^2$ (where r = radius and h = height)

Since the total surface area = 2.0 m^2 and the height $h = 82$ cm or 0.82 m, then

$$2.0 = 2\pi r(0.82) + 2\pi r^2$$
i.e. $2\pi r^2 + 2\pi r(0.82) - 2.0 = 0$

Dividing throughout by 2π gives:

$$r^2 + 0.82r - \frac{1}{\pi} = 0$$

Using the quadratic formula:

$$r = \frac{-0.82 \pm \sqrt{[(0.82)^2 - 4(1)(-\frac{1}{\pi})]}}{2(1)}$$

$$= \frac{-0.82 \pm \sqrt{1.945\ 6}}{2} = \frac{-0.82 \pm 1.394\ 8}{2}$$

$$= 0.287\ 4 \text{ or } -1.107\ 4$$

Thus the radius r of the cylinder is 0.287 4 m (the negative solution being neglected). Hence the diameter of the cylinder

$$= 2 \times 0.287\ 4$$
$$= \textbf{0.574 8 m or 57.5 cm}$$

correct to 3 significant figures.

Problem 14 The height s metres of a mass projected vertically upwards at time t seconds is $s = ut - \frac{1}{2}gt^2$. Determine how long the mass will take after being projected to reach a height of 16 m (a) on the ascent and (b) on the descent, when $u = 30$ m/s and $g = 9.81$ m/s^2.

When height $s = 16$ m, $16 = 30t - \frac{1}{2}(9.81)t^2$

i.e. $4.905t^2 - 30t + 16 = 0$

Using the quadratic formula:

$$t = \frac{-(-30) \pm \sqrt{[(-30)^2 - 4(4.905)(16)]}}{2(4.905)}$$

$$= \frac{30 \pm \sqrt{586.1}}{9.81} = \frac{30 \pm 24.21}{9.81} = 5.53 \text{ or } 0.59$$

Hence the mass will reach a height of 16 m after 0.59 s on the ascent and after 5.53 s on the descent.

Problem 15 A shed is 4.0 m long and 2.0 m wide. A concrete path of constant width is laid all the way around the shed. If the area of the path is 9.50 m^2 calculate its width to the nearest centimetre.

Figure 8.1 shows a plan view of the shed with its surrounding path of width t metres. Area of path

$$= 2(2.0 \times t) + 2t(4.0 + 2t)$$

i.e. $9.50 = 4.0t + 8.0t + 4t^2$

or $4t^2 + 12.0t - 9.50 = 0$

Hence $t = \frac{-(12.0) \pm \sqrt{[(12.0)^2 - 4(4)(-9.50)]}}{2(4)}$

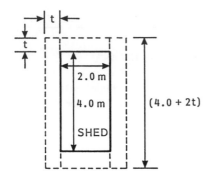

Fig. 8.1

$$= \frac{-12.0 \pm \sqrt{296.0}}{8} = \frac{-12.0 \pm 17.204\,65}{8}$$

Hence $t = 0.650\,6$ m or $-3.650\,58$ m

Neglecting the negative result which is meaningless, the width of the path, t = **0.651 m or 65 cm**, correct to the nearest centimetre.

Problem 16   If the total surface area of a solid cone is 486.2 cm^2 and its slant height is 15.3 cm, determine its base diameter.

From Chapter 15, page 140, the total surface area A of a solid cone is given by:

$$A = \pi r l + \pi r^2$$

where l is the slant height and r the base radius.

If $A = 482.2$ and $l = 15.3$ then:

$$482.2 = \pi r(15.3) + \pi r^2$$

i.e. $\pi r^2 + 15.3\pi r - 482.2 = 0$

or $r^2 + 15.3r - \dfrac{482.2}{\pi} = 0$

Using the quadratic formula,

$$r = \frac{-15.3 \pm \sqrt{\left[(15.3)^2 - 4\left(\dfrac{-482.2}{\pi}\right)\right]}}{2}$$

$$= \frac{-15.3 \pm \sqrt{848.046\,1}}{2} = \frac{-15.3 \pm 29.121\,23}{2}$$

Hence radius $r = 6.910\,6$ cm (or -22.21 cm, which is meaningless, and is thus ignored).

Thus **the diameter of the base** $= 2r = 2(6.910\,6) =$ **13.82 cm**.

Further problems on practical problems involving quadratic equations may be found in section 8.7, Problems 11 to 20, page 77.

8.6 The solution of linear and quadratic equations simultaneously

Sometimes a linear equation and a quadratic equation need to be solved simultaneously. An algebraic method of solution is shown in *Problem 17*; a graphical solution is shown in Chapter 13, page 118.

Problem 17   Determine the values of x and y which simultaneously satisfy the equations:

$y = 5x - 4 - 2x^2$ and $y = 6x - 7$

For a simultaneous solution the values of y must be equal, hence the RHS of each equation is equated. Thus

$$5x - 4 - 2x^2 = 6x - 7$$

Rearranging gives

$$5x - 4 - 2x^2 - 6x + 7 = 0$$
i.e. $-x + 3 - 2x^2 = 0$
or $2x^2 + x - 3 = 0$

Factorising gives

$$(2x + 3)(x - 1) = 0$$

i.e. $x = -\dfrac{3}{2}$ or $x = 1$

In the equation $y = 6x - 7$,

when $x = -\dfrac{3}{2}$, $y = 6\left(-\dfrac{3}{2}\right) - 7 = -16$

and when $x = 1$, $y = 6 - 7 = -1$
(Checking the result in $y = 5x - 4 - 2x^2$: when $x = -3/2$,

$$y = 5\left(-\frac{3}{2}\right) - 4 - 2\left(-\frac{3}{2}\right)^2$$

$$= -\frac{15}{2} - 4 - \frac{9}{2} = -16$$

as above; and when $x = 1$, $y = 5 - 4 - 2 = -1$ as above.)

Hence the simultaneous solutions occur when $x = -\dfrac{3}{2}$, $y = -16$ and when $x = 1$, $y = -1$.

Further problems on solving linear and quadratic equations simultaneously may be found in section 8.7, Problems 21 to 23, page 78.

8.7 Further problems on quadratic equations

Solving quadratic equations by factorization

In *Problems 1 to 4*, solve the given equations by factorization.

1 (a) $x^2 + 4x - 32 = 0$
   (b) $x^2 - 16 = 0$
   (c) $(x + 2)^2 = 16$
   [(a) 4, −8 (b) 4, −4 (c) 2, −6]
2 (a) $2x^2 - x - 3 = 0$
   (b) $6x^2 - 5x + 1 = 0$
   (c) $10x^2 + 3x - 4 = 0$
   $\left[\text{(a) } -1, 1\dfrac{1}{2} \text{ (b) } \dfrac{1}{2}, \dfrac{1}{3} \text{ (c) } \dfrac{1}{2}, -\dfrac{4}{5}\right]$
3 (a) $x^2 - 4x + 4 = 0$
   (b) $21x^2 - 25x = 4$
   (c) $8x^2 + 13x - 6 = 0$
   $\left[\text{(a) } 2 \text{ (b) } 1\dfrac{1}{3}, -\dfrac{1}{7} \text{ (c) } \dfrac{3}{8}, -2\right]$
4 (a) $5x^2 + 13x - 6 = 0$
   (b) $6x^2 - 5x - 4 = 0$
   (c) $8x^2 + 2x - 15 = 0$
   $\left[\text{(a) } \dfrac{2}{5}, -3 \text{ (b) } \dfrac{4}{3}, -\dfrac{1}{2} \text{ (c) } \dfrac{5}{4}, \dfrac{3}{2}\right]$
5 Determine the quadratic equations in x whose roots are
   (a) 3 and 1, (b) 2 and −5, (c) −1 and −4.
   [(a) $x^2 - 4x + 3 = 0$ (b) $x^2 + 3x - 10 = 0$
   (c) $x^2 + 5x + 4 = 0$]
6 Determine the quadratic equations in x whose roots are
   (a) $2\dfrac{1}{2}$ and $-\dfrac{1}{2}$, (b) 6 and −6, (c) 2.4 and −0.7.
   [(a) $4x^2 - 8x - 5 = 0$ (b) $x^2 - 36 = 0$
   (c) $x^2 - 1.7x - 1.68 = 0$]

Solving quadratic equations by 'completing the square'

In *Problems 7 and 8*, solve the given equations by completing the square, correct to 3 decimal places.

7 (a) $x^2 + 4x + 1 = 0$
   (b) $2x^2 + 5x - 4 = 0$
   (c) $3x^2 - x - 5 = 0$
   [(a) −3.732, −0.268 (b) −3.137, 0.637 (c) 1.468, −1.135]
8 (a) $5x^2 - 8x + 2 = 0$
   (b) $4x^2 - 11x + 3 = 0$
   (c) $2x^2 + 5x = 2$
   [(a) 1.290, 0.310 (b) 2.443, 0.307 (c) −2.851, 0.351]

Solving quadratic equations by formula

In *Problems 9 and 10* solve the given equations by using the quadratic formula, correct to 3 decimal places.

9 (a) $2x^2 + 5x - 4 = 0$
   (b) $5.76x^2 + 2.86x - 1.35 = 0$
   (c) $2x^2 - 7x + 4 = 0$
   [(a) 0.637, −3.137 (b) 0.296, −0.792 (c) 2.781, 0.719]
10 (a) $4x + 5 = \dfrac{3}{x}$
   (b) $(2x + 1) = \dfrac{5}{x - 3}$
   (c) $\dfrac{x + 1}{x - 1} = x - 3$
   [(a) 0.443, −1.693 (b) 3.608, −1.108 (c) 4.562, 0.438]

Practical problems involving quadratic equations

11 The angle a rotating shaft turns through in t seconds is given by $\theta = \omega t + \dfrac{1}{2}\alpha t^2$. Determine the time taken to complete 4 radians if ω is 3.0 rad/s and α is 0.60 rad/s^2.
   [1.191 s]
12 The power P developed in an electrical circuit is given by $P = 10I - 8I^2$, where I is the current in amperes. Determine the current necessary to produce a power of 2.5 watts in the circuit.
   [0.345 A or 0.905 A]
13 The area of a triangle is 47.6 cm^2 and its perpendicular height is 4.3 cm more than its base length. Determine the length of the base correct to 3 significant figures.
   [7.84 cm]
14 The sag l metres in a cable stretched between two supports, distance x m apart is given by:
   $$l = \dfrac{12}{x} + x$$
   Determine the distance between supports when the sag is 20 m.
   [0.619 m or 19.38 m]
15 The acid dissociation constant K_a of ethanoic acid is 1.8×1.0^{-5} mol dm^{-3} for a particular solution. Using the Ostwald dilution law

$$K_a = \frac{x^2}{v(1-x)}$$

determine x, the degree of ionization, given that $v = 10 \text{ dm}^3$.

[0.013 3]

16 A rectangular building is 15 m long by 11 m wide. A concrete path of constant width is laid all the way around the building. If the area of the path is 60.0 m², calculate its width correct to the nearest millimetre.

[1.066 m]

17 The total surface area of a closed cylindrical container is 20.0 m³. Calculate the radius of the cylinder if its height is 2.80 m².

[86.78 cm]

18 The bending moment M at a point in a beam is given by

$$M = \frac{3x(20-x)}{2}$$

where x metres is the distance from the point of support. Determine the value of x when the bending moment is 50 N m.

[1.835 m or 18.165 m]

19 A tennis court measures 24 m by 11 m. In the layout of a number of courts an area of ground must be allowed for at the ends and at the sides of each court. If a border of constant width is allowed around each court and the total area of the court and its border is 950 m², find the width of the borders.

[7 m]

20 Two resistors, when connected in series, have a total resistance of 40 ohms. When connected in parallel their total resistance is 8.4 ohms. If one of the resistors has a resistance R_x ohms:

(a) show that $R_x^2 - 40R_x + 420 = 0$ and
(b) calculate the resistance of each.

[12 ohms, 28 ohms]

Solving linear and quadratic equations simultaneously

In *Problems 21* to *23* determine the solutions of the simultaneous equations.

21 $y = x^2 + x + 1$
$y = 4 - x$
[$x = 1$, $y = 3$ and $x = -3$, $y = 7$]

22 $y = 15x^2 + 21x - 11$
$y = 2x - 1$
$\left[x = \frac{2}{5}, y = -\frac{1}{5} \text{ and } x = -1\frac{2}{3}, y = -4\frac{1}{3} \right]$

23 $2x^2 + y = 4 + 5x$
$x + y = 4$
[$x = 0$, $y = 4$ and $x = 3$, $y = 1$]

9

Exponential functions and Napierian logarithms

9.1 The exponential function

An exponential function is one which contains e^x, e being a constant called the exponent and having an approximate value of 2.718 3. The exponent arises from the natural laws of growth and decay and is used as a base for natural or Napierian logarithms.

9.2 Evaluating exponential functions

The value of e^x may be determined by using:

(a) the power series for e^x, or
(b) a calculator, or
(c) tables of exponential functions.

The most common method of evaluating an exponential function is by using a scientific notation calculator, this now having replaced the use of tables. However, let us first look at the power series for e^x.

The power series for e^x

The value of e^x can be calculated to any required degree of accuracy since it is defined in terms of the following **power series**:

$$e^x = 1 + x + \frac{x^2}{2!} + \frac{x^3}{3!} + \frac{x^4}{4!} + \ldots \qquad (9.1)$$

(where $3! = 3 \times 2 \times 1$ and is called 'factorial 3').
The series is valid for all values of x.
The series is said to **converge**, i.e. if all the terms are

added, an actual value for e^x (where x is a real number) is obtained. The more terms that are taken, the closer will be the value of e^x to its actual value. The value of the exponent e, correct to say 4 decimal places, may be determined by substituting $x = 1$ in the power series of equation (9.1). Thus

$$e^1 = 1 + 1 + \frac{(1)^2}{2!} + \frac{(1)^3}{3!} + \frac{(1)^4}{4!} + \frac{(1)^5}{5!} + \frac{(1)^6}{6!} + \frac{(1)^7}{7!}$$
$$+ \frac{(1)^8}{8!} + \ldots$$

$$= 1 + 1 + 0.5 + 0.166\,67 + 0.041\,67 + 0.008\,33$$
$$+ 0.001\,39 + 0.000\,20 + 0.000\,02 + \ldots$$

$$= 2.718\,28$$

i.e. $e = 2.718\,3$ correct to 4 decimal places

The value of $e^{0.05}$, correct to say 8 significant figures, is found by substituting $x = 0.05$ in the power series for e^x. Thus

$$e^{0.05} = 1 + 0.05 + \frac{(0.05)^2}{2!} + \frac{(0.05)^3}{3!} + \frac{(0.05)^4}{4!}$$
$$+ \frac{(0.05)^5}{5!} + \ldots$$

$$= 1 + 0.05 + 0.001\,25 + 0.000\,020\,833$$
$$+ 0.000\,000\,260 + 0.000\,000\,003$$

and by adding,

$$e^{0.05} = 1.051\,271\,1, \text{ correct to 8 significant figures}$$

In this example, successive terms in the series grow smaller very rapidly and it is relatively easy to determine the value of $e^{0.05}$ to a high degree of accuracy. However, when x is nearer to unity or larger than unity, a very large number of terms are required for an accurate result.

If in the series of equation (9.1), x is replaced by $-x$, then

$$e^{-x} = 1 + (-x) + \frac{(-x)^2}{2!} + \frac{(-x)^3}{3!} + \ldots$$

$$e^{-x} = 1 - x + \frac{(x)^2}{2!} - \frac{(x)^3}{3!} + \ldots$$

In a similar manner the power series for e^x may be used to evaluate any exponential function of the form ae^{kx}, where a and k are constants.

In the series of equation (9.1), let x be replaced by kx. Then

$$ae^{kx} = a\left\{1 + (kx) + \frac{(kx)^2}{2!} + \frac{(kx)^3}{3!} + \ldots\right\}$$

Thus $\quad 5e^{2x} = 5\left\{1 + (2x) + \frac{(2x)^2}{2!} + \frac{(2x)^3}{3!} + \ldots\right\}$

$$= 5\left\{1 + 2x + \frac{4x^2}{2} + \frac{8x^3}{6} + \ldots\right\}$$

i.e. $\quad 5e^{2x} = 5\left\{1 + 2x + 2x^2 + \frac{4}{3}x^3 + \ldots\right\}$

Problem 1 Determine the value of $5e^{0.5}$, correct to 5 significant figures by using the power series for e^x.

$$e^x = 1 + x + \frac{x^2}{2!} + \frac{x^3}{3!} + \ldots$$

Hence $\quad e^{0.5} = 1 + 0.5 + \frac{(0.5)^2}{(2)(1)} + \frac{(0.5)^3}{(3)(2)(1)} + \frac{(0.5)^4}{(4)(3)(2)(1)}$

$$+ \frac{(0.5)^5}{(5)(4)(3)(2)(1)} + \frac{(0.5)^6}{(6)(5)(4)(3)(2)(1)}$$

$$= 1 + 0.5 + 0.125 + 0.020\ 833 + 0.002\ 604\ 2$$
$$+ 0.000\ 260\ 4 + 0.000\ 021\ 7$$

i.e. $\quad e^{0.5} = 1.648\ 72$ correct to 6 significant figures
Hence $5e^{0.5} = 5(1.648\ 72) = \mathbf{8.243\ 6}$, correct to 5 significant figures.

Problem 2 Determine the value of $3e^{-1}$, correct to 4 decimal places, using the power series for e^x.

Substituting $x = -1$ in the power series

$$e^x = 1 + x + \frac{x^2}{2!} + \frac{x^3}{3!} + \frac{x^4}{4!} + \ldots$$

gives $\quad e^{-1} = 1 + (-1) + \frac{(-1)^2}{2!} + \frac{(-1)^3}{3!} + \frac{(-1)^4}{4!} + \ldots$

$$= 1 - 1 + 0.5 - 0.166\ 667 + 0.041\ 667$$
$$- 0.008\ 333 + 0.001\ 389 - 0.000\ 198 + \ldots$$

$$= 0.367\ 858 \text{ correct to 6 decimal places}$$

Hence $3e^{-1} = (3)(0.367\ 858) = \mathbf{1.103\ 6}$ correct to 4 decimal places.

Problem 3 Expand $e^x(x^2 - 1)$ as far as the terms in x^5.

The power series for e^x is $1 + x + \frac{x^2}{2!} + \frac{x^3}{3!} + \frac{x^4}{4!} + \ldots$

Hence

$$e^x(x^2 - 1)$$

$$= \left(1 + x + \frac{x^2}{2!} + \frac{x^3}{3!} + \frac{x^4}{4!} + \frac{x^5}{5!} + \ldots\right)(x^2 - 1)$$

$$= \left(x^2 + x^3 + \frac{x^4}{2!} + \frac{x^5}{3!} + \ldots\right)$$
$$- \left(1 + x + \frac{x^2}{2!} + \frac{x^3}{3!} + \frac{x^4}{4!} + \frac{x^5}{5!} + \ldots\right)$$

Grouping like terms gives:

$$e^x(x^2 - 1) = -1 - x + \left(x^2 - \frac{x^2}{2!}\right) + \left(x^3 - \frac{x^3}{3!}\right)$$
$$+ \left(\frac{x^4}{2!} - \frac{x^4}{4!}\right) + \left(\frac{x^5}{3!} - \frac{x^5}{5!}\right) + \ldots$$

$$= -1 - x + \frac{x^2}{2} + \frac{5}{6}x^3 + \frac{11}{24}x^4 + \frac{19}{120}x^5$$

when expanded as far as the term in x^5.

Use of a calculator

Most scientific notation calculators contain an 'e^x' function which enables all practical values of e^x and e^{-x} to be determined, correct to 8 or 9 significant figures. For example

$$e^1 = 2.718\ 281\ 8$$
$$e^{2.4} = 11.023\ 176$$
$$e^{-1.618} = 0.198\ 294\ 89$$

correct to 8 significant figures.

In practical situations the degree of accuracy given by a calculator is often far greater than is appropriate. The accepted convention is that the final result is stated to one significant figure greater than the least significant measured value. Use your calculator to check the following values:

$e^{0.12}$ = 1.127 5, correct to 5 significant figures
$e^{-1.47}$ = 0.229 93, correct to 5 decimal places
$e^{-0.431}$ = 0.649 9, correct to 4 decimal places
$e^{9.32}$ = 11 159, correct to 5 significant figures
$e^{-2.785}$ = 0.061 729 1, correct to 7 decimal places

Problem 4 Using a calculator, evaluate, correct to 5 significant figures: (a) $e^{2.731}$; (b) $e^{-3.162}$; (c) $\dfrac{5}{3}e^{5.253}$.

(a) $e^{2.731}$ = 15.348 227 . . . = **15.348**, correct to 5 significant figures.
(b) $e^{-3.162}$ = 0.042 340 97 . . . = **0.042 341**, correct to 5 significant figures.
(c) $\dfrac{5}{3}e^{5.253} = \dfrac{5}{3}(191.138\ 825 \ldots)$ = **318.56**, correct to 5 significant figures.

Problem 5 Use a calculator to determine the following, each correct to 4 significant figures:

(a) $3.72e^{0.18}$ (b) $53.2e^{-1.4}$ (c) $\dfrac{5}{122}e^{7}$

(a) $3.72e^{0.18}$ = (3.72)(1.197 217 . . .) = **4.454**, correct to 4 significant figures.
(b) $53.2e^{-1.4}$ = (53.2)(0.246 596 . . .) = **13.12**, correct to 4 significant figures.
(c) $\dfrac{5}{122}e^{7} = \left(\dfrac{5}{122}\right)(1096.633\ 1 \ldots)$ = **44.94**, correct to 4 significant figures.

Problem 6 Evaluate the following correct to 4 decimal places, using a calculator:

(a) $0.025\ 6(e^{5.21} - e^{2.49})$
(b) $5\left(\dfrac{e^{0.25} - e^{-0.25}}{e^{0.25} + e^{-0.25}}\right)$

(a) $0.025\ 6(e^{5.21} - e^{2.49})$
= 0.025 6(183.094 058 . . . − 12.061 276 1 . . .)
= **4.378 4**, correct to 4 decimal places

(b) $5\left(\dfrac{e^{0.25} - e^{-0.25}}{e^{0.25} + e^{-0.25}}\right)$

$= 5\left(\dfrac{1.284\ 025\ 41 \ldots - 0.778\ 800\ 78 \ldots}{1.284\ 025\ 41 \ldots + 0.778\ 800\ 78 \ldots}\right)$

$= 5\left(\dfrac{0.505\ 224\ 6 \ldots}{2.062\ 826\ 1 \ldots}\right)$

= **1.224 6**, correct to 4 decimal places

Problem 7 The instantaneous voltage v in a capacitive circuit is related to time t by the equation: $v = Ve^{-t/CR}$ where V, C and R are constants. Determine v, correct to 4 significant figures, when $t = 30 \times 10^{-3}$ seconds, $C = 10 \times 10^{-6}$ farads, $R = 47 \times 10^{3}$ ohms and $V = 200$ volts.

$v = Ve^{-t/CR} = 200e^{(-30\times10^{-3})/(10\times10^{-6}\times47\times10^{3})}$

Using a calculator, $v = 200e^{-0.063\ 829\ 7 \ldots}$
$= 200(0.938\ 164\ 6 \ldots)$
= **187.6 volts**

Further problems on evaluating exponential functions may be found in section 9.7, Problems 1 to 11.

9.3 Graphs of exponential functions

(i) Values of e^{x} and e^{-x}, obtained from a calculator, correct to 2 decimal places, over the range $x = -3$ to $x = 3$, are shown in the table below.

x	−3.0	−2.5	−2.0	−1.5	−1.0	−0.5	0
e^{x}	0.05	0.08	0.14	0.22	0.37	0.61	1.00
e^{-x}	20.09	12.18	7.39	4.48	2.72	1.65	1.00

x	0.5	1.0	1.5	2.0	2.5	3.0
e^{x}	1.65	2.72	4.48	7.39	12.18	20.09
e^{-x}	0.61	0.37	0.22	0.14	0.08	0.05

Fig. 9.1 shows graphs of $y = e^{x}$ and $y = e^{-x}$.
(ii) A similar table may be drawn up for $y = 5e^{1/2x}$, a graph of which is shown in *Fig. 9.2*. The gradient of the curve at any point, dy/dx, is obtained by drawing a tangent to the curve at that point and measuring the gradient of the tangent. For example: when $x = 0$, $y = 5$ and

$$\dfrac{dy}{dx} = \dfrac{BC}{AB} = \dfrac{(6.2 - 3.7)}{1} = 2.5$$

and when $x = 2$, $y = 13.6$ and

$$\dfrac{dy}{dx} = \dfrac{EF}{DE} = \dfrac{(16.8 - 10)}{1} = 6.8$$

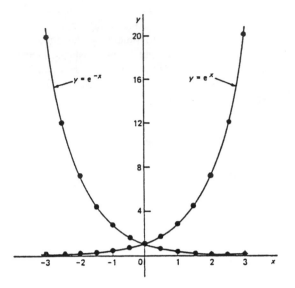

Fig. 9.1

These two results each show that $\dfrac{dy}{dx} = \dfrac{1}{2}y$ and further

determinations of the gradients of $y = 5e^{\frac{1}{2}x}$ would give the same result for each. In general, for all natural growth and decay laws of the form $y = Ae^{kx}$, where k is a positive constant for growth laws (as in *Fig. 9.2*) and a negative constant for decay curves, $dy/dx = ky$, i.e. **the rate of change of the variable, y, is proportional to the variable itself.**

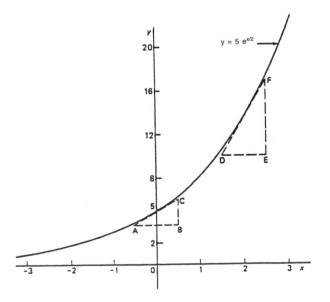

Fig. 9.2

(iii) For any natural law of growth and decay of the form $dy/dx = ky$, the solution is always $y = Ae^{kx}$.

> *Problem 8*   Plot a graph of $y = 2e^{0.3x}$ over a range of $x = -3$ to $x = 3$. Hence determine the value of y when $x = 2.2$ and the value of x when $y = 1.6$.

A table of values is drawn up as shown below.

x	-3	-2	-1	0	1	2	3
$0.3x$	-0.9	-0.6	-0.3	0	0.3	0.6	0.9
$e^{0.3x}$	0.407	0.549	0.741	1.00	1.350	1.822	2.460
$2e^{0.3x}$	0.81	1.10	1.48	2.00	2.70	3.64	4.92

A graph of $y = 2e^{0.3x}$ is shown plotted in *Fig. 9.3*. When $x = 2.2$, $y = \mathbf{3.87}$ and when $y = 1.6$, $x = \mathbf{-0.74}$.

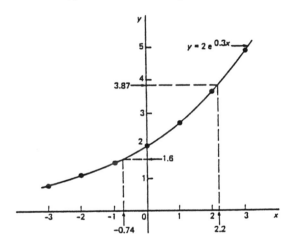

Fig. 9.3

> *Problem 9*   Plot the graph of $y = \dfrac{1}{3}e^{-2x}$ over the range $x = -1.5$ to $x = 1.5$. Determine, from the graph, the value of y when $x = -1.2$ and the value of x when $y = 1.4$.

A table of values is drawn up as shown below.

x	-1.5	-1.0	-0.5	0	0.5	1.0	1.5
$-2x$	3	2	1	0	-1	-2	-3
e^{-2x}	20.086	7.389	2.718	1.00	0.368	0.135	0.050
$\dfrac{1}{3}e^{-2x}$	6.70	2.46	0.91	0.33	0.12	0.05	0.02

A graph of $y = \dfrac{1}{3}e^{-2x}$ is shown in *Fig. 9.4*.

When $x = -1.2$, $y = \mathbf{3.67}$ and when $y = 1.4$, $x = \mathbf{-0.72}$.

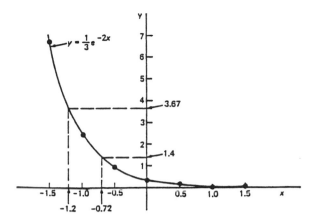

Fig. 9.4

Problem 10 A natural law of growth is of the form $y = 4e^{0.2x}$. Plot a graph depicting this law for values of x from $x = -3$ to $x = 3$. From the graph determine (a) the value of y when x is 2.2, (b) the value of x when y is 3.4 and (c) the rate of change of y with respect to x (i.e. dy/dx) at $x = -2$.

A table of values is drawn up as shown below.

x	-3	-2	-1	0	1	2	3
$0.2x$	-0.6	-0.4	-0.2	0	0.2	0.4	0.6
$e^{0.2x}$	0.549	0.670	0.819	1.00	1.221	1.492	1.822
$4e^{0.2x}$	2.20	2.68	3.28	4.00	4.88	5.97	7.29

A graph of $y = 4e^{0.2x}$ is shown in *Fig. 9.5*. From the graph:

(a) when $x = 2.2$, $y =$ **6.2**,

(b) when $y = 3.4$, $x =$ **−0.8**,

(c) at $x = -2$, gradient, i.e. $\dfrac{dy}{dx} = \dfrac{BC}{AB} = \dfrac{1.08}{2} =$ **0.54**

(From para. (iii), when $y = Ae^{kx}$ then $dy/dx = ky$. In this case $A = 4$, $k = 0.2$ thus $y = 4e^{0.2x}$. Hence, when $x = -2$, $dy/dx = (0.2)4e^{0.2(-2)} =$ **0.54**.)

Problem 11 The decay of voltage, v volts, across a capacitor at time t seconds is given by $v = 250e^{-t/3}$. Draw a graph showing the natural decay curve over the first 6 s. From the graph find (a) the voltage after 3.4 s, and (b) the time when the voltage is 150 V. (c) Determine the rate of change of voltage after 2 s and after 4 s.

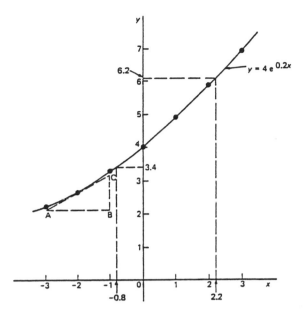

Fig. 9.5

A table of values is drawn up as shown below.

t	0	1	2	3
$e^{-t/3}$	1.00	0.716 5	0.513 4	0.367 9
$v = 250e^{-t/3}$	250.0	179.1	128.4	91.97

t	4	5	6
$e^{-t/3}$	0.263 6	0.188 9	0.135 3
$v = 250e^{-t/3}$	65.90	47.22	33.83

The natural decay curve of $v = 250e^{-t/3}$ is shown in *Fig. 9.6*.

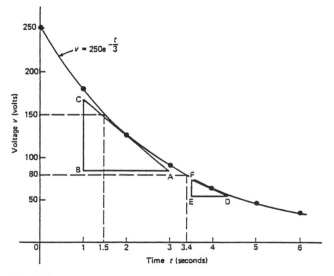

Fig. 9.6

From the graph:

(a) when time $t = 3.4$ s, **voltage $v = 80$ volts**, and
(b) when voltage $v = 150$ volts, **time $t = 1.5$ s**,
(c) Rate of change of voltage (i.e. dv/dt after 2 s is given by BC/AB

$$\frac{BC}{AB} = \frac{(170 - 85)}{(1 - 3)} = \frac{-85}{2} = \mathbf{-42.5 \ V/s}$$

After 4s $\quad \dfrac{dv}{dt} = \dfrac{EF}{DE} = \dfrac{(77 - 55)}{(3.5 - 4.5)} = \dfrac{-22}{1} = \mathbf{-22 \ V/s}$

Problem 12   A law of natural growth is of the form $dy/dx = 8y$. Determine the solution of this equation.

An equation of the form $dy/dx = ky$ has the solution $y = Ae^{kx}$, where A and k are constants (see para. (iii)). With the equation $dy/dx = 8y$, $k = 8$ thus the solution is $\mathbf{y = Ae^{8x}}$.

Problem 13   The rate of change of charge on a capacitor dQ/dt is proportional to the charge Q and is given by $R\dfrac{dQ}{dt} + \dfrac{Q}{C} = 0$ where R and C are constants. Solve the equation for Q.

Rearranging $R\dfrac{dQ}{dt} + \dfrac{Q}{C} = 0$ gives $R\dfrac{dQ}{dt} = -\dfrac{Q}{C}$

Thus $\dfrac{dQ}{dt} = -\dfrac{Q}{CR}$ i.e. $\dfrac{dQ}{dt} = \left(-\dfrac{1}{CR}\right) Q$ which is of the form

$\dfrac{dy}{dx} = ky$ where $k = -\dfrac{1}{CR}$

The solution is $Q = Ae^{(-1/CR)t} = Ae^{-t/CR}$.

Problem 14   The rate of decay of radioactive atoms, dN/dt, is proportional to the number of radioactive atoms N and is given by $dN/dt + 0.8 \times 10^4 N = 0$. Solve the equation for N.

Rearranging $\dfrac{dN}{dt} + 0.8 \times 10^4 N = 0$ gives $\dfrac{dN}{dt} = -0.8 \times 10^4 N$

which is of the form $\dfrac{dy}{dx} = ky$ where $k = -0.8 \times 10^4$.

Hence the solution is $N = Ae^{-0.8 \times 10^4 t}$.

Further problems on graphs of exponential functions may be found in section 9.7, Problems 12 to 22, page 88.

9.4   Napierian logarithms

(i) A **logarithm** of a number is the power to which a base has to be raised to be equal to the number. Thus, if $y = a^x$ then $x = \log_a y$.

(ii) Logarithms having a base of 10 are called **common logarithms** and the common logarithm of x is written as $\lg x$.

(iii) Logarithms having a base of e are called **hyperbolic, Napierian** or **natural logarithms** and the Napierian logarithm of x is written as $\log_e x$, or more commonly, $\ln x$.

9.5   Evaluating Napierian logarithms

The value of a Napierian logarithm may be determined by using:

(a) a calculator, or
(b) a relationship between common and Napierian logarithms, or
(c) Napierian logarithm tables.

The most common method of evaluating a Napierian logarithm is by a scientific notation calculator, this now having replaced the use of four-figure tables, and also the relationship between common and Napierian logarithms,

$$\log_e y = 2.302\ 6 \log_{10} y$$

Most scientific notation calculators contain a '$\ln x$' function which displays the value of the Napierian logarithm of a number when the appropriate key is pressed.

Using a calculator,

$\ln 4.692 = 1.545\ 858\ 9 \ldots = 1.545\ 9$, correct to 4 decimal places

and $\ln 35.78 = 3.577\ 389\ 07 \ldots = 3.577\ 4$, correct to 4 decimal places

Use your calculator to check the following values:

$\ln 1.732 = 0.549\ 28$, correct to 5 significant figures
$\ln 1 = 0$
$\ln 0.52 = -0.653\ 9$, correct to 4 decimal places
$\ln 593 = 6.385\ 2$, correct to 5 significant figures
$\ln 1750 = 7.467\ 4$, correct to 4 decimal places
$\ln 0.17 = -1.772$, correct to 4 significant figures
$\ln 0.000\ 32 = -8.047\ 19$, correct to 6 significant figures
$\ln e^3 = 3$
$\ln e^1 = 1$

From the last two examples we can conclude that

$$\log_e e^x = x$$

This is useful when solving equations involving exponential functions.
For example, to solve $e^{3x} = 8$, take Napierian logarithms of both sides, which gives

$$\ln e^{3x} = \ln 8$$
i.e. $\quad 3x = \ln 8$

from which $x = \dfrac{1}{3} \ln 8 = \mathbf{0.693\ 1}$, correct to 4 decimal places.

Problem 15 Using a calculator evaluate correct to 5 significant figures: (a) ln 47.291; (b) ln 0.062 13; (c) 3.2 ln 762.923.

(a) ln 47.291 = 3.856 320 0 ... = **3.856 3**, correct to 5 significant figures.
(b) ln 0.062 13 = −2.778 526 3 ... = **−2.778 5**, correct to 5 significant figures.
(c) 3.2 ln 762.923 = 3.2(6.637 157 1 ...) = **21.239**, correct to 5 significant figures.

Problem 16 Use a calculator to evaluate the following, each correct to 5 significant figures:

(a) $\dfrac{1}{4} \ln 4.729\ 1$ (b) $\dfrac{\ln 7.869\ 3}{7.869\ 3}$ (c) $\dfrac{5.29 \ln 24.07}{e^{-0.176\ 2}}$

(a) $\dfrac{1}{4} \ln 4.729\ 1 = \dfrac{1}{4}(1.553\ 734\ 9 \ldots) = \mathbf{0.388\ 43}$, correct to 5 significant figures

(b) $\dfrac{\ln 7.869\ 3}{7.869\ 3} = \dfrac{2.062\ 969\ 11 \ldots}{7.869\ 3} = \mathbf{0.262\ 15}$, correct to 5 significant figures

(c) $\dfrac{5.29 \ln 24.07}{e^{-0.176\ 2}} = \dfrac{5.29(3.180\ 966\ 25 \ldots)}{(0.838\ 450\ 27 \ldots)} = \mathbf{20.070}$, correct to 5 significant figures

---

Problem 17 Evaluate the following:

(a) $\dfrac{\ln e^{2.5}}{\lg 10^{0.5}}$

(b) $\dfrac{4e^{2.23} \lg 2.23}{\ln 2.23}$ (correct to 3 decimal places)

(a) $\dfrac{\ln e^{2.5}}{\lg 10^{0.5}} = \dfrac{2.5}{0.5} = \mathbf{5}$

(b) $\dfrac{4e^{2.23} \lg 2.23}{\ln 2.23}$

$= \dfrac{4(9.299\ 866\ 07 \ldots)(0.348\ 304\ 86 \ldots)}{(0.802\ 001\ 58 \ldots)}$

$= \mathbf{16.156}$, correct to 3 decimal places

Problem 18 Solve the equation $7 = 4e^{-3x}$ to find x, correct to 4 significant figures.

Rearranging $7 = 4e^{-3x}$ gives:

$$\dfrac{7}{4} = e^{-3x}$$

Taking the reciprocal of both sides gives:

$$\dfrac{4}{7} = \dfrac{1}{e^{-3x}} = e^{3x}$$

Taking Napierian logarithms of both sides gives:

$$\ln\left(\dfrac{4}{7}\right) = \ln\left(e^{3x}\right)$$

Since $\log_e e^\alpha = \alpha$, then $\ln\left(\dfrac{4}{7}\right) = 3x$.

Hence $x = \dfrac{1}{3} \ln\left(\dfrac{4}{7}\right) = \dfrac{1}{3}(-0.559\ 62) = \mathbf{-0.186\ 5}$, correct to 4 significant figures.

Problem 19 Given $20 = 60(1 - e^{-t/2})$ determine the value of t, correct to 3 significant figures.

Rearranging $20 = 60(1 - e^{-t/2})$ gives:

$$\dfrac{20}{60} = 1 - e^{-t/2}$$

and

$$e^{-t/2} = 1 - \dfrac{20}{60} = \dfrac{2}{3}$$

Taking the reciprocal of both sides gives:

$$e^{t/2} = \dfrac{3}{2}$$

Taking Napierian logarithms of both sides gives:

$$\ln e^{t/2} = \ln \frac{3}{2}$$

i.e. $\dfrac{t}{2} = \ln \dfrac{3}{2}$

from which, $t = 2 \ln \dfrac{3}{2} = \mathbf{0.881}$, correct to 3 significant figures.

Problem 20   Solve the equation $3.72 = \ln \left(\dfrac{5.14}{x} \right)$ to find x.

From the definition of a logarithm, since $3.72 = \ln \left(\dfrac{5.14}{x} \right)$

then $e^{3.72} = \dfrac{5.14}{x}$

Rearranging gives: $x = \dfrac{5.14}{e^{3.72}} = 5.14 \, e^{-3.72}$

i.e. $x = \mathbf{0.124\ 6}$, correct to 4 significant figures.

Further problems on evaluating Napierian logarithms may be found in section 9.7, Problems 23 to 32, page 88.

9.6   Laws of growth and decay

The laws of exponential growth and decay are of the form $y = Ae^{kx}$ and $y = A(1 - e^{kx})$, where A and k are constants. The laws occur frequently in engineering and science and examples of quantities related by a natural law include:

(i)   Linear expansion $l = l_0 e^{\alpha \theta}$
(ii)   Change in electrical resistance
       with temperature $R_\theta = R_0 e^{\alpha \theta}$
(iii) Tension in belts $T_1 = T_0 e^{\mu \alpha}$
(iv)   Newton's law of cooling $\theta = \theta_0 e^{-kt}$
(v)   Biological growth $y = y_0 e^{kt}$
(vi)   Discharge of a capacitor $q = Q e^{-t/CR}$
(vii) Atmospheric pressure $p = p_0 e^{-h/c}$
(viii) Radioactive decay $N = N_0 e^{-\lambda t}$
(ix)   Decay of current in an inductive
       circuit $i = I e^{-Rt/L}$
(x)   Growth of current in a capacitive
       circuit $i = I(1 - e^{-t/CR})$

Problem 21   The resistance R of an electrical conductor at temperature $\theta°C$ is given by $R = R_0 e^{\alpha \theta}$, where α is a constant and $R_0 = 5 \times 10^3$ ohms. Determine the value of α when $R = 6 \times 10^3$ ohms and $\theta = 1500°C$. Also, find the temperature when the resistance R is 5.4×10^3 ohms.

Transposing $R = R_0 e^{\alpha \theta}$ gives $\dfrac{R}{R_0} = e^{\alpha \theta}$

Taking Napierian logarithms of both sides gives:

$$\ln \left(\frac{R}{R_0} \right) = \ln e^{\alpha \theta} = \alpha \theta$$

Hence

$$\alpha = \frac{1}{\theta} \ln \left(\frac{R}{R_0} \right) = \frac{1}{1500} \ln \left(\frac{6 \times 10^3}{5 \times 10^3} \right)$$

$$= \frac{1}{1500} (0.182\ 3)$$

Hence $\alpha = \mathbf{1.215 \times 10^{-4}}$.

From above, $\ln \left(\dfrac{R}{R_0} \right) = \alpha \theta$ hence $\theta = \dfrac{1}{\alpha} \ln \left(\dfrac{R}{R_0} \right)$

When $R = 5.4 \times 10^3$, $\alpha = 1.215 \times 10^{-4}$ and $R_0 = 5 \times 10^3$,

$$\theta = \frac{1}{1.215 \times 10^{-4}} \ln \left(\frac{5.4 \times 10^3}{5 \times 10^3} \right)$$

$$= \frac{10^4}{1.215} (7.696 \times 10^{-2}) = \mathbf{633.4°C}$$

Problem 22   In an experiment involving Newton's law of cooling, the temperature $\theta(°C)$ is given by $\theta = \theta_0 e^{-kt}$. Find the value of constant k when $\theta_0 = 56.6°C$, $\theta = 16.5°C$ and $t = 83.0$ seconds.

Transposing $\theta = \theta_0 e^{-kt}$ gives $\dfrac{\theta}{\theta_0} = e^{-kt}$ from which

$$\frac{\theta_0}{\theta} = \frac{1}{e^{-kt}} = e^{kt}$$

Taking Napierian logarithms of both sides gives:

$$\ln \left(\frac{\theta_0}{\theta} \right) = kt$$

from which,

$$k = \frac{1}{t} \ln \left(\frac{\theta_0}{\theta} \right) = \frac{1}{83.0} \ln \left(\frac{56.6}{16.5} \right) = \frac{1}{83.0}(1.232\ 6)$$

Hence $k = \mathbf{1.485 \times 10^{-2}}$.

Problem 23 The current i amperes flowing in a capacitor at time t seconds is given by $i = 8.0(1 - e^{-t/CR})$, where the circuit resistance R is 25×10^3 ohms and capacitance C is 16×10^{-6} farads. Determine (a) the current i after 0.5 seconds and (b) the time for the current to reach 6.0 A.

(a) Current $i = 8.0(1 - e^{-t/CR})$

$= 8.0[1 - e^{-0.5/(16\times10^{-4})(25\times10^3)}] = 8.0(1 - e^{-1.25})$
$= 8.0(1 - 0.286\ 5) = 8.0(0.713\ 5)$
$= \textbf{5.708 amperes}$

(b) Transposing $i = 8.0(1 - e^{-t/CR})$ gives $\dfrac{i}{8.0} = 1 - e^{-t/CR}$

from which, $e^{-t/CR} = 1 - \dfrac{i}{8.0} = \dfrac{8.0 - i}{8.0}$

Taking the reciprocal of both sides gives: $e^{t/CR} = \dfrac{8.0}{8.0 - i}$

Taking Napierian logarithms of both sides gives:

$$\frac{t}{CR} = \ln\left(\frac{8.0}{8.0 - i}\right)$$

Hence $t = CR \ln\left(\dfrac{8.0}{8.0 - i}\right)$

$= (16 \times 10^{-6})(25 \times 10^3) \ln\left(\dfrac{8.0}{8.0 - 6.0}\right)$

when $i = 6.0$ amperes, i.e.

$t = \dfrac{400}{10^3} \ln\left(\dfrac{8.0}{2.0}\right) = 0.4 \ln 4.0 = 0.4(1.386\ 3)$

$= \textbf{0.554 5 s}$

Problem 24 The temperature θ_2 of a winding which is being heated electrically at time t is given by: $\theta_2 = \theta_1(1 - e^{-t/\tau})$ where θ_1 is the temperature (in degrees Celsius) at time $t = 0$ and τ is a constant.

Calculate (a) θ_1, correct to the nearest degree, when θ_2 is 50°C, t is 30 s and τ is 60 s,
(b) the time t, correct to 1 decimal place, for θ_2 to be half the value of θ_1.

(a) Transposing the formula to make θ_1 the subject gives:

$\theta_1 = \dfrac{\theta_2}{(1 - e^{-t/\tau})} = \dfrac{50}{1 - e^{-30/60}} = \dfrac{50}{1 - e^{-1/2}} = \dfrac{50}{0.393\ 5}$

i.e. $\boldsymbol{\theta_1 = 127°C}$, correct to the nearest degree.
(b) Transposing to make t the subject of the formula gives:

$$\frac{\theta_2}{\theta_1} = 1 - e^{-t/\tau}$$

from which $e^{-t/\tau} = 1 - \dfrac{\theta_2}{\theta_1}$

Hence $-\dfrac{t}{\tau} = \ln\left(1 - \dfrac{\theta_2}{\theta_1}\right)$

i.e. $t = -\tau \ln\left(1 - \dfrac{\theta_2}{\theta_1}\right)$

Since $\theta_2 = \dfrac{1}{2}\theta_1$,

$t = -60 \ln\left(1 - \dfrac{1}{2}\right) = -60 \ln 0.5 = 41.59$ s

Hence the time for the temperature to fall to one half of its original value is 41.6 s, correct to 1 decimal place.

Further problems on the laws of growth and decay may be found in section 9.7, Problems 33 to 43, page 89.

9.7 Further problems on exponential functions and Napierian logarithms

Evaluating exponential functions

In *Problems 1* and *2* use a calculator to evaluate the given functions correct to 4 significant figures.

1 (a) $e^{4.4}$ (b) $e^{-0.25}$ (c) $e^{0.92}$
   [(a) 81.45 (b) 0.778 8 (c) 2.509]
2 (a) $e^{-1.8}$ (b) $e^{-0.78}$ (c) e^{10}
   [(a) 0.165 3 (b) 0.458 4 (c) 22 030]
3 Evaluate, correct to 5 significant figures,

   (a) $3.5e^{2.8}$ (b) $-\dfrac{6}{5}e^{-1.5}$ (c) $2.16e^{5.7}$
   [(a) 57.556 (b) −0.267 76 (c) 645.55]
4 Use a calculator to evaluate the following, correct to 5 significant figures:
   (a) $e^{1.629}$ (b) $e^{-2.748\ 3}$ (c) $0.62e^{4.178}$
   [(a) 5.098 8 (b) 0.064 037 (c) 40.446]

In *Problems 5* and *6*, evaluate correct to 5 decimal places.

5 (a) $\dfrac{1}{7}e^{3.462\ 9}$ (b) $8.52e^{-1.265\ 1}$ (c) $\dfrac{5e^{2.692\ 1}}{3e^{1.117\ 1}}$
   [(a) 4.558 48 (b) 2.404 44 (c) 8.051 24]
6 (a) $\dfrac{5.682\ 3}{e^{-2.134\ 7}}$ (b) $\dfrac{e^{2.112\ 7} - e^{-2.112\ 7}}{2}$ (c) $\dfrac{4(e^{-1.729\ 5} - 1)}{e^{3.681\ 7}}$
   [(a) 48.041 06 (b) 4.074 82 (c) −0.082 86]

7 The length of a bar, l, at temperature θ is given by $l = l_0 e^{\alpha\theta}$, where l_0 and α are constants. Evaluate l, correct to 4 significant figures, where $l_0 = 2.587$, $\theta = 321.7$ and $\alpha = 1.771 \times 10^{-4}$.
[2.739]

8 Evaluate $5.6e^{-1}$, correct to 4 decimal places, using the power series for e^x.
[2.060 1]

9 Use the power series for e^x to determine, correct to 4 significant figures,
(a) e^2 (b) $e^{-0.3}$ and check your result by using a calculator.
[(a) 7.389 (b) 0.740 8]

10 Expand $(1 - 2x)e^{2x}$ as far as the term in x^4.
$$\left[1 - 2x^2 - \frac{8x^3}{3} - 2x^4\right]$$

11 Expand $(2e^{x^2})(x^{1/2})$ to six terms.
$$\left[2x^{1/2} + 2x^{5/2} + x^{9/2} + \frac{1}{3}x^{13/2} + \frac{1}{12}x^{17/2} + \frac{1}{60}x^{21/2}\right]$$

Graphs of exponential functions

12 Plot a graph of $y = 3e^{0.2x}$ over the range $x = -3$ to $x = 3$. Hence determine the value of y when $x = 1.4$ and the value of x when $y = 4.5$.
[3.97, 2.03]

13 Plot a graph of $y = \frac{1}{2} e^{-1.5x}$ over a range $x = -1.5$ to $x = 1.5$ and hence determine the value of y when $x = -0.8$ and the value of x when $y = 3.5$.
[1.66, -1.30]

14 Plot a graph of $y = 2.5e^{-0.15x}$ over a range $x = -8$ to $x = 8$. Determine from the graph the value of y when $x = -6.2$ and the value of x when $y = 5.4$.
[6.34, -5.13]

15 Draw a graph of $y = 2(2e^{-x} - 3e^{2x})$ over a range of $x = -3$ to $x = 3$. Determine the value of y when $x = -2.2$ and the value of x when $y = 17.4$.
[36.0, -1.49]

16 In a chemical reaction the amount of starting material C cm^3 left after t minutes is given by $C = 40e^{-0.006t}$. Plot a graph of C against t and determine (a) the concentration C after 1 hour, (b) the time taken for the concentration to decrease by half. (c) Determine the rate of change of C with t after 40 min.
[(a) 27.9 cm^3 (b) 115.5 min (c) -0.189 cm^3/min]

17 The rate at which a body cools is given by $\theta = 250e^{-0.05t}$, where the excess of temperature of a body above its surroundings at time t minutes is θ°C. Plot a graph showing this natural decay curve for the first hour of cooling and hence determine the rate of cooling after (a) 15 minutes, (b) 45 minutes.
[(a) -5.90°C/min (b) -1.32°C/min]

18 The tensions in two sides of a belt, T and T_0 newtons, passing round a pulley wheel and in contact with the pulley for an angle θ radians is given by $T = T_0 e^{0.3\theta}$. Plot a graph depicting this relationship over a range $\theta = 0$ to $\theta = 2.0$ radians, given $T_0 = 50$ N. From the graph determine the value of $dT/d\theta$ when $\theta = 1.2$ radians.
[21.5 N/rad]

19 The voltage drop, v volts, across an inductor is related to time, t ms, by $v = 30 \times 10^3 e^{-t/10}$. Plot a graph of v against t from $t = 0$ to $t = 10$ ms. Use the graph to determine the rate of change of voltage with time (i.e. dv/dt) when $t = 5.5$ ms.
[-1731 V/ms]

20 Determine the solution of the following equations:
(a) $\dfrac{dT}{d\theta} = 0.4T$ (b) $\dfrac{dm}{di} - 2m = 0$
(c) $\dfrac{dv}{dx} + 9.81v = 0$ (d) $\dfrac{1}{5}\dfrac{dm}{d\theta} - \dfrac{1}{2}m = 0$
[(a) $T = Ae^{0.4\theta}$ (b) $m = Ae^{2i}$ (c) $v = Ae^{-9.81x}$ (d) $m = Ae^{5/2\theta}$]

21 The change of length l of a bar of metal with respect to temperature θ is directly proportional to its length and may be represented by $(dl/d\theta) - \alpha l = 0$, where α is a constant equal to 2.5×10^{-6}. Solve the differential equation for l.
[$l = Ae^{2.5\times10^{-6}\theta}$]

22 The rate of change of voltage across an electrical circuit, dV/dt, is directly proportional to the applied voltage V such that $7.5V - 5.0(dV/dt) = 0$. Solve the equation for V.
[$V = Ae^{1.5t}$]

Evaluating Napierian logarithms

In *Problems 23* to *25* use a calculator to evaluate the given functions, correct to 4 decimal places.

23 (a) $\ln 1.73$ (b) $\ln 5.413$ (c) $\ln 9.412$
[(a) 0.548 1 (b) 1.688 8 (c) 2.242 0]

24 (a) $\ln 17.3$ (b) $\ln 541.3$ (c) $\ln 941 2$
[(a) 2.850 7 (b) 6.294 0 (c) 9.149 7]

25 (a) $\ln 0.173$ (b) $\ln 0.005 413$ (c) $\ln 0.094 12$
[(a) -1.754 5 (b) -5.219 0 (c) -2.363 2]

In *Problems 26* and *27*, evaluate correct to 5 significant figures.

26 (a) $\dfrac{1}{6} \ln 5.293 2$ (b) $\dfrac{\ln 82.473}{4.829}$ (c) $\dfrac{5.62 \ln 321.62}{e^{1.294 2}}$
[(a) 0.277 74 (b) 0.913 74 (c) 8.894 1]

27 (a) $\dfrac{2.946 \ln e^{1.76}}{\lg 10^{1.41}}$ (b) $\dfrac{5e^{-0.162 9}}{2 \ln 0.001 65}$
(c) $\dfrac{\ln 4.862 9 - \ln 2.471 1}{5.173}$
[(a) 3.677 3 (b) -0.331 54 (c) 0.130 87]

In *Problems 28* to *32* solve the given equations, each correct to 4 significant figures.

28 $1.5 = 4e^{2t}$
    [−0.490 4]

29 $7.83 = 2.91e^{-1.7x}$
    [−0.582 2]

30 $16 = 24(1 - e^{-t/2})$
    [2.197]

31 $5.17 = \ln\left(\dfrac{x}{4.64}\right)$
    [816.2]

32 $3.72 \ln\left(\dfrac{1.59}{x}\right) = 2.43$
    [0.827 4]

Laws of growth and decay

33 Two quantities x and y are related by the equation $y = ae^{-kx}$, where a and k are constants. (a) Determine the value of y when $a = 2.114$, $k = -3.20$ and $x = 1.429$. (b) Determine the value of x when $y = 115.4$, $a = 17.8$ and $k = 4.65$.
    [(a) 204.7 (b) −0.402 0]

34 The pressure p pascals at height h metres above ground level is given by $p = p_0 e^{-h/C}$, where p_0 is the pressure at ground level and C is a constant. When p_0 is 1.012×10^5 Pa and the pressure at a height of 1420 m is 9.921×10^4 Pa, determine the value of C.
    [71 500]

35 The length l metres of a metal bar at temperature $t°C$ is given by $l = l_0 e^{\alpha t}$, where l_0 and α are constants. Determine (a) the value of α when $l = 1.993$ m, $l_0 = 1.894$ m and $t = 250°C$, and (b) the value of l_0 when $l = 2.416$, $t = 310°C$ and $\alpha = 1.682 \times 10^{-4}$.
    [(a) 2.038×10^{-4} (b) 2.293 m]

36 The temperature $\theta_2°C$ of an electrical conductor at time t seconds is given by $\theta_2 = \theta_1(1 - e^{-t/T})$, where θ_1 is the initial temperature and T seconds is a constant. Determine (a) θ_1 when $\theta_2 = 50°C$, $t = 30$ s and $T = 80$ s, and (b) the time t for θ_2 to fall to half the value of θ_1 if T remains at 80 s.
    [(a) 159.9°C (b) 55.45 s]

37 Quantities x and y are related by $y = 8.317(1 - e^{cx/t})$, where c and t are constants. Determine (a) the value of y when $c = 2.9 \times 10^{-3}$, $x = 841.2$ and $t = 4.379$, and (b) the value of t when $y = -83.68$, $x = 841.2$ and $c = 2.9 \times 10^{-2}$.
    [(a) −6.201 (b) 10.15]

38 The voltage drop, v volts, across an inductor L henrys at time t seconds is given by $v = 200e^{-Rt/L}$, where $R =$ 150 Ω and $L = 12.5 \times 10^{-3}$ H. Determine (a) the voltage when $t = 160 \times 10^{-6}$ s, and (b) the time for the voltage to reach 85 V.
    [(a) 29.32 volts (b) 71.31×10^{-6} s]

39 A belt is in contact with a pulley for a sector of $\theta = 1.12$ radians and the coefficient of friction between these two surfaces is $\mu = 0.26$.
    (a) Determine the tension on the taut side of the belt, T newtons, when the tension on the slack side is given by $T_0 = 22.7$ newtons, given that these quantities are related by the law $T = T_0 e^{\mu\theta}$.
    (b) It is required that the transmitted force $(T - T_0)$ be increased to 24.0 newtons. Assuming that T_0 remains at 22.7 newtons and θ at 1.12 radians, find the coefficient of friction.
    [(a) 30.4 N (b) 0.644]

40 The instantaneous current i at time t is given by:

$$i = 10e^{-t/CR}$$

when a capacitor is being charged. The capacitance C is 7×10^{-6} farads and the resistance R is 0.3×10^6 ohms. Determine:
    (a) the instantaneous current when t is 2.5 seconds, and
    (b) the time for the instantaneous current to fall to 5 amperes.
    Sketch a curve of current against time from $t = 0$ to $t = 6$ seconds.
    [(a) 3.04 A (b) 1.46 s]

41 The amount of product x (in mol/cm^3) found in a chemical reaction starting with 2.5 mol/cm^3 of reactant is given by $x = 2.5(1 - e^{-4t})$ where t is the time, in minutes, to form product x. Plot a graph at 30 second intervals up to 2.5 minutes and determine x after 1 minute.
    [2.45 mol/cm^3]

42 The current i flowing in a capacitor at time t is given by:
    $$i = 12.5(1 - e^{-t/CR})$$
    where resistance R is 30 kilohms and the capacitance C is 20 microfarads. Determine
    (a) the current flowing after 0.5 seconds, and
    (b) the time for the current to reach 10 amperes.
    [(a) 7.07 A (b) 0.966 s]

43 The amount A after n years of a sum invested P is given by the compound interest law: $A = Pe^{rn/100}$ when the per unit interest rate r is added continuously. Determine, correct to the nearest pound, the amount after 8 years for a sum of £1500 invested if the interest rate is 6% per annum.
    [£2424]

10

Straight line graphs

10.1 Introduction to graphs

A **graph** is a pictorial representation of information showing how one quantity varies with another related quantity.

The most common method of showing a relationship between two sets of data is to use **Cartesian** or **rectangular axes** as shown in *Fig. 10.1*.

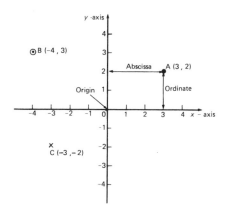

Fig. 10.1

The points on a graph are called **coordinates**. Point A in *Fig. 10.1* has the coordinates (3,2), i.e. 3 units in the x direction and 2 units in the y direction. Similarly, point B has coordinates (-4,3) and C has coordinates (-3,-2). The origin has coordinates (0,0).

The horizontal distance of a point from the vertical axis is called the **abscissa** and the vertical distance from the horizontal axis is called the **ordinate**.

10.2 The straight line graph

Let a relationship between two variables x and y be $y = 3x + 2$.

When $x = 0$, $y = 3(0) + 2 = 2$. When $x = 1$, $y = 3(1) + 2 = 5$. When $x = 2$, $y = 3(2) + 2 = 8$, and so on.

Thus coordinates (0,2), (1,5) and (2,8) have been produced from the equation by selecting arbitrary values of x, and are shown plotted in *Fig. 10.2*. When the points are joined together a **straight-line graph** results.

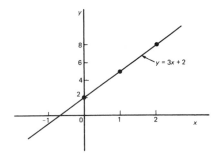

Fig. 10.2

The **gradient** or **slope** of a straight line is the ratio of the change in the value of y to the change in the value of x between any two points on the line. If, as x increases ($\rightarrow$), y also increases ($\uparrow$), then the gradient is positive. In *Fig. 10.3(a)* the gradient of AC

$$= \frac{\text{change in } y}{\text{change in } x} = \frac{\text{CB}}{\text{BA}} = \frac{7 - 3}{3 - 1} = \frac{4}{2} = 2$$

If as x increases ($\rightarrow$), y decreases ($\downarrow$), then the gradient is negative.

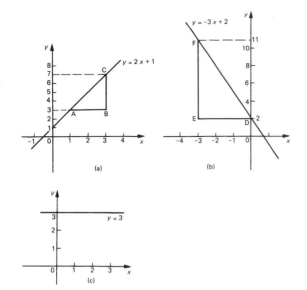

Fig. 10.3

In *Fig. 10.3(b)*, the gradient of DF

$$= \frac{\text{change in } y}{\text{change in } x} = \frac{FE}{ED} = \frac{11 - 2}{-3 - 0} = \frac{9}{-3} = -3$$

Figure 10.3(c) shows a straight line graph $y = 3$. Since the straight line is horizontal the gradient is zero.

The value of y when $x = 0$ is called the **y-axis intercept**. In *Fig. 10.3(a)* the y-axis intercept is 1 and in *Fig. 10.3(b)* is 2.

If the equation of a graph is of the form $y = mx + c$, where m and c are constants, the graph will always be a straight line, m representing the gradient and c the y-axis intercept. Thus $y = 5x + 2$ represents a straight line of gradient 5 and y-axis intercept 2. Similarly, $y = -3x - 4$ represents a straight line of gradient −3 and y-axis intercept −4.

Summary of general rules to be applied when drawing graphs

(i)   Give the graph a title clearly explaining what is being illustrated.
(ii) Choose scales such that the graph occupies as much space as possible on the graph paper being used.
(iii) Choose scales so that interpolation is made as easy as possible. Usually scales such as 1 cm = 1 unit, or 1 cm = 2 units, or 1 cm = 10 units are used. Awkward scales such as 1 cm = 3 units or 1 cm = 7 units should not be used.
(iv) The scales need not start at zero, particularly when starting at zero produces an accumulation of points within a small area of the graph paper.
(v)   The coordinates, or points, should be clearly marked.

This may be done either by a cross, or by a dot and circle, or just by a dot (see *Fig. 10.1*).

(vi) A statement should be made next to each axis explaining the numbers represented with their appropriate units.
(vii) Sufficient numbers should be written next to each axis without cramping.

Problem 1   Plot the graph $y = 4x + 3$ in the range $x = -3$ to $x = +4$. From the graph, find (a) the value of y when $x = 2.2$, and (b) the value of x when $y = -3$.

Whenever an equation is given and a graph is required, a table giving corresponding values of the variable is necessary. The table is achieved as follows:

When $x = -3$, $y = 4x + 3 = 4(-3) + 3 = -12 + 3 = -9$.
When $x = -2$, $y = 4(-2) + 3 = -8 + 3 = -5$, and so on.

Such a table is shown below:

x	−3	−2	−1	0	1	2	3	4
y	−9	−5	−1	3	7	11	15	19

The coordinates (−3,−9), (−2,−5), (−1,−1), and so on, are plotted and joined together to produce the straight line shown in *Fig. 10.4*. (Note that the scales used on the x and y axes do not have to be the same.) From the graph:

(a) when $x = 2.2$, **$y = 11.8$**, and
(b) when $y = -3$, **$x = -1.5$**.

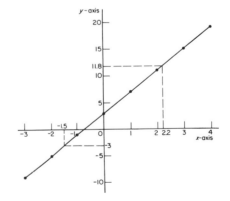

Fig. 10.4   *Graph of* y = 4x + 3

Problem 2   Plot the following graphs on the same axes between the range $x = -4$ to $x = +4$, and determine the gradients of each.

(a) $y = x$ (b) $y = x + 2$ (c) $y = x + 5$ (d) $y = x - 3$

A table of coordinates is produced for each graph.

(a) $y = x$

x	-4	-3	-2	-1	0	1	2	3	4
y	-4	-3	-2	-1	0	1	2	3	4

(b) $y = x + 2$

x	-4	-3	-2	-1	0	1	2	3	4
y	-2	-1	0	1	2	3	4	5	6

(c) $y = x + 5$

x	-4	-3	-2	-1	0	1	2	3	4
y	1	2	3	4	5	6	7	8	9

(d) $y = x - 3$

x	-4	-3	-2	-1	0	1	2	3	4
y	-7	-6	-5	-4	-3	-2	-1	0	1

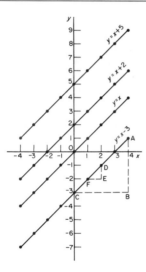

Fig. 10.5   *Graphs of* y = x + 5, y = x + 2, y = x *and* y = x − 3

The coordinates are plotted and joined for each graph. The results are shown in *Fig. 10.5*. Each of the straight lines produced are parallel to each other, i.e. the slope or gradient is the same for each.

To find the gradient of any straight line, say, $y = x - 3$, a horizontal and vertical component needs to be constructed. In *Fig. 10.5*, AB is constructed vertically at $x = 4$ and BC constructed horizontally at $y = -3$. The gradient of

$$AC = \frac{AB}{BC} = \frac{1 - (-3)}{4 - 0} = \frac{4}{4} = 1$$

i.e. the gradient of the straight line $y = x - 3$ is 1. The actual positioning of AB and BC is unimportant for the gradient is also given by

$$\frac{DE}{EF} = \frac{-1 - (-2)}{2 - 1} = \frac{1}{1} = 1$$

The slope or gradient of each of the straight lines in *Fig. 10.5* is thus 1 since they are all parallel to each other.

Problem 3   Plot the following graphs on the same axes between the values $x = -3$ to $x = +3$ and determine the gradient and y-axis intercept of each.

(a) $y = 3x$ (b) $y = 3x + 7$
(c) $y = -4x + 4$ (d) $y = -4x - 5$

A table of coordinates is drawn up for each equation.

(a) $y = 3x$

x	-3	-2	-1	0	1	2	3
y	-9	-6	-3	0	3	6	9

(b) $y = 3x + 7$

x	-3	-2	-1	0	1	2	3
y	-2	1	4	7	10	13	16

(c) $y = -4x + 4$

x	-3	-2	-1	0	1	2	3
y	16	12	8	4	0	-4	-8

(d) $y = -4x - 5$

x	-3	-2	-1	0	1	2	3
y	7	3	-1	-5	-9	-13	-17

Each of the graphs is plotted as shown in *Fig. 10.6*, and each is a straight line. $y = 3x$ and $y = 3x + 7$ are parallel to each other and thus have the same gradient. The gradient of AC is given by

$$AC = \frac{BC}{AB} = \frac{16 - 7}{3 - 0} = \frac{9}{3} = 3$$

Hence the gradient of both $y = 3x$ and $y = 3x + 7$ is 3. $y = -4x + 4$ and $y = -4x - 5$ are parallel to each other and thus have the same gradient. The gradient of DF is given by

$$DF = \frac{EF}{ED} = \frac{-5 - (-17)}{0 - 3} = \frac{12}{-3} = -4$$

Hence the gradient of both $y = -4x + 4$ and $y = -4x - 5$ is −4.

The y-axis intercept means the value of y where the straight line cuts the y-axis. From *Fig. 10.6*,

Fig. 10.6 *Graphs of y = 3x, y = 3x + 7, y = −24x + 4 y = −4x − 5*

$y = 3x$ cuts the y-axis at $y = 0$
$y = 3x + 7$ cuts the y-axis at $y = +7$
$y = −4x + 4$ cuts the y-axis at $y = +4$
and $y = −4x − 5$ cuts the y-axis at $y = −5$

Some general conclusions can be drawn from the graphs shown in *Figs 10.4, 10.5* and *10.6*.
When an equation is of the form $y = mx + c$, where m and c are constants, then

(i) a graph of y against x produces a straight line,
(ii) m represents the slope or gradient of the line, and
(iii) c represents the y-axis intercept.

Thus, given an equation such as $y = 3x + 7$, it may be deduced 'on sight' that its gradient is +3 and its y-axis intercept is +7, as shown in *Fig. 10.6*. Similarly, if $y = −4x − 5$, then the gradient is −4 and the y-axis intercept is −5, as shown in *Fig. 10.6*.
    When plotting a graph of the form $y = mx + c$, only two coordinates need be determined. When the coordinates are plotted a straight line is drawn between the two points. Normally, three coordinates are determined, the third one acting as a check.

Problem 4 The following equations represent straight lines. Determine, without plotting graphs, the gradient and y-axis intercept for each.

(a) $y = 3$ (b) $y = 2x$
(c) $y = 5x − 1$ (d) $2x + 3y = 3$

(a) $y = 3$ (which is of the form $y = 0x + 3$) represents a horizontal straight line intercepting the y-axis at **3**. Since the line is horizontal its **gradient is zero.**

(b) $y = 2x$ is of the form $y = mx + c$, where c is zero. Hence **gradient = 2** and **y-axis intercept = 0** (i.e. the origin).
(c) $y = 5x − 1$ is of the form $y = mx + c$. Hence **gradient = 5** and **y-axis intercept = −1**.
(d) $2x + 3y = 3$ is not in the form $y = mx + c$ as it stands. Transposing to make y the subject gives $3y = 3 − 2x$, i.e.

$$y = \frac{3 − 2x}{3} = \frac{3}{3} − \frac{2x}{3}$$

i.e. $y = −\dfrac{2x}{3} + 1$ which is of the form $y = mx + c$.

Hence **gradient** $= −\dfrac{2}{3}$ and **y-axis intercept = +1**.

Problem 5 Without plotting graphs, determine the gradient and y-axis intercept values of the following equations:

(a) $y = 7x − 3$ (b) $3y = −6x + 2$
(c) $y − 2 = 4x + 9$ (d) $\dfrac{y}{3} = \dfrac{x}{3} − \dfrac{1}{5}$
(e) $2x + 9y + 1 = 0$.

(a) $y = 7x − 3$ is of the form $y = mx + c$, hence **gradient, m = 7** and **y-axis intercept, c = −3**.
(b) Rearranging $3y = −6x + 2$ gives $y = \dfrac{6x}{3} + \dfrac{2}{3}$, i.e.

$y = −2x + \dfrac{2}{3}$ which is of the form $y = mx + c$. Hence

gradient m = −2 and **y-axis intercept, c** $= \dfrac{2}{3}$.

(c) Rearranging $y − 2 = 4x + 9$ gives $y = 4x + 11$, hence **gradient = 4** and **y-axis intercept = 11**.
(d) Rearranging

$\dfrac{y}{3} = \dfrac{x}{2} − \dfrac{1}{5}$ gives $y = 3\left(\dfrac{x}{2} − \dfrac{1}{5}\right) = \dfrac{3}{2}x − \dfrac{3}{5}$

Hence **gradient** $= \dfrac{3}{2}$ and **y-axis intercept** $= −\dfrac{3}{5}$.

(e) Rearranging $2x + 9y + 1 = 0$ gives $9y = −2x − 1$, i.e.

$$y = −\frac{2}{9}x − \frac{1}{9}$$

Hence **gradient** $= −\dfrac{2}{9}$ and **y-axis intercept** $= −\dfrac{1}{9}$.

Problem 6 Determine the gradient of the straight line graph passing through the coordinates (a) (−2,5) and (3,4) (b) (−2,−3) and (−1,3).

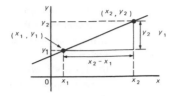

Fig. 10.7

A straight line graph passing through coordinates (x_1, y_1) and (x_2, y_2) has a gradient given by

$$m = \frac{y_2 - y_1}{x_2 - x_1} \quad \text{(see } Fig. \text{ } 10.7)$$

(a) A straight line passes through $(-2,5)$ and $(3,4)$, hence $x_1 = -2$, $y_1 = 5$, $x_2 = 3$ and $y_2 = 4$; hence gradient m

$$= \frac{y_2 - y_1}{x_2 - x_1} = \frac{4 - 5}{3 - (-2)} = -\frac{1}{5}$$

(b) A straight line passes through $(-2,-3)$ and $(-1,3)$, hence $x_1 = -2$, $y_1 = -3$, $x_2 = -1$ and $y_2 = 3$; hence gradient m

$$= \frac{y_2 - y_1}{x_2 - x_1} = \frac{3 - (-3)}{-1 - (-2)} = \frac{3 + 3}{-1 + 2} = \frac{6}{1} = 6$$

Problem 7 Plot the graphs $3x + y + 1 = 0$ and $2y - 5 = x$ on the same axes and find their point of intersection.

Rearranging $3x + y + 1 = 0$ gives $y = -3x - 1$.

Rearranging $2y - 5 = x$ gives $2y = x + 5$ and $y = \frac{1}{2}x + 2\frac{1}{2}$

Since both equations are of the form $y = mx + c$ both are straight lines.

Knowing an equation is a straight line means that only two coordinates need be plotted and a straight line drawn through them. A third coordinate is usually determined to act as a check. A table of values is produced for each equation as shown below

x	1	0	−1
$-3x - 1$	−4	−1	2

x	2	0	−3
$\frac{1}{2}x + 2\frac{1}{2}$	$3\frac{1}{2}$	$2\frac{1}{2}$	1

The graphs are plotted as shown in *Fig. 10.8*.

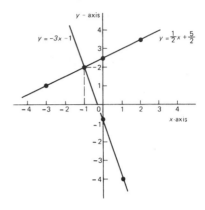

Fig. 10.8

The two straight lines are seen to intersect at $(-1,2)$.

Further problems on straight line graphs may be found in section 10.4, Problems 1 to 11, page 97.

10.3 Practical problems involving straight line graphs

When a set of coordinate values are given or are obtained experimentally and it is believed that they follow a law of the form $y = mx + c$, then if a straight line can be drawn reasonably close to most of the coordinate values when plotted, this verifies that a law of the form $y = mx + c$ exists. From the graph, constants m (i.e. gradient) and c (i.e. y-axis intercept) can be determined. This technique is called **determination of law** (see Chapter 11).

Problem 8 The temperature in degrees Celsius and the corresponding values in degrees Fahrenheit are shown in the table below. Construct rectangular axes, choose a suitable scale and plot a graph of degrees Celsius (on the horizontal axis) against degrees Fahrenheit (on the vertical scale).

°C	10	20	40	60	80	100
°F	50	68	104	140	176	212

From the graph find (a) the temperature in degrees Fahrenheit at 55°C, (b) the temperature in degrees Celsius at 167°F, (c) the Fahrenheit temperature at 0°C, and (d) the Celsius temperature at 230°F.

The coordinates (10,50), (20,68), (40,104), and so on are plotted as shown in *Fig. 10.9*. When the coordinates are joined, a straight line is produced. Since a straight line results there is a linear relationship between degrees Celsius and degrees Fahrenheit.

(a) To find the Fahrenheit temperature at 55°C a vertical line AB is constructed from the horizontal axis to meet the straight line at B. The point where the horizontal line BD meets the vertical axis indicates the equivalent Fahrenheit temperature.
Hence 55°C is equivalent to 131°F.
This process of finding an equivalent value in between the given information in the above table is called **interpolation**.

(b) To find the Celsius temperature at 167°F, a horizontal line EF is constructed as shown in *Fig. 10.9*. The point where the vertical line FG cuts the horizontal axis indicates the equivalent Celsius temperature. **Hence 167°F is equivalent to 75°C.**

(c) If the graph is assumed to be linear even outside of the given data, then the graph may be extended at both ends (shown by broken lines in *Fig. 10.9*).
From *Fig. 10.9*, **0°C corresponds to 32°F.**

(d) **230°F is seen to correspond to 110°C.**
The process of finding equivalent values outside of the given range is called **extrapolation**.

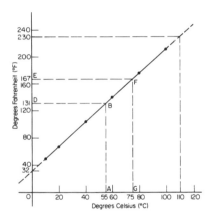

Fig. 10.9 *Graph of degrees Celsius against degrees Fahrenheit*

Problem 9   In an experiment on Charles's law, the value of the volume of gas, V m^3, was measured for various temperatures T°C. Results are shown below.

V m^3	25.0	25.8	26.6	27.4	28.2	29.0
T°C	60	65	70	75	80	85

Plot a graph of volume (vertical) against temperature (horizontal) and from it find (a) the temperature when the volume is 28.6 m^3, and (b) the volume when the temperature is 67°C.

If a graph is plotted with both the scales starting at zero then the result is as shown in *Fig. 10.10*. All of the points

lie in the top right-hand corner of the graph, making interpolation difficult. A more accurate graph is obtained if the temperature axis starts at 55°C and the volume axis starts at 24.5 m^3. The axes corresponding to these values is shown by the broken lines in *Fig. 10.10* and are called **false axes**, since the origin is not now at zero. A magnified version of this relevant part of the graph is shown in *Fig. 10.11*. From the graph:

(a) when the volume is 28.6 m^3, the equivalent temperature is **82.5°C**, and

(b) when the temperature is 67°C, the equivalent volume is **26.1 m^3**.

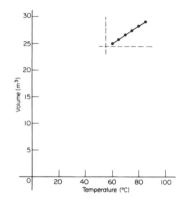

Fig. 10.10 *Graph of volume against temperature with a zero origin*

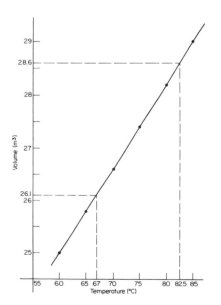

Fig. 10.11 *Graph of volume against temperature with a non-zero origin*

Problem 10 In an experiment demonstrating Hooke's law, the strain in an aluminium wire was measured for various stresses. The results were:

Stress N/mm^2	4.9	8.7	15.0
Strain	0.000 07	0.000 13	0.000 21

Stress N/mm^2	18.4	24.2	27.3
Strain	0.000 27	0.000 34	0.000 39

Plot a graph of stress (vertically) against strain (horizontally). Find:

(a) Young's Modulus of Elasticity for aluminium, which is given by the gradient of the graph,
(b) the value of the strain at a stress of 20 N/mm^2, and
(c) the value of the stress when the strain is 0.000 20.

The coordinates (0.000 07,4.9), (0.000 13,8.7), and so on, are plotted as shown in *Fig. 10.12*. The graph produced is the best straight line which can be drawn corresponding to these points. (With experimental results it is unlikely that all the points will lie exactly on a straight line.) The graph, and each of its axes, are labelled. Since the straight line passes through the origin, then stress is directly proportional to strain for the given range of values.

(a) The gradient of the straight line,

$$AC = \frac{AB}{BC} = \frac{28 - 7}{0.000\ 40 - 0.000\ 10} = \frac{21}{0.000\ 30}$$

$$= \frac{21}{3 \times 10^{-4}} = \frac{7}{10^{-4}} = 7 \times 10^4 = 70\ 000\ \text{N/mm}^2$$

Thus Young's Modulus of Elasticity for aluminium is 70 000 N/mm^2. Since $1\ \text{m}^2 = 10^6\ \text{mm}^2$, 70 000 N/mm^2 is equivalent to $70\ 000 \times 10^6\ \text{N/m}^2$, i.e. **$70 \times 10^9\ \text{N/m}^2$ (or Pascals).**

From *Fig. 10.12*:
(b) the value of the strain at a stress of 20 N/mm^2 is **0.000 285,** and
(c) the value of the stress when the strain is 0.000 20 is **14 N/mm^2.**

Problem 11 The following values of resistance R ohms and corresponding voltage V volts are obtained from a test on a filament lamp.

R ohms	30	48.5	73	107	128
V volts	16	29	52	76	94

Choose suitable scales and plot a graph with R representing the vertical axis and V the horizontal axis. Determine (a) the slope of the graph, (b) the R axis intercept value, (c) the equation of the graph, (d) the value of resistance when the voltage is 60 V, and (e) the value of the voltage when the resistance is 40 ohms. (f) If the graph were to continue in the same manner, what value of resistance would be obtained at 110 V?

The coordinates (16,30), (29,48.5), and so on, are shown plotted in *Fig. 10.13* where the best straight line is drawn through the points.

(a) The slope or gradient of the straight line,

$$AC = \frac{AB}{BC} = \frac{135 - 10}{100 - 0} = \frac{125}{100} = \mathbf{1.25}$$

(Note that the vertical line AB and the horizontal line BC may be constructed anywhere along the length of the straight line. However, calculations are made easier

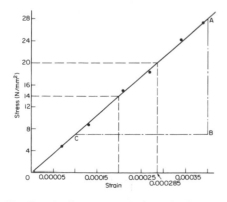

Fig. 10.12 *Graph of stress against strain for aluminium*

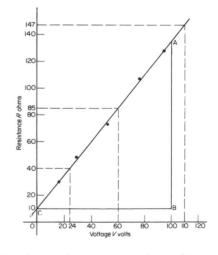

Fig. 10.13 *Graph of resistance against voltage*

if the horizontal line BC is carefully chosen, in this case, 100.)
(b) The *R*-axis intercept is at **R = 10 ohms** (by extrapolation).
(c) The equation of a straight line is $y = mx + c$, when y is plotted on the vertical axis and x on the horizontal axis. m represents the gradient and c the y-axis intercept. In this case, R corresponds to y, V corresponds to x, $m = 1.25$ and $c = 10$. Hence the equation of the graph is $R = (1.25V + 10)\ \Omega$.

From *Fig. 10.13*,
(d) when the voltage is 60 V, the resistance is **85 Ω**,
(e) when the resistance is 40 ohms, the voltage is **24 V**, and
(f) by extrapolation, when the voltage is 110 V, the resistance is **147 Ω**.

Problem 12 Experimental tests to determine the breaking stress σ of rolled copper at various temperatures t gave the following results.

Stress σ N/cm^2	8.46	8.02	7.75	7.35	7.06	6.63
Temperature $t°$C	70	200	280	410	500	640

Show that the values obey the law $\sigma = at + b$, where a and b are constants and determine approximate values for a and b. Use the law to determine the stress at 250°C and the temperature when the stress is 7.54 N/cm^2.

The coordinates (70,8.46), (200,8.04), and so on, are plotted as shown in *Fig. 10.14*. Since the graph is a straight line then the values obey the law $\sigma = at + b$, and the gradient of the straight line is

$$a = \frac{AB}{BC} = \frac{8.36 - 6.76}{100 - 600} = \frac{1.60}{-500} = -0.003\ 2$$

Vertical axis intercept, $b = 8.68$.
Hence the law of the graph is $\sigma = -0.003\ 2t + 8.68$.
When the temperature is 250°C, stress σ is given by

$$\sigma = -0.003\ 2(250) + 8.68 = 7.88\ \text{N/cm}^2$$

Rearranging $\sigma = -0.003\ 2t + 8.68$ gives

$$0.003\ 2t = 8.68 - \sigma, \text{ i.e. } t = \frac{8.68 - \sigma}{0.003\ 2}$$

Hence when the stress $\sigma = 7.54$ N/cm^2, temperature

$$t = \frac{8.68 - 7.54}{0.003\ 2} = 356.3°\text{C}$$

Further practical problems involving straight line graphs may be found in section 10.4, Problems 12 to 22, page 98.

10.4 Further problems on straight line graphs

The straight line graph

1 Corresponding values obtained experimentally for two quantities are:

x	−2.0	−0.5	0	1.0	2.5	3.0	5.0
y	−13.0	−5.5	−3.0	2.0	9.5	12.0	22.0

Use a horizontal scale for x of 1 cm = $\frac{1}{2}$unit and a vertical scale for y of 1 cm = 2 units and draw a graph of x against y. Label the graph and each of its axes. By interpolation, find from the graph the value of y when x is 3.5.
[14.5]

2 The equation of a line is $4y = 2x + 5$. A table of corresponding values is produced and is shown below. Complete the table and plot a graph of y against x. Find the gradient of the graph.

x	−4	−3	−2	−1	0	1	2	3	4
y		−0.50			1.25				3.25

$\left[\frac{1}{2}\right]$

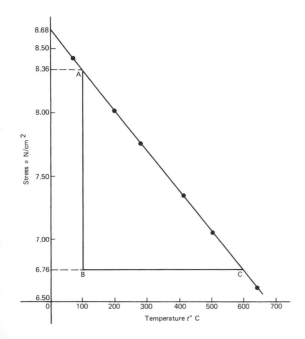

Fig. 10.14 *Graph of stress against temperature*

3   Determine the gradient and intercept on the y-axis for each of the following equations:
   (a) $y = 4x - 2$ (b) $y = -x$
   (c) $y = -3x - 4$ (d) $y = 4$
   [(a) 4,–2 (b) –1,0 (c) –3,–4 (d) 0,4]

4   Find the gradient and intercept on the y-axis for each of the following equations:
   (a) $2y - 1 = 4x$ (b) $6x - 2y = 5$

   (c) $3(2y - 1) = \dfrac{x}{4}$

   $\left[\text{(a) } 2\dfrac{1}{2} \text{ (b) } 3,-2\dfrac{1}{2} \text{ (c) } \dfrac{1}{24},\dfrac{1}{2} \right]$

Determine the gradient and y-axis intercept for each of the equations in *Problems 5* and *6* and sketch the graphs.

5   (a) $y = 6x - 3$ (b) $y = -2x + 4$
   (c) $y = 3x$ (d) $y = 7$
   [(a) 6,–3 (b) –2,4 (c) 3,0 (d) 0,7]

6   (a) $2y + 1 = 4x$ (b) $2x + 3y + 5 = 0$

   (c) $3(2y - 4) = \dfrac{x}{3}$ (d) $5x - \dfrac{y}{2} - \dfrac{7}{3} = 0$

   $\left[\text{(a) } 2,-\dfrac{1}{2} \text{ (b) } -\dfrac{2}{3},-1\dfrac{2}{3} \text{ (c) } \dfrac{1}{18},2 \text{ (d) } 10,-4\dfrac{2}{3} \right]$

7   Determine the gradient of the straight line graphs passing through the coordinates
   (a) (2,7) and (–3,4)
   (b) (–4,–1) and (–5,3)

   (c) $\left(\dfrac{1}{4}, -\dfrac{3}{4} \right)$ and $\left(-\dfrac{1}{2}, \dfrac{5}{8} \right)$

   $\left[\text{(a) } \dfrac{3}{5} \text{ (b) } -4 \text{ (c) } -1\dfrac{5}{6} \right]$

8   State which of the following equations will produce graphs which are parallel to one another:
   (a) $y - 4 = 2x$ (b) $4x = -(y + 1)$

   (c) $x = \dfrac{1}{2}(y + 5)$ (d) $1 + \dfrac{1}{2}y = \dfrac{3}{2}x$

   (e) $2x = \dfrac{1}{2}(7 - y)$

   [(a) and (c), (b) and (e)]

9   Draw a graph of $y - 3x + 5 = 0$ over a range of $x = -3$ to $x = 4$. Hence determine (a) the value of y when $x = 1.3$ and (b) the value of x when y is –9.2.
   [(a) –1.1 (b) –1.4]

10   Draw on the same axes the graphs of $y = 3x - 5$ and $3y + 2x = 7$. Find the coordinates of the point of intersection. Check the result obtained by solving the two simultaneous equations algebraically.
   [(2,1)]

11   Plot the graphs $y = 2x + 3$ and $2y = 15 - 2x$ on the same axes and determine their point of intersection.

   $\left[\left(1\dfrac{1}{2}, 6 \right) \right]$

Practical problems involving straight line graphs

12   The resistance R ohms of a copper winding is measured at various temperatures $t°C$ and the results are as follows:

R ohms	112	120	126	131	136
$t°C$	20	36	48	58	64

Plot a graph of R (vertically) against t (horizontally) and find from it (a) the temperature when the resistance is 122 Ω and (b) the resistance when the temperature is 52°C.
[(a) 40°C (b) 128 Ω]

13   The speed of a motor varies with armature voltage as shown by the following experimental results:

n (rev/min)	285	517	615	750	917	1050
V (volts)	60	95	110	130	155	175

Plot a graph of speed (horizontally) against voltage (vertically) and draw the best straight line through the points. Find from the graph (a) the speed at a voltage of 145 V, and (b) the voltage at a speed of 400 rev/min.
[(a) 850 rev/min (b) 77.5 V]

14   The following table gives the force F newtons which, when applied to a lifting machine, overcomes a corresponding load of L newtons.

Force F newtons	25	47	64	120	149	187
Load L newtons	50	140	210	430	550	700

Choose suitable scales and plot a graph of F (vertically) against L (horizontally). Draw the best straight line through the points. Determine from the graph (a) the gradient, (b) the F-axis intercept, (c) the equation of the graph, (d) the force applied when the load is 310 N, and (e) the load that a force of 160 N will overcome. (f) If the graph were to continue in the same manner, what value of force will be needed to overcome a 800 N load?
$\left[\begin{array}{l} \text{(a) } 0.25 \text{ (b) } 12 \text{ (c) } F = 0.25L + 12 \\ \text{(d) } 89.5 \text{ N (e) } 592 \text{ N (f) } 212 \text{ N} \end{array} \right]$

15   The following table gives the results of tests carried out to determine the breaking stress σ of rolled copper at various temperatures, t:

Stress σ (N/cm^2)	8.51	8.07	7.80	7.47	7.23	6.78
Temperature t (°C)	75	220	310	420	500	650

Plot a graph of stress (vertically) against temperature (horizontally). Draw the best straight line through the plotted coordinates. Determine the slope of the graph and the vertical axis intercept.
[–0.003,8.73]

16 The velocity v of a body after varying time intervals t was measured as follows:

t (seconds)	2	5	8	11	15	18
v (m/s)	16.9	19.0	21.1	23.2	26.0	28.1

Plot v vertically and t horizontally and draw a graph of velocity against time. Determine from the graph (a) the velocity after 10 s, (b) the time at 20 m/s and (c) the equation of the graph.
[(a) 22.5 m/s (b) 6.43 s (c) $v = 0.7t + 15.5$]

17 The mass m of a steel joist varies with length l as follows:

mass, m (kg)	80	100	120	140	160
length, l (m)	3.00	3.74	4.48	5.23	5.97

Plot a graph of mass (vertically) against length (horizontally). Determine the equation of the graph.
[$m = 26.9l - 0.63$]

18 The crushing strength of mortar varies with the percentage of water used in its preparation, as shown below.

Crushing strength, F (tonnes)	1.64	1.36	1.07	0.78	0.50	0.22
% of water used, $w\%$	6	9	12	15	18	21

Plot a graph of F (vertically) against w (horizontally).
(a) Interpolate and determine the crushing strength when 10% of water is used.
(b) Assuming the graph continues in the same manner extrapolate and determine the percentage of water used when the crushing strength is 0.15 tonnes.
(c) What is the equation of the graph?
[(a) 1.26 t (b) 21.68% (c) $F = -0.095w + 2.21$]

19 The velocity v of a body after varying time intervals t was measured as follows:

t seconds	2	5	7	10	14	17
v m/s	15.5	17.3	18.5	20.3	22.7	24.5

Plot a graph with velocity vertical and time horizontal. Determine from the graph (a) the gradient, (b) the vertical axis intercept, (c) the equation of the graph, (d) the velocity after 12.5 s, and (e) the time when the velocity is 18 m/s.
[(a) 0.6 (b) 14.3 (c) $v = 0.6t + 14.3$ (d) 21.8 m/s (e) 6.17 s]

20 In an experiment demonstrating Hooke's law, the strain in a copper wire was measured for various stresses. The results were:

Stress (pascals)	10.6×10^6	18.2×10^6	24.0×10^6
Strain	0.000 11	0.000 19	0.000 25

Stress (pascals)	30.7×10^6	39.4×10^6
Strain	0.000 32	0.000 41

Plot a graph of stress (vertically) against strain (horizontally).
Determine (a) Young's modulus of elasticity for copper, which is given by the gradient of the graph, (b) the value of strain at a stress of 21×10^6 Pa, (c) the value of stress when the strain is 0.000 30.
[(a) 96×10^9 Pa (b) 0.000 22 (c) 28.8×10^6 Pa]

21 An experiment with a set of pulley blocks gave the following results:

Effort, E (newtons)	19.0	11.0	13.6	17.4	20.8	23.6
Load, L (newtons)	15	25	38	57	74	88

Plot a graph of effort (vertically) against load (horizontally) and determine (a) the gradient, (b) the vertical axis intercept, (c) the law of the graph, (d) the effort when the load is 30 N and (e) the load when the effort is 19 N.
[(a) $\frac{1}{5}$ (b) 6 (c) $E = \frac{1}{5}L + 6$ (d) 12 N (e) 65 N]

22 The variation of pressure p in a vessel with temperature T is believed to follow a law of the form $p = aT + b$, where a and b are constants. Verify this law for the results given below and determine the approximate values of a and b. Hence determine the pressures at temperatures of 285 K and 310 K and the temperature at a pressure of 250 kPa.

pressure, p kPa	244	247	252	258	262	267
temperature, T K	273	277	282	289	294	300

[$a = 0.85$, $b = 12$, 254.3 kPa, 275.5 kPa, 280 K]

11

Reduction of non-linear laws to linear form

11.1 Determination of law

Frequently, the relationship between two variables, say x and y, is not a linear one, i.e. when x is plotted against y a curve results. In such cases the non-linear equation may be modified to the linear form, $y = mx + c$, so that the constants, and thus the law relating the variables can be determined. This technique is called '**determination of law**'.

Some examples of the reduction of equations to linear form include:

(i) $y = ax^2 + b$ compares with $Y = mX + c$, where $m = a$, $c = b$ and $X = x^2$. Hence y is plotted vertically against x^2 horizontally to produce a straight line graph of gradient 'a' and y-axis intercept 'b'.

(ii) $y = \dfrac{a}{x} + b$

y is plotted vertically against $1/x$ horizontally to produce a straight line graph of gradient 'a' and y-axis intercept 'b'.

(iii) $y = ax^2 + bx$

Dividing both sides by x gives $y/x = ax + b$.
Comparing with $Y = mX + c$ shows that y/x is plotted vertically against x horizontally to produce a straight line graph of gradient 'a' and y/x axis intercept 'b'.

(iv) $y = ax^n$

Taking logarithms to a base of 10 of both sides gives:

$$\lg y = \lg (ax^n) = \lg a + \lg x^n$$

i.e. $\qquad\qquad \lg y = n \lg x + \lg a$
which compares with $Y = mX + c$
and shows that $\lg y$ is plotted vertically against $\lg x$ horizontally to produce a straight line graph of gradient n and $\lg y$-axis intercept $\lg a$.

(v) $y = ab^x$

Taking logarithms to a base of 10 of both sides gives:

$$\lg y = \lg (ab^x)$$
i.e. $\lg y = \lg a + \lg b^x$
i.e. $\lg y = x \lg b + \lg a$

or $\qquad\qquad \lg y = (\lg b)x + \lg a$
which compares with $Y = mX + c$
and shows that $\lg y$ is plotted vertically against x horizontally to produce a straight line graph of gradient $\lg b$ and $\lg y$-axis intercept $\lg a$.

(vi) $y = ae^{bx}$

Taking logarithms to a base of e of both sides gives:

$$\ln y = \ln (ae^{bx})$$
i.e. $\ln y = \ln a + \ln e^{bx}$
i.e. $\ln y = \ln a + bx \ln e$

i.e. $\qquad\qquad \ln y = bx + \ln a$
which compares with $Y = mX + c$
and shows that $\ln y$ is plotted vertically against x horizontally to produce a straight line graph of gradient b and $\ln y$-axis intercept $\ln a$.

11.2 Worked problems on reducing non-linear laws to linear form

---

Problem 1 Experimental values of x and y, shown below, are believed to be related by the law $y = ax^2 + b$. By plotting a suitable graph verify this law and determine approximate values of a and b.

x	1	2	3	4	5
y	9.8	15.2	24.2	36.5	53.0

If y is plotted against x a curve results and it is not possible to determine the values of constants a and b from the curve. Comparing $y = ax^2 + b$ with $Y = mX + c$ shows that y is to be plotted vertically against x^2 horizontally. A table of values is drawn up as shown below.

x	1	2	3	4	5
x^2	1	4	9	16	25
y	9.8	15.2	24.2	36.5	53.0

A graph of y against x^2 is shown in *Fig. 11.1*, with the best straight line drawn through the points. Since a straight line graph results, the law is verified.

From the graph, gradient $a = \dfrac{AB}{BC} = \dfrac{53 - 17}{25 - 5} = \dfrac{36}{20} = \mathbf{1.8}$

and the y-axis intercept, $b = \mathbf{8.0}$.
Hence the law of the graph is $y = \mathbf{1.8}x^2 + \mathbf{8.0}$.

Problem 2 Values of load L newtons and distance d metres obtained experimentally are shown in the following table.

Load, L N	32.3	29.6	27.0	23.2
distance, d m	0.75	0.37	0.24	0.17

Load, L N	18.3	12.8	10.0	6.4
distance, d m	0.12	0.09	0.08	0.07

Verify that load and distance are related by a law of the form $L = \dfrac{a}{d} + b$ and determine approximate values of a and b. Hence calculate the load when the distance is 0.20 m and the distance when the load is 20 N.

Comparing $L = \dfrac{a}{d} + b$ i.e. $L = a\dfrac{1}{d} + b$ with $Y = mX + c$

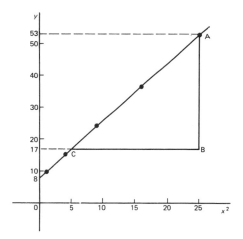

Fig. 11.1 *Graph of y against x^2*

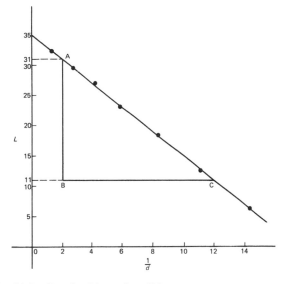

Fig. 11.2 *Graph of L against 1/d*

shows that L is to be plotted vertically against $1/d$ horizontally. Another table of values is drawn up as shown below.

L	32.3	29.6	27.0	23.2	18.3	12.8	10.0	6.4
d	0.75	0.37	0.24	0.17	0.12	0.09	0.08	0.07
$1/d$	1.33	2.70	4.17	5.88	8.33	11.1	12.5	14.3

A graph of L against $1/d$ is shown in *Fig. 11.2*. A straight line can be drawn through the points, which verifies that load and distance are related by a law of the form

$$L = \frac{a}{d} + b$$

Gradient of straight line, $a = \dfrac{AB}{BC} = \dfrac{31 - 11}{2 - 12} = \dfrac{20}{-10} = \mathbf{-2}$

L-axis intercept, $b = \mathbf{35}$.

Hence the law of the graph is $L = -\dfrac{2}{d} + \mathbf{35}$

When the distance $d = 0.20$ m, load L

$$= \frac{-2}{0.20} + 35 = \mathbf{25.0\ N}$$

Rearranging $L = -\dfrac{2}{d} + 35$ gives

$$\frac{2}{d} = 35 - L \text{ and } d = \frac{2}{35 - L}$$

Hence when load $L = 20$ N, distance, d

$$= \frac{2}{35 - 20} = \frac{2}{15} = \mathbf{0.133\ m}$$

Problem 3   The solubility s of potassium chlorate is shown by the following table:

$t°C$	10	20	30	40	50	60	80	100
s	4.9	7.6	11.1	15.4	20.4	26.4	40.6	58.0

The relationship between s and t is thought to be of the form $s = 3 + at + bt^2$. Plot a graph to test the supposition and use the graph to find approximate values of a and b. Hence calculate the solubility of potassium chlorate at 70°C.

Rearranging $s = 3 + at + bt^2$ gives $s - 3 = at + bt^2$ and $\left(\dfrac{s - 3}{t}\right) = a + bt$ or $\left(\dfrac{s - 3}{t}\right) = bt + a$ which is of the form $Y = mX + c$, showing that $\left(\dfrac{s - 3}{t}\right)$ is to be plotted vertically and t horizontally. Another table of values is drawn up as shown below.

t	10	20	30	40
s	4.9	7.6	11.1	15.4
$\left(\dfrac{s - 3}{t}\right)$	0.19	0.23	0.27	0.31

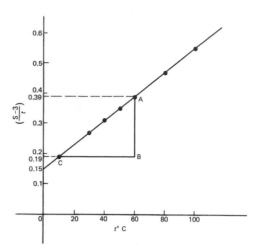

Fig. 11.3

t	50	60	80	100
s	20.4	26.4	40.6	58.0
$\left(\dfrac{s - 3}{t}\right)$	0.35	0.39	0.47	0.55

A graph of $\left(\dfrac{s - 3}{t}\right)$ against t is shown plotted in *Fig. 11.3*. A straight line fits the points which shows that s and t are related by $s = 3 + at + bt^2$.

Gradient of straight line, $b = \dfrac{AB}{BC} = \dfrac{0.39 - 0.19}{60 - 10}$

$$= \frac{0.20}{50} = \mathbf{0.004}$$

Vertical axis intercept, $a = \mathbf{0.15}$.
Hence the law of the graph is $s = 3 + \mathbf{0.15}t + \mathbf{0.004}t^2$.
The solubility of potassium chlorate at 70°C is given by

$$s = 3 + 0.15(70) + 0.004(70)^2 = 3 + 10.5 + 19.6$$
$$= \mathbf{33.1}$$

Problem 4   The current flowing in, and the power dissipated by, a resistor are measured experimentally for various values and the results are as shown below.

Current, I amperes	2.2	3.6	4.1	5.6	6.8
Power, P watts	116	311	403	753	1110

Show that the law relating current and power is of the form $P = RI^n$, where R and n are constants, and determine the law.

Taking logarithms to a base of 10 of both sides of $P = RI^n$ gives:

$$\lg P = \lg (RI^n) = \lg R + \lg I^n = \lg R + n \lg I$$

i.e. $\lg P = n \lg I + \lg R$, which is of the form $Y = mX + c$, showing that $\lg P$ is to be plotted vertically against $\lg I$ horizontally. A table of values for $\lg I$ and $\lg P$ is drawn up as shown below.

I	2.2	3.6	4.1	5.6	6.8
$\lg I$	0.342	0.556	0.613	0.748	0.833
P	116	311	403	753	1110
$\lg P$	2.064	2.493	2.605	2.877	3.045

A graph of $\lg P$ against $\lg I$ is shown in *Fig. 11.4* and since a straight line results the law $P = RI^n$ is verified. Gradient of straight line,

$$n = \frac{AB}{BC} = \frac{2.98 - 2.18}{0.8 - 0.4} = \frac{0.80}{0.4} = 2$$

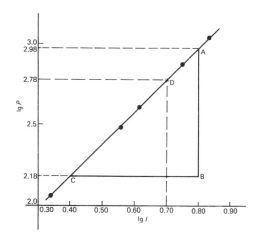

Fig. 11.4

It is not possible to determine the vertical axis intercept on sight since the horizontal axis scale does not start at zero. Selecting any point from the graph, say point D, where $\lg I = 0.70$ and $\lg P = 2.78$, and substituting values into $\lg P = n \lg I + \lg R$ gives

$$2.78 = (2)(0.70) + \lg R$$

from which $\lg R = 2.78 - 1.40 = 1.38$
Hence $R = $ antilog 1.38 $(= 10^{1.38}) = 24.0$
Hence the law of the graph is $P = 24.0\,I^2$.

Problem 5 The periodic time, T, of oscillation of a pendulum is believed to be related to its length, l, by a law of the form $T = kl^n$, where k and n are constants. Values of T were measured for various lengths of the pendulum and the results are as shown below.

Periodic time, T s	1.0	1.3	1.5	1.8	2.0	2.3
Length, l m		0.25	0.42	0.56	0.81	1.0	1.32

Show that the law is true and determine the approximate values of k and n. Hence find the periodic time when the length of the pendulum is 0.75 m.

From para. (iv) of section 11.1, if $T = kl^n$ then

$$\lg T = n \lg l + \lg k$$

and comparing with

$$Y = mX + c$$

shows that $\lg T$ is plotted vertically against $\lg l$ horizontally. A table of values for $\lg T$ and $\lg l$ is drawn up as shown below.

T	1.0	1.3	1.5	1.8	2.0	2.3
$\lg T$	0	0.114	0.176	0.255	0.301	0.362
l	0.25	0.42	0.56	0.81	1.0	1.32
$\lg l$	-0.602	-0.377	-0.252	-0.092	0	0.121

A graph of $\lg T$ against $\lg l$ is shown in *Fig. 11.5* and the law $T = kl^n$ is true since a straight line results. From the graph, gradient of straight line,

$$n = \frac{AB}{BC} = \frac{0.25 - 0.05}{-0.10 - (-0.50)} = \frac{0.20}{0.40} = \frac{1}{2}$$

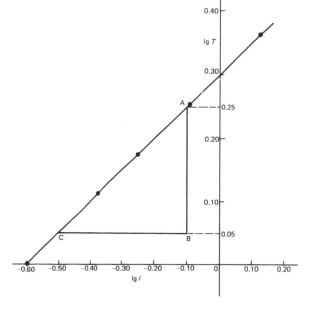

Fig. 11.5

Vertical axis intercept, lg k = 0.30. Hence k = antilog 0.30 (= $10^{0.30}$) = **2.0**.

Hence the law of the graph is $T = 2.0l^{1/2}$ or $T = 2.0\sqrt{l}$.

When length l = 0.75 m then $T = 2.0\sqrt{(0.75)}$ = **1.73 s**.

Problem 6   Quantities x and y are believed to be related by a law of the form $y = ab^x$, where a and b are constants. Values of x and corresponding values of y are:

x	0	0.6	1.2	1.8	2.4	3.0
y	5.0	9.67	18.7	36.1	69.8	135.0

Verify the law and determine the approximate values of a and b. Hence determine (a) the value of y when x is 2.1 and (b) the value of x when y is 100.

From para. (v) of section 11.1, if $y = ab^x$ then

$$\lg y = (\lg b)x + \lg a$$

and comparing with

$$Y = mX + c$$

shows that $\lg y$ is plotted vertically and x horizontally. Another table is drawn up as shown below.

x	0	0.6	1.2	1.8	2.4	3.0
y	5.0	9.67	18.7	36.1	69.8	135.0
$\lg y$	0.70	0.99	1.27	1.56	1.84	2.13

A graph of $\lg y$ against x is shown in *Fig. 11.6* and since a straight line results, the law $y = ab^x$ is verified. Gradient of straight line,

$$\lg b = \frac{AB}{BC} = \frac{2.13 - 1.17}{3.0 - 1.0} = \frac{0.96}{2.0} = 0.48$$

Hence b = antilog 0.48 (= $10^{0.48}$) = **3.0**, correct to 2 significant figures.
Vertical axis intercept, $\lg a$ = 0.70, from which

a = antilog 0.70 (= $10^{0.70}$)
$\quad$ = **5.0**, correct to 2 significant figures.

Hence the law of the graph is $y = 5.0 (3.0)^x$.

(a) When x = 2.1, $y = 5.0(3.0)^{2.1}$ = **50.2**.
(b) When y = 100, 100 = $5.0(3.0)^x$, from which 100/5.0 = $(3.0)^x$, i.e. 20 = $(3.0)^x$

Taking logarithms of both sides gives $\lg 20 = \lg (3.0)^x = x \lg 3.0$.

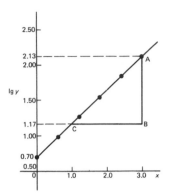

Fig. 11.6

Hence $x = \dfrac{\lg 20}{\lg 3.0} = \dfrac{1.301\ 0}{0.477\ 1} = \mathbf{2.73}$

Problem 7   The current i mA flowing in a capacitor which is being discharged varies with time t ms as shown below.

i mA	203	61.14	22.49	6.13	2.49	0.615
t ms	100	160	210	275	320	390

Show that these results are related by a law of the form $i = Ie^{t/T}$, where I and T are constants. Determine the approximate values of I and T.

Taking Napierian logarithms of both sides of $i = Ie^{t/T}$ gives

$$\ln i = \ln (Ie^{t/T}) = \ln I + \ln e^{t/T}$$

i.e. $\ln i = \ln I + \dfrac{t}{T}$ (since $\ln e$ = 1)

or $\ln i = \left(\dfrac{1}{T}\right)t + \ln I$

which compares with $y = mx + c$, showing that $\ln i$ is plotted vertically against t horizontally. (For methods of evaluating Napierian logarithms see Chapter 9.) Another table of values is drawn up as shown below.

t	100	160	210	275	320	390
i	203	61.14	22.49	6.13	2.49	0.615
$\ln i$	5.31	4.11	3.11	1.81	0.91	-0.49

A graph of $\ln i$ against t is shown in *Fig. 11.7* and since a straight line results the law $i = Ie^{t/T}$ is verified. Gradient of straight line,

$$\frac{1}{T} = \frac{AB}{BC} = \frac{5.30 - 1.30}{100 - 300} = \frac{4.0}{-200} = -0.02$$

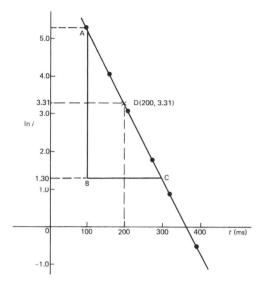

Fig. 11.7

Hence

$$T = \frac{1}{-0.02} = -50$$

Selecting any point on the graph, say point D, where $t = 200$ and $\ln i = 3.31$, and substituting into

$$\ln i = \left(\frac{1}{T}\right)t + \ln I$$

gives

$$3.31 = \left(-\frac{1}{50}\right)(200) + \ln I$$

from which

$$\ln I = 3.31 + 4.0 = 7.31$$

and

$$I = \text{antilog } 7.31 \ (= e^{7.31}) = 1495 \text{ or } \mathbf{1500} \text{ correct to 3 significant figures.}$$

Hence the law of the graph is $i = 1500e^{-t/50}$.

11.3 Further problems on reducing non-linear laws to linear form

In *Problems 1* to *4*, x and y are two related variables and all other letters denote constants. For the stated laws to be verified it is necessary to plot graphs of the variables in a modified form. State for each (a) what should be plotted on the vertical axis, (b) what should be plotted on the horizontal axis, (c) the gradient and (d) the vertical axis intercept.

1 (i) $y = d + cx^2$
   (ii) $y - a = b\sqrt{x}$
   $$\begin{bmatrix} \text{(i) (a) } y \text{ (b) } x^2 \text{ (c) } c \text{ (d) } d \\ \text{(ii) (a) } y \text{ (b) } \sqrt{x} \text{ (c) } b \text{ (d) } a \end{bmatrix}$$

2 (i) $y - e = \dfrac{f}{x}$
   (ii) $y - cx = bx^2$
   $$\begin{bmatrix} \text{(i) (a) } y \text{ (b) } \dfrac{1}{x} \text{ (c) } f \text{ (d) } e \\ \text{(ii) (a) } \dfrac{y}{x} \text{ (b) } x \text{ (c) } b \text{ (d) } ca \end{bmatrix}$$

3 (i) $y = \dfrac{a}{x} + bx$
   (ii) $y = ba^x$
   $$\begin{bmatrix} \text{(i) (a) } \dfrac{y}{x} \text{ (b) } \dfrac{1}{x^2} \text{ (c) } a \text{ (d) } b \\ \text{(ii) (a) } \lg y \text{ (b) } x \text{ (c) } \lg a \text{ (d) } \lg b \end{bmatrix}$$

4 (i) $y = kx^l$
   (ii) $\dfrac{y}{m} = e^{nx}$
   $$\begin{bmatrix} \text{(i) (a) } \lg y \text{ (b) } \lg x \text{ (c) } l \text{ (d) } \lg k \\ \text{(ii) (a) } \ln y \text{ (b) } x \text{ (c) } n \text{ (d) } \ln m \end{bmatrix}$$

5 In an experiment the resistance of wire is measured for wires of different diameters with the following results.

R ohms	1.64	1.14	0.89	0.76	0.63
d mm	1.10	1.42	1.75	2.04	2.56

It is thought that R is related to d by the law $R = (a/d^2) + b$, where a and b are constants. Verify this and find the approximate values for a and b. Determine the cross-sectional area needed for a resistance reading of 0.50 ohms.
[$a = 1.5$, $b = 0.4$, 11.78 mm²]

6 Corresponding experimental values of two quantities x and y are given below.

x	1.5	3.0	4.5	6.0	7.5	9.0
y	11.5	25.0	47.5	79.0	119.5	169.0

By plotting a suitable graph verify that y and x are connected by a law of the form $y = kx^2 + c$, where k and c are constants. Determine the law of the graph and hence find the value of x when y is 60.0.
[$y = 2x^2 + 7$, 5.15]

7   Experimental results of the safe load, L kN, applied to girders of varying spans, d m, are shown below.

Span, d m	2.0	2.8	3.6	4.2	4.8
Load, L kN	475	339	264	226	198

It is believed that the relationship between load and span is $L = c/d$, where c is a constant. Determine (a) the value of constant c and (b) the safe load for a span of 3.0 m.
[(a) 950 (b) 317 kN]

8   The following results give corresponding values of two quantities x and y which are believed to be related by a law of the form $y = ax^2 + bx$ where a and b are constants.

y	33.86	55.54	72.80	84.10	111.4	168.1
x	3.4	5.2	6.5	7.3	9.1	12.4

Verify the law and determine approximate values of a and b. Hence determine (i) the value of y when x is 8.0 and (ii) the value of x when y is 146.5.
[$a = 0.4$, $b = 8.6$ (i) 94.4 (ii) 11.2]

9   The luminosity I of a lamp varies with the applied voltage V and the relationship between I and V is thought to be $I = kV^n$. Experimental results obtained are:

I candelas	1.92	4.32	9.72	15.87	23.52	30.72
V volts	40	60	90	115	140	160

Verify that the law is true and determine the law of the graph. Determine also the luminosity when 75 V is applied across the lamp.
[$I = 0.001\,2\,V^2$, 6.75 candelas]

10   The head of pressure h and the flow velocity v are measured and are believed to be connected by the law $v = ah^b$, where a and b are constants. The results are as shown below.

h	10.6	13.4	17.2	24.6	29.3
v	9.77	11.00	12.44	14.88	16.24

Verify that the law is true and determine values of a and b.
[$a = 3.0$, $b = 0.5$]

11   Experimental values of x and y are measured as follows.

x	0.4	0.9	1.2	2.3	3.8
y	8.35	13.47	17.94	51.32	215.20

The law relating x and y is believed to be of the form $y = ab^x$, where a and b are constants. Determine the approximate values of a and b. Hence find the value of y when x is 2.0 and the value of x when y is 100.
[$a = 5.7$, $b = 2.6$, 38.53, 3.0]

12   The activity of a mixture of radioactive isotope is believed to vary according to the law $R = R_0 t^{-c}$, where R_0 and c are constants. Experimental results are shown below.

R	9.72	2.65	1.15	0.47	0.32	0.23
t	2	5	9	17	22	28

Verify that the law is true and determine approximate values of R_0 and c.
[$R_0 = 26.0$, $c = 1.42$]

13   Determine the law of the form $y = ae^{kx}$ which relates the following values.

y	0.030\,6	0.285	0.841	5.21	173.2	1181
x	−4.0	5.3	9.8	17.4	32.0	40.0

[$y = 0.08e^{0.24x}$]

14   The tension T in a belt passing round a pulley wheel and in contact with the pulley over an angle of θ radians is given by $T = T_0 e^{\mu\theta}$, where T_0 and μ are constants. Experimental results obtained are:

T newtons	47.9	52.8	60.3	70.1	80.9
θ radians	1.12	1.48	1.97	2.53	3.06

Determine approximate values of T_0 and μ. Hence find the tension when θ is 2.25 radians and the value of θ when the tension is 50.0 newtons.
[$T_0 = 35.4$ N, $\mu = 0.27$, 65.0 N, 1.28 radians]

12

Graphs with logarithmic scales

12.1 Logarithmic scales

Graph paper is available where the scale markings along the horizontal and vertical axes are proportional to the logarithms of the numbers. Such graph paper is called **log-log graph paper**.

A **logarithmic scale** is shown in *Fig. 12.1* where the distance between, say 1 and 2, proportional to $\lg 2 - \lg 1$, i.e. 0.301 0 of the total distance from 1 to 10. Similarly, the distance between 7 and 8 is proportional to $\lg 8 - \lg 7$, i.e. 0.057 99 of the total distance from 1 to 10. Thus the distance between markings progressively decreases as the numbers increase from 1 to 10.

With log-log graph paper the scale markings are from 1 to 9, and this pattern can be repeated several times. The number of times the pattern of markings is repeated on an axis signifies the number of **cycles**. When the vertical axis has, say, 3 sets of values from 1 to 9 and the horizontal axis has 2 sets of values from 1 to 9, then this log-log graph paper is called 'log 3 cycle × 2 cycle' (see *Fig. 12.2*). Many different arrangements are available ranging from 'log 1 cycle × 1 cycle' through to 'log 5 cycle × 5 cycle'.

To depict a set of values, say, from 0.4 to 161, on an axis of log-log graph paper, 4 cycles are required, from 0.1 to 1, 1 to 10, 10 to 100 and 100 to 1000.

12.2 Graphs of the form $y = ax^n$

Taking logarithms to a base of 10 of both sides of $y = ax^n$ gives:

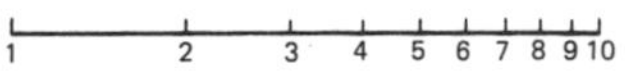

Fig. 12.1

$$\lg y = \lg (ax^n) = \lg a + \lg x^n$$

i.e. $\qquad \lg y = n \lg x + \lg a$

which compares with $\quad Y = mX + c$

Thus, by plotting $\lg y$ vertically against $\lg x$ horizontally, a straight line results, i.e. the equation $y = ax^n$ is reduced to linear form. With log-log graph paper available x and y may be plotted directly, without having first to determine their logarithms, as shown in Chapter 11.

Problem 1 Experimental values of two related quantities x and y are shown below:

x	0.41	0.63	0.92	1.36	2.17	3.95
y	0.45	1.21	2.89	7.10	20.79	82.46

The law relating x and y is believed to be $y = ax^b$, where a and b are constants. Verify that this law is true and determine the approximate values of a and b.

If $y = ax^b$ then $\lg y = b \lg x + \lg a$, from above, which is of the form $Y = mX + c$, showing that to produce a straight line graph $\lg y$ is plotted vertically against $\lg x$ horizontally. x and y may be plotted directly on to log-log graph paper as shown in *Fig. 12.2*. The values of y range from 0.45 to 82.46 and 3 cycles are needed (i.e. 0.1 to 1, 1 to 10 and 10 to 100). The values of x range from 0.41 to 3.95 and 2 cycles are needed (i.e. 0.1 to 1 and 1 to 10). Hence 'log 3 cycle × 2 cycle' is used as shown in *Fig. 12.2* where the axes are marked and the points plotted. Since the points lie on a straight line the law $y = ax^b$ is verified.

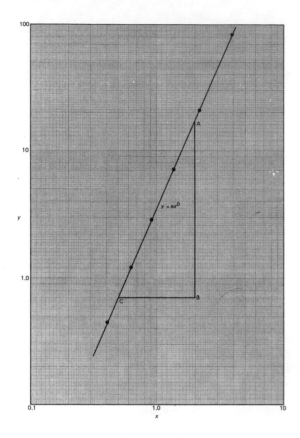

Fig. 12.2   *Graph to verify a law of the form $y = ax^b$*

To evaluate constants a and b:

Method 1. Any two points on the straight line, say points A and C, are selected, and AB and BC are measured (say in centimetres)

Then, gradient, $b = \dfrac{AB}{BC} = \dfrac{11.5 \text{ units}}{5 \text{ units}} = 2.3$

Since $\lg y = b \lg x + \lg a$, when $x = 1$, $\lg x = 0$ and $\lg y = \lg a$. The straight line crosses the ordinate $x = 1.0$ at $y = 3.5$. Hence $\lg a = \lg 3.5$, i.e. $a = 3.5$.

Method 2. Any two points on the straight line, say points A and C, are selected. A has coordinates $(2, 17.25)$ and C has coordinates $(0.5, 0.7)$

Since $y = ax^b$ then $17.25 = a(2)^b$ (1)
and $0.7 = a(0.5)^b$ (2)

i.e. two simultaneous equations are produced and may be solved for a and b.
Dividing equation (1) by equation (2) to eliminate a gives

$$\frac{17.25}{0.7} = \frac{(2)^b}{(0.5)^b} = \left(\frac{2}{0.5}\right)^b$$

i.e. $24.643 = (4)^b$.
Taking logarithms of both sides gives $\lg 24.643 = b \lg 4$, i.e.

$$b = \frac{\lg 24.643}{\lg 4} = 2.3, \text{ correct to 2 significant figures}$$

Substituting $b = 2.3$ in equation (1) gives: $17.25 = a(2)^{2.3}$, i.e.

$$a = \frac{17.25}{(2)^{2.3}} = \frac{17.25}{4.925} = 3.5, \text{ correct to 2 significant figures}$$

Hence the law of the graph is $y = 3.5x^{2.3}$.

Problem 2   The power dissipated by a resistor was measured for varying values of current flowing in the resistor and the results are as shown:

Current, I amperes	1.4	4.7	6.8	9.1	11.2	13.1
Power, P watts	49	552	1156	2070	3136	4290

Prove that the law relating current and power is of the form $P = RI^n$, where R and n are constants, and determine the law. Hence calculate the power when the current is 12 amperes and the current when the power is 1000 W.

Since $P = RI^n$ then $\lg P = n \lg I + \lg R$, which is of the form $Y = mX + c$, showing that to produce a straight line graph $\lg P$ is plotted vertically against $\lg I$ horizontally. Power values range from 49 to 4290, hence 3 cycles of log-log graph paper are needed (10 to 100, 100 to 1000 and 1000 to 10 000).

Current values range from 1.4 to 11.2, hence 2 cycles of log-log graph paper are needed (1 to 10 and 10 to 100) Thus 'log 3 cycles × 2 cycles' is used as shown in *Fig. 12.3* (or, if not available, graph paper having a larger number of cycles per axis can be used). The coordinates are plotted and a straight line results which proves that the law relating current and power is of the form $P = RI^n$.
Gradient of straight line

$$n = \frac{AB}{BC} = \frac{14 \text{ units}}{7 \text{ units}} = 2$$

At point C, $I = 2$ and $P = 100$. Substituting these values into $P = RI^n$ gives: $100 = R (2)^2$. Hence $R = 100/(2)^2 = 25$ which may have been found from the intercept on the $I = 1.0$ axis in *Fig. 12.3*. **Hence the law of the graph is $P = 25I^2$.**
When current $I = 12$, power $P = 25(12)^2 = $ **3600 watts** (which may be read from the graph).

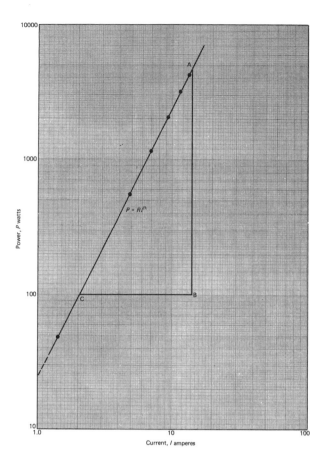

Fig. 12.3 *Variation of power with current*

When power

$$P = 1000, \; 1000 = 25I^2$$

Hence $\quad I^2 = \dfrac{1000}{25} = 40$

from which, $I = \sqrt{40} = \mathbf{6.32\ A}$

Problem 3 The pressure p and volume v of a gas are believed to be related by a law of the form $p = cv^n$, where c and n are constants. Experimental values of p and corresponding values of v obtained in a laboratory are:

p pascals	2.28×10^5	8.04×10^5	2.03×10^6
v m^3	3.2×10^{-2}	1.3×10^{-2}	6.7×10^{-3}

p pascals	5.05×10^6	1.82×10^7
v m^3	3.5×10^{-3}	1.4×10^{-3}

Verify that the law is true and determine approximate values of c and n.

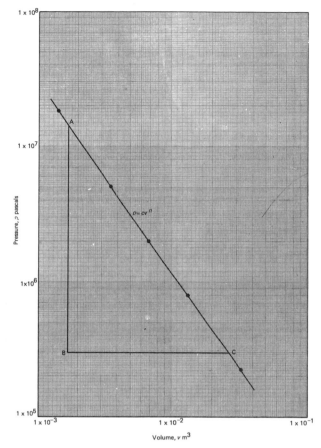

Fig. 12.4 *Variation of pressure with volume*

Since $p = cv^n$, then $\lg p = n \lg v + \lg c$, which is of the form $Y = mX + c$, showing that to produce a straight line graph $\lg p$ is plotted vertically against $\lg v$ horizontally. The co-ordinates are plotted on 'log 3 cycle $\times$ 2 cycle' graph paper as shown in *Fig. 12.4*. With the data expressed in standard form, the axes are marked in standard form also. Since a straight line results the law $p = cv^n$ is verified.

The straight line has a negative gradient and the value of the gradient is given by

$$\frac{AB}{BC} = \frac{14 \text{ units}}{10 \text{ units}} = 1.4. \text{ Hence } n = \mathbf{-1.4}$$

Selecting any point on the straight line, say point C, having coordinates $(2.63 \times 10^{-2}, 3 \times 10^5)$, and substituting these values in $p = cv^n$ gives:

$$3 \times 10^5 = c(2.63 \times 10^{-2})^{-1.4}$$

Hence $c = \dfrac{3 \times 10^5}{(2.63 \times 10^{-2})^{-1.4}} = \dfrac{3 \times 10^5}{(0.026\ 3)^{-1.4}} = \dfrac{3 \times 10^5}{1.63 \times 10^2}$

$\qquad = \mathbf{1840}$, correct to 3 significant figures

Hence the law of the graph is $p = 1840v^{-1.4}$ or $pv^{1.4} = 1840$.

Further problems on graphs of the form $y = ax^n$ may be found in section 12.5, Problems 1 to 3, page 112.

12.3   Graphs of the form $y = ab^x$

Taking logarithms to a base of 10 of both sides of $y = ab^x$ gives:

$$\lg y = \lg (ab^x) = \lg a + \lg b^x = \lg a + x \lg b$$

i.e. $\qquad\qquad \lg y = (\lg b)x + \lg a$

which compares with $\quad Y = mX + c$

Thus, by plotting $\lg y$ vertically against x horizontally a straight line results, i.e. the graph $y = ab^x$ is reduced to linear form. In this case, graph paper having a linear horizontal scale and a logarithmic vertical scale may be used. This type of graph paper is called **log-linear graph paper**, and is specified by the number of cycles on the logarithmic scale. For example, graph paper having 3 cycles on the logarithmic scale is called 'log 3 cycle × linear' graph paper.

Problem 4   Experimental values of quantities x and y are believed to be related by a law of the form $y = ab^x$, where a and b are constants. The values of x and corresponding values of y are:

x	0.7	1.4	2.1	2.9	3.7	4.3
y	18.4	45.1	111	308	858	1850

Verify the law and determine the approximate values of a and b. Hence evaluate (i) the value of y when x is 2.5 and (ii) the value of x when y is 1200.

Since $y = ab^x$ then $\lg y = (\lg b)x + \lg a$ (from above), which is of the form $Y = mX + c$, showing that to produce a straight line graph $\lg y$ is plotted vertically against x horizontally. Using log-linear graph paper, values of x are marked on the horizontal scale to cover the range 0.7 to 4.3. Values of y range from 18.4 to 1850 and 3 cycles are needed (i.e. 10 to 100, 100 to 1000 and 1000 to 10 000). Thus 'log 3 cycles × linear' graph paper is used as shown in *Fig. 12.5*. A straight line is drawn through the coordinates, hence the law $y = ab^x$ is verified.

Gradient of straight line, $\lg b = AB/BC$. Direct measurement (say in centimetres) is not made with log-linear graph paper since the vertical scale is logarithmic and the horizontal scale is linear. Hence

$$\frac{AB}{BC} = \frac{\lg 1000 - \lg 100}{3.82 - 2.02} = \frac{3-2}{1.80} = \frac{1}{1.80} = 0.555\,6$$

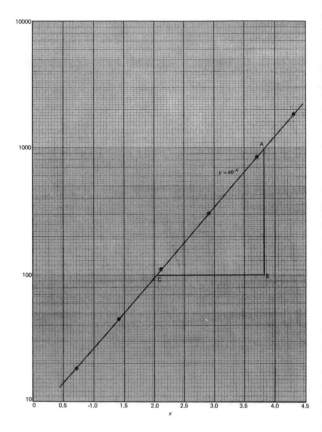

Fig. 12.5   *Graph to verify a law of the form* $y = ab^x$

Here $b = $ antilog $0.555\,6 (= 10^{0.555\,6}) = $ **3.6**, correct to 2 significant figures.

Point A has coordinates (3.82,1000). Substituting these values into $y = ab^x$ gives: $1000 = a(3.6)^{3.82}$, i.e.

$$a = \frac{1000}{(3.6)^{3.82}} = \textbf{7.5}, \text{ correct to 2 significant figures}$$

Hence the law of the graph is $y = 7.5(3.6)^x$.

(i)   When $x = 2.5$, $y = 7.5(3.6)^{2.5} = \textbf{184}$.
(ii)   When $y = 1200$, $1200 = 7.5(3.6)^x$, hence

$$(3.6)^x = \frac{1200}{7.5} = 160$$

Taking logarithms gives $x \lg 3.6 = \lg 160$

i.e. $\qquad\qquad\qquad x = \dfrac{\lg 160}{\lg 3.6} = \dfrac{2.204\,1}{0.556\,3}$

$$= \textbf{3.96}$$

A further problem on graphs of the form $y = ab^x$ may be found in section 12.5, Problem 4, page 112.

12.4   Graphs of the form $y = ae^{kx}$

Taking logarithms to a base of e of both sides of $y = ae^{kx}$ gives:

$$\ln y = \ln (ae^{kx}) = \ln a + \ln e^{kx} = \ln a + kx \ln e$$

i.e. $\ln y = kx + \ln a$ (since $\ln e = 1$)

which compares with $Y = mX + c$

Thus, by plotting $\ln y$ vertically against x horizontally, a straight line results, i.e. the equation $y = ae^{kx}$ is reduced to linear form. Since $\ln y = 2.302\,6 \lg y$, i.e. $\ln y = $ (a constant)($\lg y$), the same log-linear graph paper can be used for Napierian logarithms as for logarithms to a base of 10.

Problem 5   The data given below is believed to be related by a law of the form $y = ae^{kx}$, where a and b are constants. Verify that the law is true and determine approximate values of a and b. Also determine the value of y when x is 3.8 and the value of x when y is 85.

x	−1.2	0.38	1.2	2.5	3.4	4.2	5.3
y	9.3	22.2	34.8	71.2	117	181	332

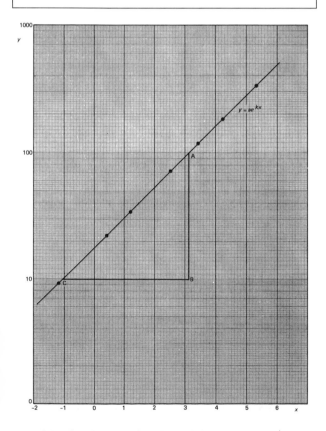

Fig. 12.6   *Graph to verify a law of the form* $y = ae^{kx}$

Since $y = ae^{kx}$ then $\ln y = kx + \ln a$ (from above), which is of the form $Y = mX + c$, showing that to produce a straight line graph $\ln y$ is plotted vertically against x horizontally. The value of y ranges from 9.3 to 332 hence 'log 3 cycle × linear' graph paper is used. The plotted coordinates are shown in *Fig. 12.6* and since a straight line passes through the points the law $y = ae^{kx}$ is verified.

Gradient of straight line,

$$k = \frac{AB}{BC} = \frac{\ln 100 - \ln 10}{3.12 - (-1.08)} = \frac{2.302\,6}{4.20}$$

$$= 0.55, \text{ correct to 2 significant figures}$$

Since $\ln y = kx + \ln a$, when $x = 0$, $\ln y = \ln a$, i.e. $y = a$. The vertical axis intercept value at $x = 0$ is 18, hence $a = 18$.

The law of the graph is thus $y = 18e^{0.55x}$.

When x is 3.8, $y = 18e^{0.55(3.8)} = 18e^{2.09} = 18(8.084\,9) = $ **146**.

When y is 85, $85 = 18e^{0.55x}$. Hence $e^{0.55x} = \dfrac{85}{18} = 4.722\,2$

and $0.55x = \ln 4.722\,2 = 1.552\,3$. Hence $x = \dfrac{1.552\,3}{0.55} = $ **2.82**

Problem 6   The voltage, v volts, across an inductor is believed to be related to time, t ms, by the law $v = Ve^{t/T}$, where V and T are constants. Experimental results obtained are:

v volts	883	347	90	55.5	18.6	5.2
t ms	10.4	21.6	37.8	43.6	56.7	72.0

Show that the law relating voltage and time is as stated and determine the approximate values of V and T. Find also the value of voltage after 25 ms and the time when the voltage is 30.0 V.

Since $v = Ve^{t/T}$ then

$$\ln v = \frac{1}{T}t + \ln V$$

which is of the form $Y = mX + c$.

Using 'log 3 cycle × linear' graph paper, the points are plotted as shown in *Fig. 12.7*.

Since the points are joined by a straight line the law $v = Ve^{t/T}$ is verified.

Gradient of straight line,

$$\frac{1}{T} = \frac{AB}{BC} = \frac{\ln 100 - \ln 10}{36.5 - 64.2} = \frac{2.302\,6}{-27.7}$$

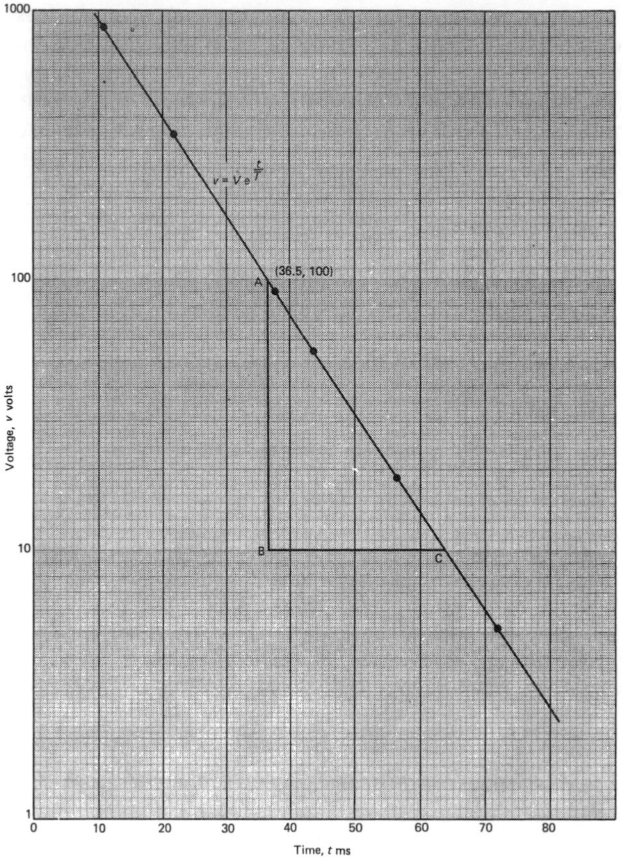

Fig. 12.7   *Variation of voltage with time*

Hence

$$T = \frac{-27.7}{2.302\ 6} = \textbf{-12.0}, \text{ correct to 3 significant figures}$$

Since the straight line does not cross the vertical axis at $t = 0$ in *Fig. 12.7*, the value of V is determined by selecting any point, say A, having coordinates (36.5,100) and substituting these values into $v = Ve^{t/T}$. Thus

$$100 = Ve^{36.5/-12.0}$$

i.e. $V = \dfrac{100}{e^{-36.5/12.0}} = \textbf{2090 volts}$, correct to 3 significant figures

Hence the law of the graph is $v = 2090e^{-t/12.0}$.
When time $t = 25$ ms, voltage $v = 2090e^{-25/12.0} = \textbf{260 V}$.
When the voltage is 30.0 volts, $30.0 = 2090e^{t/12.0}$, hence

$$e^{-t/12.0} = \frac{30.0}{2090} \text{ and } e^{t/12.0} = \frac{2090}{30.0} = 69.67$$

Taking Napierian logarithms gives: $\dfrac{t}{12.0} = \ln 69.67 = 4.243\ 8$

from which, time $t = (12.0)(4.243\ 8) = \textbf{50.9 ms}$.

___

Further problems on graphs of the form $y = ae^{kx}$ may be found in section 12.5, Problems 5 and 6, page 113.

12.5   Further problems on graphs having logarithmic scales

Graphs of the form $y = ax^n$

1 Quantities x and y are believed to be related by a law of the form $y = ax^n$, where a and n are constants. Experimental values of x and corresponding values of y are:

x	0.8	2.3	5.4	11.5	21.6	42.9
y	8	54	250	974	3028	10 410

Show that the law is true and determine the values of a and n. Hence determine the value of y when x is 7.5 and the value of x when y is 5000.
[$a = 12$, $n = 1.8$, 451, 28.5]

2 Show from the following results of voltage V and admittance Y of an electrical circuit that the law connecting the quantities is of the form $V = kY^n$, and determine the values of k and n.

Voltage, V volts	2.88	2.05	1.60	1.22	0.96
Admittance, Y siemens	0.52	0.73	0.94	1.23	1.57

[$k = 1.5$, $n = -1$]

3 Quantities x and y are believed to be related by a law of the form $y = mn^x$. The values of x and corresponding values of y are:

x	0	0.5	1.0	1.5	2.0	2.5	3.0
y	1.0	3.2	10	31.6	100	316	1000

Verify the law and find the values of m and n.
[$m = 1$, $n = 10$]

Graphs of the form $y = ab^x$

4 Experimental values of p and corresponding values of q are shown below.

p	-13.2	-27.9	-62.2	-383.2	-1581	-2931
q	0.30	0.75	1.23	2.32	3.17	3.54

Show that the law relating p and q is $p = ab^q$, where a and b are constants. Determine (i) values of a and b, and state the law, (ii) the value of p when q is 2.0 and (iii) the value of q when p is -2000.

$$\left[\begin{array}{l} \text{(i) } a = -8, b = 5.3, p = -8(5.3)^q \\ \text{(ii) } -224.7 \text{ (iii) } 3.31 \end{array} \right]$$

Graphs of the form $y = ae^{kx}$

5 Atmospheric pressure p is measured at varying altitudes h and the results are as shown below:

Altitude, h m	500	1500	3000	5000	8000
pressure, p cm	73.39	68.42	61.60	53.56	43.41

Show that the quantities are related by the law $p = ae^{kh}$, where a and k are constants. Determine the values of a

and k and state the law. Find also the atmospheric pressure at 10 000 m.

$[a = 76, k = -7 \times 10^{-5}, p = 76e^{-7\times10^5 h}, 37.74 \text{ cm}]$

6 At particular times, t minutes, measurements are made of the temperature, $\theta°C$, of a cooling liquid and the following results are obtained:

Temperature $\theta°C$	92.2	55.9	33.9	20.6	12.5
Time t minutes	10	20	30	40	50

Prove that the quantities follow a law of the form $\theta = \theta_0 e^{kt}$, where θ_0 and k are constants, and determine the approximate values of θ_0 and k.
$[\theta_0 = 152, k = -0.05]$

7–12 See *Problems 9* to *14* on page 106 which may all be solved using logarithmic graph paper.

13

Graphical solution of equations

13.1 Graphical solution of simultaneous equations

Linear simultaneous equations in two unknowns may be solved graphically by:

(i) plotting the two straight lines on the same axes, and
(ii) noting their point of intersection.

The coordinates of the point of intersection give the required solution.

> *Problem 1* Solve graphically the simultaneous equations $2x - y = 4$
> $$x + y = 5$$

Rearranging each equation into $y = mx + c$ form gives:

$$y = 2x - 4 \tag{1}$$
$$y = -x + 5 \tag{2}$$

Only three coordinates need be calculated for each graph since both are straight lines.

x	0	1	2
$y = 2x - 4$	−4	−2	0

x	0	1	2
$y = -x + 5$	5	4	3

Each of the graphs is plotted as shown in *Fig. 13.1*. The point of intersection is at (3,2) and since this is the only point which lies simultaneously on both lines then $x = 3$, $y = 2$ is the solution of the simultaneous equations.

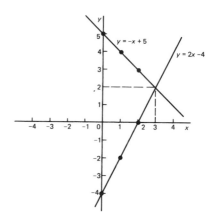

Fig. 13.1

> *Problem 2* Solve graphically the equations
> $$1.20x + y = 1.80$$
> $$x - 5.0y = 8.50$$

Rearranging each equation into $y = mx + c$ form gives:

$$y = -1.20x + 1.80 \tag{1}$$

$$y = \frac{x}{5.0} - \frac{8.5}{5.0}$$

i.e. $y = 0.20x - 1.70 \tag{2}$

Three coordinates are calculated for each equation as shown below.

x	0	1	2
$y = -1.20x + 1.80$	1.80	0.60	−0.60

x	0	1	2
$y = 0.20x - 1.70$	-1.70	-1.50	-1.30

The two lines are plotted as shown in *Fig. 13.2*. The point of intersection is (2.50,−1.20). Hence the solution of the simultaneous equations is $x = 2.50$, $y = -1.20$. (It is sometimes useful initially to sketch the two straight lines to determine the region where the point of intersection is. Then, for greater accuracy, a graph having a smaller range of values can be drawn to 'magnify' the point of intersection.)

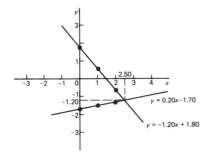

Fig. 13.2

Further problems on the graphical solution of simultaneous equations may be found in section 13.5, Problems 1 to 6, page 120.

13.2 Graphical solution of quadratic equations

A general **quadratic equation** is of the form $y = ax^2 + bx + c$, where a, b and c are constants and a is not equal to zero.

A graph of a quadratic equation always produces a shape called a **parabola**.

The gradient of the curve between 0 and A and between B and C in *Fig. 13.3* is positive, whilst the gradient between A and B is negative. Points such as A and B are called **turning points**. At A the gradient is zero and, as x increases, the gradient of the curve changes from positive just before A to negative just after. Such a point is called a **maximum value**. At B the gradient is also zero and, as x

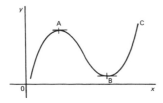

Fig. 13.3

increases, the gradient of the curve changes from negative just before B to positive just after. Such a point is called a **minimum value**. (More on maximum and minimum values may be found in Chapter 29.)

Quadratic graphs

(i) $y = ax^2$

Graphs of $y = x^2$, $y = 3x^2$ and $y = \frac{1}{2}x^2$ are shown in *Fig. 13.4*. All have minimum values at the origin (0,0).

Graphs of $y = -x^2$, $y = -3x^2$ and $y = -\frac{1}{2}x^2$ are shown in *Fig. 13.5*. All have maximum values at the origin (0,0).

When $y = ax^2$, (a) curves are symmetrical about the y-axis,

(b) the magnitude of 'a' affects the gradient of the curve,

and (c) the sign of 'a' determines whether it has a maximum or minimum value.

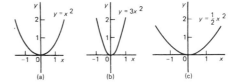

Fig. 13.4

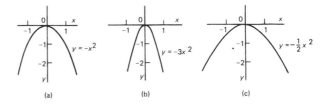

Fig. 13.5

(ii) $y = ax^2 + c$
Graphs of $y = x^2 + 3$, $y = x^2 - 2$, $y = -x^2 + 2$ and $y = -2x^2 - 1$ are shown in *Fig. 13.6*.
When $y = ax^2 + c$: (a) curves are symmetrical about the y-axis,

(b) the magnitude of 'a' affects the gradient of the curve,

and (c) the constant 'c' is the y-axis intercept.

(iii) $y = ax^2 + bx + c$.
Whenever 'b' has a value other than zero the curve is displaced to the right or left of the y-axis. When b/a is positive, the curve is displaced $b/2a$ to the left of the

y-axis, as shown in *Fig. 13.7(a)*. When *b/a* is negative the curve is displaced *b/2a* to the right of the *y*-axis, as shown in *Fig. 13.7(b)*.

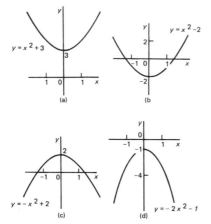

(a)   (b)

(c)   (d)

Fig. 13.6

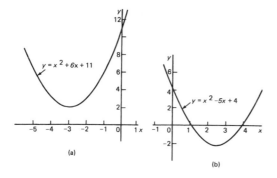

(a)   (b)

Fig. 13.7

Quadratic equations of the form $ax^2 + bx + c = 0$ may be solved graphically by:

(i) plotting the graph $y = ax^2 + bx + c$, and
(ii) noting the points of intersection on the *x*-axis (i.e. where $y = 0$).

The *x* values of the points of intersection give the required solutions since at these points both $y = 0$ and $ax^2 + bx + c = 0$. The number of solutions, or roots of a quadratic equation, depends on how many times the curve cuts the *x*-axis and there can be no real roots (as in *Fig. 13.7(a)*) or one root (as in *Figs 13.4* and *13.5*) or two roots (as in *Fig. 13.7(b)*).

Problem 3   Solve the quadratic equation $4x^2 + 4x - 15 = 0$ graphically given that the solutions lie in the range $x = -3$ to $x = 2$. Determine also the coordinates and nature of the turning point of the curve.

Let $y = 4x^2 + 4x - 15$. A table of values is drawn up as shown below.

x				-3	-2	-1	0	1	2
$4x^2$				36	16	4	0	4	16
$4x$				-12	-8	-4	0	4	8
-15				-15	-15	-15	-15	-15	-15
$y = 4x^2 + 4x - 15$				9	-7	-15	-15	-7	9

A graph of $y = 4x^2 + 4x - 15$ is shown in *Fig. 13.8*. The only points where $y = 4x^2 + 4x - 15$ and $y = 0$ are the points marked A and B. This occurs at $x = -2.5$ **and** $x = 1.5$ and these are the solutions of the quadratic equation $4x^2 + 4x - 15 = 0$. (By substituting $x = -2.5$ and $x = 1.5$ into the original equation the solutions may be checked.) The curve has a turning point at $(-0.5, -16)$ and the nature of the point is a **minimum.**

An alternative graphical method of solving $4x^2 + 4x - 15 = 0$ is to rearrange the equation as $4x^2 = -4x + 15$ and then plot two separate graphs – in this case $y = 4x^2$ and $y = -4x + 15$. Their points of intersection give the roots of equation $4x^2 = -4x + 15$, i.e. $4x^2 + 4x - 15 = 0$. This is shown in *Fig. 13.9*, where the roots are $x = -2.5$ and $x = 1.5$ as before.

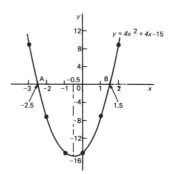

Fig. 13.8

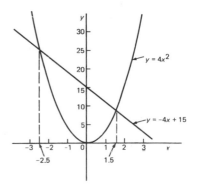

Fig. 13.9

Problem 4 Solve graphically the quadratic equation $-5x^2 + 9x + 7.2 = 0$ given that the solutions lie between $x = -1$ and $x = 3$. Determine also the coordinates of the turning point and state its nature.

Let $y = -5x^2 + 9x + 7.2$. A table of values is drawn up as shown below. A graph of $y = -5x^2 + 9x + 7.2$ is shown plotted in *Fig. 13.10*. The graph crosses the x-axis (i.e. where $y = 0$) at $x = -0.6$ **and** $x = 2.4$ and these are the solutions of the quadratic equation $-5x^2 + 9x + 7.2 = 0$. The turning point is a **maximum** having coordinates **(0.9,11.25)**.

x	-1	-0.5	0	1
$-5x^2$	-5	-1.25	0	-5
$+9x$	-9	-4.5	0	9
$+7.2$	7.2	7.2	7.2	7.2
$y = -5x^2 + 9x + 7.2$	-6.8	1.45	7.2	11.2

x	2	2.5	3
$-5x^2$	-20	-31.25	-45
$+9x$	18	22.5	27
$+7.2$	7.2	7.2	7.2
$y = -5x^2 + 9x + 7.2$	5.2	-1.55	-10.8

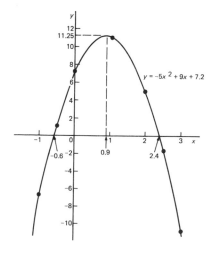

Fig. 13.10

Problem 5 Plot a graph of $y = 2x^2$ and hence solve the equations

(a) $2x^2 - 8 = 0$ and (b) $2x^2 - x - 3 = 0$.

A graph of $y = 2x^2$ is shown in *Fig. 13.11*.

(a) Rearranging $2x^2 - 8 = 0$ gives $2x^2 = 8$ and the solution of this equation is obtained from the points of intersection of $y = 2x^2$ and $y = 8$, i.e. at coordinates $(-2,8)$ and $(2,8)$, shown as A and B, respectively, in *Fig. 13.11*. Hence the solutions of $2x^2 - 8 = 0$ are $x = -2$ **and** $x = +2.$

(b) Rearranging $2x^2 - x - 3 = 0$ gives $2x^2 = x + 3$ and the solution of this equation is obtained from the points of intersection of $y = 2x^2$ and $y = x + 3$, i.e. at C and D in *Fig. 13.11*. Hence the solutions of $2x^2 - x - 3 = 0$ are $x = -1$ **and** $x = 1.5.$

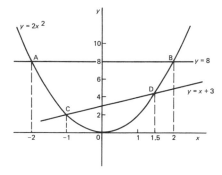

Fig. 13.11

Problem 6 Plot the graph of $y = -2x^2 + 3x + 6$ for values of x from $x = -2$ to $x = 4$. Use the graph to find the roots of the following equations

(a) $-2x^2 + 3x + 6 = 0$ (b) $-2x^2 + 3x + 2 = 0$
(c) $-2x^2 + 3x + 9 = 0$ (d) $-2x^2 + x + 5 = 0.$

A table of values is drawn up as shown below.

x	-2	-1	0	1	2	3	4
$-2x^2$	-8	-2	0	-2	-8	-18	-32
$+3x$	-6	-3	0	3	6	9	12
$+6$	6	6	6	6	6	6	6
y	-8	1	6	7	4	-3	-14

A graph of $y = -2x^2 + 3x + 6$ is shown in *Fig. 13.12*.

(a) The parabola $y = -2x^2 + 3x + 6$ and the straight line $y = 0$ intersect at A and B, where $x = -1.13$ and $x = 2.63$ and these are the roots of the equation $-2x^2 + 3x + 6 = 0$.

(b) Comparing $y = -2x^2 + 3x + 6$ (1)
with $0 = -2x^2 + 3x + 2$ (2)
shows that if 4 is added to both sides of equation (2), the right-hand side of both equations will be the same.

Hence $4 = -2x^2 + 3x + 6$. The solution of this equation is found from the points of intersection of the line $y = 4$ and the parabola $y = -2x^2 + 3x + 6$, i.e. points C and D in *Fig. 13.12*. Hence the roots of $-2x^2 + 3x + 2 = 0$ are $x = -0.5$ **and** $x = 2$.

(c) $-2x^2 + 3x + 9 = 0$ may be rearranged as $-2x^2 + 3x + 6 = -3$, and the solution of this equation is obtained from the points of intersection of the line $y = -3$ and the parabola $y = -2x^2 + 3x + 6$, i.e. at points E and F in *Fig. 13.12*. Hence the roots of $-2x^2 + 3x + 9 = 0$ are $x = -1.5$ **and** $x = 3$.

(d) Comparing $y = -2x^2 + 3x + 6$ (3)
with $0 = -2x^2 + x + 5$ (4)

shows that if $2x + 1$ is added to both sides of equation (4) the right-hand side of both equations will be the same. Hence equation (4) may be written as $2x + 1 = -2x^2 + 3x + 6$. The solution of this equation is found from the points of intersection of the line $y = 2x + 1$ and the parabola $y = -2x^2 + 3x + 6$, i.e. points G and H in *Fig. 13.12*. Hence the roots of $-2x^2 + x + 5 = 0$ are $x = -1.35$ **and** $x = 1.85$.

Further problems on the graphical solution of quadratic equations may be found in section 13.5, Problems 7 to 16, page 120.

13.3   Graphical solution of linear and quadratic equations simultaneously

The solution of **linear and quadratic equations simultaneously** may be achieved graphically by: (i) plotting the straight line and parabola on the same axes, and (ii) noting the points of intersection. The coordinates of the points of intersection give the required solutions.

> *Problem 7*   Determine graphically the values of x and y which simultaneously satisfy the equations
>
> $y = 2x^2 - 3x - 4$
> and $y = 2 - 4x$

$y = 2x^2 - 3x - 4$ is a parabola and a table of values is drawn up as shown below.

x	-2	-1	0	1	2	3
$2x^2$	8	2	0	2	8	18
$-3x$	6	3	0	-3	-6	-9
-4	-4	-4	-4	-4	-4	-4
y	10	1	-4	-5	-2	5

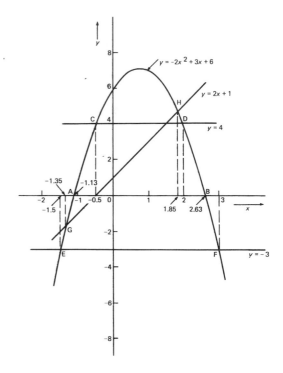

Fig. 13.12

$y = 2 - 4x$ is a straight line and only three coordinates need be calculated.

x	0	1	2
y	2	-2	-6

The two graphs are shown plotted in *Fig. 13.13* and the points of intersection, shown as A and B, are at coordinates $(-2, 10)$ and $(1\frac{1}{2}, -4)$. Hence the simultaneous solutions occur when $x = -2$, $y = 10$ and when $x = 1\frac{1}{2}$, $y = -4$. (These solutions may be checked by substituting into each of the original equations.)

Further problems on the graphical solution of linear and quadratic equations simultaneously may be found in section 13.5. Problems 17 and 18, page 120.

13.4   Graphical solution of cubic equations

A **cubic equation** of the form $ax^3 + bx^2 + cx + d = 0$ may be solved graphically by: (i) plotting the graph $y = ax^3 + bx^2 + cx + d$, and (ii) noting the points of intersection on the x-axis (i.e. where $y = 0$). The x-values of the points of intersection give the required solution since at these points both $y = 0$ and $ax^3 + bx^2 + cx + d = 0$.

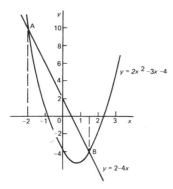

Fig. 13.13

The number of solutions, or roots of a cubic equation depends on how many times the curve cuts the *x*-axis and there can be one, two or three possible roots, as shown in *Fig. 13.14*.

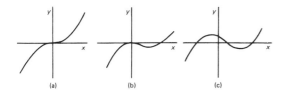

Fig. 13.14

Problem 8 Solve graphically the cubic equation $4x^3 - 8x^2 - 15x + 9 = 0$ given that the roots lie between $x = -2$ and $x = 3$. Determine also the coordinates of the turning points and distinguish between them.

Let $y = 4x^3 - 8x^2 - 15x + 9$. A table of values is drawn up as shown below.

x	-2	-1	0	1	2	3
$4x^3$	-32	-4	0	4	32	108
$-8x^2$	-32	-8	0	-8	-32	-72
$-15x$	30	15	0	-15	-30	-45
$+9$	9	9	9	9	9	9
y	-25	12	9	-10	-21	0

A graph of $y = 4x^3 - 8x^2 - 15x + 9$ is shown in *Fig. 13.15*. The graph crosses the *x*-axis (where $y = 0$) at $x = -1\frac{1}{2}$, $x = \frac{1}{2}$ and $x = 3$ and these are the solutions to the cubic equation $4x^3 - 8x^2 - 15x + 9 = 0$. The turning points occur at $(-0.6, 14.2)$, which is a **maximum**, and $(2, -21)$, which is a **minimum**.

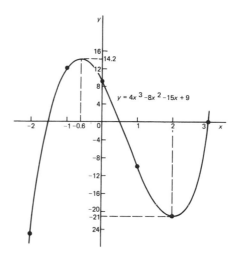

Fig. 13.15

Problem 9 Plot the graph of $y = 2x^3 - 7x^2 + 4x + 4$ for values of *x* between $x = -1$ and $x = 3$. Hence determine the roots of the equation $2x^3 - 7x^2 + 4x + 4 = 0$.

A table of values is drawn up as shown below.

x	-1	0	1	2	3
$2x^3$	-2	0	2	16	54
$-7x^2$	-7	0	-7	-28	-63
$+4x$	-4	0	4	8	12
$+4$	4	4	4	4	4
y	-9	4	3	0	7

A graph of $y = 2x^3 - 7x^2 + 4x + 4$ is shown in *Fig. 13.16*. The graph crosses the *x*-axis at $x = -0.5$ and touches the *x*-axis at $x = 2$.

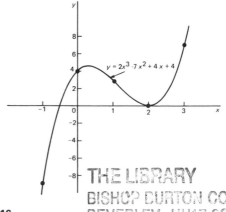

Fig. 13.16

Hence the solutions of the equation $2x^3 - 7x^2 + 4x + 4 = 0$ are $x = -0.5$ and $x = 2$.

Further problems on the graphical solutions of cubic equations may be found in section 13.5, Problems 19 to 25.

13.5 Further problems on the graphical solution of equations

Simultaneous equations

In *Problems 1* to *5*, solve the simultaneous equations graphically.

1 $x + y = 2$
   $3y - 2x = 1$
   $[x = 1, y = 1]$
2 $y = 5 - x$
   $x - y = 2$
   $\left[x = 3\frac{1}{2}, y = 1\frac{1}{2}\right]$
3 $3x + 4y = 5$
   $2x - 5y + 12 = 0$
   $[x = -1, y = 2]$
4 $1.4x - 7.06 = 3.2y$
   $2.1x - 6.7y = 12.87$
   $[x = 2.3, y = -1.2]$
5 $3x - 2y = 0$
   $4x + y + 11 = 0$
   $[x = -2, y = -3]$
6 The friction force F newtons and load L newtons are connected by a law of the form $F = aL + b$, where a and b are constants. When $F = 4$ newtons, $L = 6$ newtons and when $F = 2.4$ newtons, $L = 2$ newtons. Determine graphically the values of a and b.
   $[a = 0.4, b = 1.6]$

Quadratic equations

7 Sketch the following graphs and state the nature and coordinates of their turning points.
   (a) $y = 4x^2$ (b) $y = 2x^2 - 1$
   (c) $y = -x^2 + 3$ (d) $y = -\frac{1}{2}x^2 - 1$
   $\left[\begin{array}{l}\text{(a) Minimum (0,0) (b) Minimum (0,-1)}\\\text{(c) Maximum (0,3) (d) Maximum (0,-1)}\end{array}\right]$

Solve graphically the quadratic equations in *Problems 8* to *11* by plotting the curves between the given limits. Give answers correct to 1 decimal place.

8 $4x^2 - x - 1 = 0$; $x = -1$ to $x = 1$
   $[-0.4$ or $0.6]$
9 $x^2 - 3x = 27$; $x = -5$ to $x = 8$
   $[-3.9$ or $6.9]$

10 $2x^2 - 6x - 9 = 0$; $x = -2$ to $x = 5$
    $[-1.1$ or $4.1]$
11 $2x(5x - 2) = 39.6$; $x = -2$ to $x = 3$
    $[-1.8$ or $2.2]$
12 Solve the quadratic equation $2x^2 + 7x + 6 = 0$ graphically, given that the solutions lie in the range $x = -2$ to $x = 3$. Determine also the nature and coordinates of its turning point.
    $\left[x = -1\frac{1}{2} \text{ or } 2, \text{ Minimum at } \left(-1\frac{3}{4}, -\frac{1}{8}\right)\right]$
13 Solve graphically the quadratic equation $10x^2 - 9x - 11.2 = 0$, given that the roots lie between $x = -1$ and $x = 2$.
    $[x = -0.7$ or $1.6]$
14 Plot a graph of $y = 3x^2$ and hence solve the equations (a) $3x^2 - 8 = 0$ and (b) $3x^2 - 2x - 1 = 0$.
    $\left[\text{(a) } \pm 1.63 \text{ (b) } 1 \text{ or } -\frac{1}{3}\right]$
15 Plot the graphs $y = 2x^2$ and $y = 3 - 4x$ on the same axes and find the coordinates of the points of intersection. Hence determine the roots of the equation $2x^2 + 4x - 3 = 0$.
    $[(-2.58,13.31), (0.58,0.67); x = -2.58 \text{ or } 0.58]$
16 Plot a graph of $y = 10x^2 - 13x - 30$ for values of x between $x = -2$ and $x = 3$. Solve the equation $10x^2 - 13x - 30 = 0$ and from the graph determine (a) the value of y when x is 1.3, (b) the value of x when y is 10 and (c) the roots of the equation $10x^2 - 15x - 18 = 0$.
    $\left[\begin{array}{l}x = -1.2 \text{ or } 2.5\\\text{(a) } -30 \text{ (b) } 2.75 \text{ and } -1.45\\\text{(c) } x = 2.29 \text{ or } -0.79\end{array}\right]$

Linear and quadratic equations simultaneously

17 Determine graphically the values of x and y which simultaneously satisfy the equations $y = 2(x^2 - 2x - 4)$ and $y + 4 = 3x$.
    $[x = 4, y = 8 \text{ and } x = -\frac{1}{2}, y = -5\frac{1}{2}]$
18 Plot the graph of $y = 4x^2 - 8x - 21$ for values of x from -2 to $+4$. Use the graph to find the roots of the following equations:
    (a) $4x^2 - 8x - 21 = 0$ (b) $4x^2 - 8x - 16 = 0$
    (c) $4x^2 - 6x - 18 = 0$.
    $\left[\begin{array}{l}\text{(a) } x = -1.5 \text{ or } 3.5 \text{ (b) } x = -1.24 \text{ or } 3.24\\\text{(c) } x = -1.5 \text{ or } 3.0\end{array}\right]$

Cubic equations

19 Plot the graph $y = 4x^3 + 4x^2 - 11x - 6$ between $x = -2$ and $x = 2$ and use the graph to solve the cubic equation $4x^3 + 4x^2 - 11x - 6 = 0$.
    $[x = -2.0, -0.5 \text{ or } 1.5]$
20 By plotting a graph of $y = x^3 - 2x^2 - 5x + 6$ between

$x = -3$ and $x = 4$ solve the equation $x^3 - 2x^2 - 5x + 6 = 0$. Determine also the coordinates of the turning points and distinguish between them.

[($x = -2$, 1 or 3, Minimum at $(2.12, -4.10)$, Maximum at $(-0.79, 8.21)$)]

In *Problems 21* to *24*, solve graphically the cubic equations given, each correct to 2 significant figures.

21 $x^3 - 1 = 0$
   [$x = 1$]

22 $x^3 - x^2 - 5x + 2 = 0$
   [$x = -2.0$, 0.38 or 2.6]

23 $x^3 - 2x^2 = 2x - 2$
   [$x = 0.69$ or 2.5]

24 $2x^3 - 2x^2 - 9.08x + 8.28 = 0$
   [$x = -2.3$, 1.0 or 1.8]

25 Show that the cubic equation $8x^3 + 36x^2 + 54x + 27 = 0$ has only one real root and determine its value.
   [$x = -1.5$]

14

Geometry

14.1 Angular measurement

Geometry is a part of mathematics in which the properties of points, lines, surfaces and solids are investigated.

An **angle** is the amount of rotation between two straight lines.

Angles may be measured in either **degrees** or **radians** (see section 14.8).

1 revolution = 360 degrees, thus 1 degree = $\frac{1}{360}$ th of one revolution. Also 1 minute = $\frac{1}{60}$ th of a degree and 1 second = $\frac{1}{60}$th of a minute. 1 minute is written as 1′ and 1 second is written as 1″. **Thus 1° = 60′ and 1′ = 60″**.

Problem 1 Add 14°53′ and 37°18′.

14°53′ 53′ + 19′ = 72′. Since 60′ = 1°, 72′ = 1°12′.
37°19′ Thus the 12′ is placed in the minutes column
52°12′ and 1° is carried in the degrees column. Then
1° 14° + 37° + 1° (carried) = 52°.

Thus **14°53′ + 37°19′ = 52°12′**.

Problem 2 Subtract 15°47′ from 28°13′.

27°
2̶8̶°13′ 13′ 47′ cannot be done. Hence 1° or 60′ is
15°47′ 'borrowed' from the degrees column, which
12°26′ leaves 27° in that column. Now (60′ + 13′) −

47′ = 26′, which is placed in the minutes column. 27° − 15° = 12°, which is placed in the degrees column.

Thus **28°13′ − 15°47′ = 12°26′**.

Problem 3 Determine (a) 13°42′51″ + 48°22′17″, (b) 37°12′8″ − 21°17′25″.

(a)   13°42′51″ (b)   36°11′
    48°22′17″   3̶7̶°1̶2̶′ 8″
Adding: 62° 5′ 8″   21°17′25″
    1° 1′ Subtracting: 15°54′43″

Problem 4 Convert (a) 24°42′, (b) 78°15′26″ to degrees and decimals of a degree.

(a) Since 1 minute = $\frac{1}{60}$th of a degree, 42′ = 42°/60 = 0.70°.
 Hence **24°42′ = 24.70°**.

(b) Since 1 second = $\frac{1}{60}$th of a minute, 26″ = 26′/60
 = 0.433 3′.
 Hence 78°15′26″ = 78°15.433 3′
 15.433 3′ = 15.433 3°/60 = 0.257 2°, correct to 4 decimal places.
 Hence **78°15′26″ = 78.257 2°**, correct to 4 decimal places.

Problem 5 Convert 45.371° into degrees, minutes and seconds.

Since $1° = 60'$, $0.371° = (0.371 \times 60)' = 22.26'$.
Since $1' = 60''$, $0.26' = (0.26 \times 60)'' = 15.6'' = 16''$ to the nearest second.
Hence **45.371° = 45°22'16''**.

Further problems on angular measurement may be found in section 14.9, Problems 1 to 4, page 132.

14.2 Types and properties of angles

(a) (i) Any angle between 0° and 90° is called an **acute angle**.
   (ii) An angle equal to 90° is called a **right angle**.
   (iii) Any angle between 90° and 180° is called an **obtuse angle**.
   (iv) Any angle greater than 180° and less than 360° is called a **reflex angle**.
(b) (i) An angle of 180° lies on a straight line.
   (ii) If two angles add up to 90° they are called **complementary angles**.
   (iii) If two angles add up to 180° they are called **supplementary angles.**
   (iv) **Parallel lines** are straight lines which are in the same plane and never meet. (Such lines are denoted by arrows, as in *Fig. 14.1*.)
   (v) A straight line which crosses two parallel lines is called a **transversal** (see MN in *Fig. 14.1*).

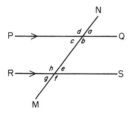

Fig. 14.1

(c) With reference to *Fig. 14.1*:
   (i) $a = c$, $b = d$, $e = g$ and $f = h$. Such pairs of angles are called **vertically opposite angles**.
   (ii) $a = e$, $b = f$, $c = g$ and $d = h$. Such pairs of angles are called **corresponding angles**.
   (iii) $c = e$ and $b = h$. Such pairs of angles are called **alternate angles**.
   (iv) $b + e = 180°$ and $c + h = 180°$. Such pairs of angles are called **interior angles**.

Problem 6 State the general name given to the following angles:

(a) 159° (b) 63° (c) 90° (d) 227°

(a) 159° lies between 90° and 180° and is therefore called an **obtuse angle**.
(b) 63° lies between 0° and 90° and is therefore called an **acute angle**.
(c) 90° is called a **right angle**.
(d) 227° is greater than 180° and less than 360° and is therefore called a **reflex angle**.

Problem 7 Find the angles complementary to
(a) 41° (b) 58°39'.

(a) The complement of 41° is $(90° - 41°)$, i.e. **49°**.
(b) The complement of 58°39' is $(90° - 58°39')$, i.e. **31°21'**.

Problem 8 Find the angles supplementary to
(a) 27° (b) 111°11'.

(a) The supplement of 27° is $(180° - 27°)$, i.e. **153°**.
(b) The supplement of 111°11' is $(180° - 111°11')$, i.e. **68°49'**.

Problem 9 Two straight lines AB and CD intersect at 0. If ∠AOC is 43°, find ∠AOD, ∠DOB and ∠BOC.

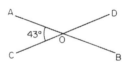

Fig. 14.2

From *Fig. 14.2*, ∠AOD is supplementary to ∠AOC. Hence ∠AOD = $180° - 43° = $ **137°**. When two straight lines intersect the vertically opposite angles are equal. Hence **∠DOB = 43°** and **∠BOC = 137°**.

Problem 10 Determine angle β in *Fig. 14.3*.

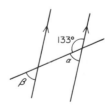

Fig. 14.3

$\alpha = 180° - 133° = 47°$ (i.e. supplementary angles).
$\alpha = \beta = 47°$ (corresponding angles between parallel lines).

Problem 11 Determine the value of angle θ in *Fig. 14.4*.

Let a straight line FG be drawn through E such that FG is parallel to AB and CD. ∠BAE = ∠AEF (alternate angles between parallel lines AB and FG), hence ∠AEF = 23°37′. ∠ECD = ∠FEC (alternate angles between parallel lines FG and CD), hence ∠FEC = 35°49′.

Angle θ = ∠AEF + ∠FEC = 23°37′ + 35°49′ = **59°26′**.

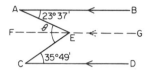

Fig. 14.4

Problem 12   Determine angles *c* and *d* in *Fig. 14.5*.

---

b = 46° (corresponding angles between parallel lines). Also *b* + *c* + 90° = 180° (angles on a straight line). Hence 46° + *c* + 90° = 180°, from which *c* = **44°**. *b* and *d* are supplementary, hence *d* = 180° − 46° = **134°**. Alternatively, 90° + *c* = *d* (vertically opposite angles).

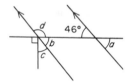

Fig. 14.5

Further problems on types and properties of angles may be found in section 14.9. Problems 5 to 11, page 132.

14.3 Properties of triangles

A **triangle** is a figure enclosed by three straight lines. The sum of the three angles of a triangle is equal to 180°.

Types of triangles:

(i)   An **acute-angled triangle** is one in which all the angles are acute, i.e. all the angles are less than 90°.

(ii) A **right-angled triangle** is one which contains a right angle.

(iii) An **obtuse-angled triangle** is one which contains an obtuse angle, i.e. one angle which lies between 90° and 180°.

(iv) An **equilateral triangle** is one in which all the sides and all the angles are equal (i.e. each 60°).

(v)   An **isosceles triangle** is one in which two angles and two sides are equal.

(vi) A **scalene triangle** is one with unequal angles and therefore unequal sides.

With reference to *Fig. 14.6*:

(i)   Angles *A*, *B* and *C* are called **interior angles** of the triangle.

(ii) Angle θ is called an **exterior angle** of the triangle and is equal to the sum of the two opposite interior angles, i.e. θ = *A* + *C*.

(iii) *a* + *b* + *c* is called the **perimeter** of the triangle.

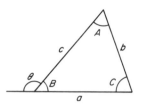

Fig. 14.6

Problem 13   Name the types of triangles shown in *Fig. 14.7*.

(a) Equilateral triangle.
(b) Acute-angled scalene triangle.
(c) Right-angled triangle.
(d) Obtuse-angled scalene triangle.
(e) Isosceles triangle.

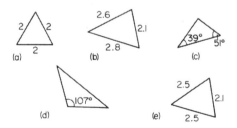

Fig. 14.7

Problem 14   Determine the value of angles θ and α in *Fig. 14.8*.

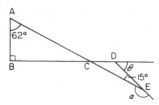

Fig. 14.8

In triangle ABC, ∠A + ∠B + ∠C = 180° (angles in a triangle add up to 180°), hence ∠C = 180° − 90° − 62° = 28°. Thus ∠DCE = 28° (vertically opposite angles).

$\angle\theta = \angle DCE + \angle DEC$ (exterior angle of a triangle is equal to the sum of the two opposite interior angles). Hence $\angle\theta = 28° + 15° = \mathbf{43°}$.
$\angle\alpha$ and $\angle DEC$ are supplementary, thus $\alpha = 180° - 15° = \mathbf{165°}$.

Problem 15 ABC is an isosceles triangle in which the unequal angle BAC is 56°. AB is extended to D as shown in *Fig. 14.9*. Determine the angle DBC.

Since the three interior angles of a triangle add up to 180° then $56° + \angle B + \angle C = 180°$, i.e. $\angle B + \angle C = 180° - 56° = 124°$.

Triangle ABC is isosceles hence $\angle B = \angle C = 124°/2 = 62°$.

$\angle DBC = \angle A + \angle C$ (exterior angle equals sum of two interior opposite angles), i.e. $\angle DBC = 56° + 62° = \mathbf{118°}$.
Alternatively, $\angle DBC + \angle ABC = 180°$ (i.e. supplementary angles).)

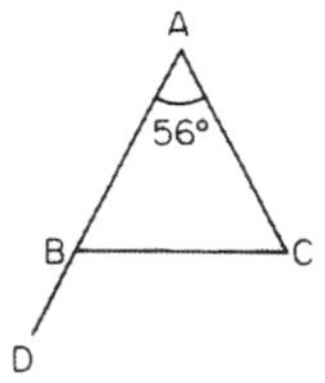

Fig. 14.9

Problem 16 Find angles a, b, c, d and e in *Fig. 14.10*.

$= 62°$ and $c = \mathbf{55°}$ (alternate angles between parallel lines).
$55° + b + 62° = 180°$ (angles in a triangle add up to 180°), hence $b = 180° - 55° - 62° = \mathbf{63°}$.
$= \mathbf{d} = \mathbf{63°}$ (alternate angles between parallel lines).
$+ 55° + 63° = 180°$ (angles in a triangle add up to 180°), hence $e = 180° - 55° - 63° = \mathbf{62°}$.
Check: $e = a = 62°$ (corresponding angles between parallel lines).)

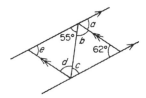

Fig. 14.10

Further problems on properties of triangles may be found in section 14.9, Problems 12 to 16, page 132.

14.4 The theorem of Pythagoras

With reference to *Fig. 14.11*, the side opposite the right angle (side b) is called the **hypotenuse**. The **theorem of Pythagoras** states: 'In any right-angled triangle, the square on the hypotenuse is equal to the sum of the squares on the other two sides.' Hence $b^2 = a^2 + c^2$.

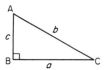

Fig. 14.11

Problem 17 In *Fig. 14.12*, find the length of BC.

By Pythagoras' theorem: $a^2 = b^2 + c^2$
i.e. $a^2 = 4^2 + 3^2 = 16 + 9 = 25$
Hence $a = \sqrt{25} = \pm 5$
(−5 has no meaning in this context and is thus ignored).
Thus **BC = 5 cm**.

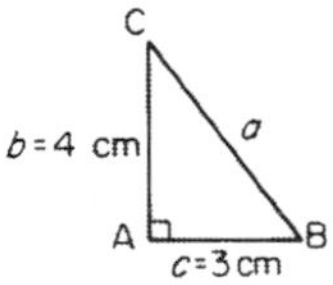

Fig. 14.12

Problem 18 In *Fig. 14.13*, find the length of EF.

By Pythagoras' theorem: $e^2 = d^2 + f^2$
Hence $13^2 = d^2 + 5^2$
$169 = d^2 + 25$
$d^2 = 169 - 25 = 144$
Thus $d = \sqrt{144} = 12$ cm
Thus **EF = 12 cm**.

Fig. 14.13

Problem 19 Two aircraft leave an airfield at the same time. One travels due north at an average speed of 300 km/h and the other due west at an average speed of 220 km/h. Calculate their distance apart after 4 hours.

After 4 hours, the first aircraft has travelled $4 \times 300 = 1200$ km, due north, and the second aircraft has travelled $4 \times 220 = 880$ km due west, as shown in *Fig. 14.14*. Distance apart after 4 hours = BC.
From Pythagoras' theorem:

$$BC^2 = 1200^2 + 880^2$$
$$= 1\,440\,000 + 774\,400 \text{ and } BC = \sqrt{(2\,214\,400)}$$

Hence distance apart after 4 hours = 1488 km.

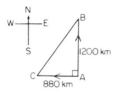

Fig. 14.14

Further problems on the theorem of Pythagoras may be found in section 14.9, Problems 17 to 23, page 133.

14.5 Congruent triangles

Two triangles are said to be **congruent** if they are equal in all respects, i.e. three angles and three sides in one triangle are equal to three angles and three sides in the other triangle. Two triangles are congruent if:

(i)   the three sides of one are equal to the three sides of the other (SSS),
(ii) they have two sides of the one equal to two sides of the other, and if the angles included by these sides are equal (SAS),
(iii) two angles of the one are equal to two angles of the other and any side of the first is equal to the corresponding side of the other (ASA), or
(iv) their hypotenuses are equal and if one other side of one is equal to the corresponding side of the other (RHS).

> **Problem 20** State which of the pairs of triangles shown in *Fig. 14.15* are congruent and name their correct sequence.

(a) Congruent ABC, FDE (Angle, side, angle, i.e. ASA).
(b) Congruent GIH, JLK (Side, angle, side, i.e. SAS).
(c) Congruent MNO, RQP (Right-angle, hypotenuse, side, i.e. RHS).
(d) Not necessarily congruent. It is not indicated that any side coincides.
(e) Congruent ABC, FED (Side, side, side, i.e. SSS).

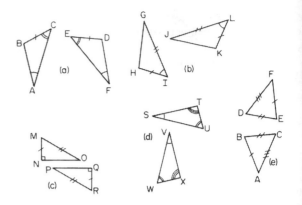

Fig. 14.15

> **Problem 21** In *Fig. 14.16*, triangle PQR is isosceles with Z the mid-point of PQ. Prove that triangle PXZ and QYZ are congruent, and that triangles RXZ and RYZ are congruent. Determine the values of angles RPZ and RXZ.

Since triangle PQR is isosceles PR = RQ and thus $\angle$QPR $= \angle$RQP.
$\angle$RXZ $= \angle$QPR $+ 28°$ and $\angle$RYZ $= \angle$RQP $+ 28°$ (exterior angles of a triangle equal the sum of the two interior opposite angles). Hence $\angle$RXZ $= \angle$RYZ.
$\angle$PXZ $= 180° - \angle$RXZ and $\angle$QYZ $= 180° - \angle$RYZ. Thus $\angle$PXZ $= \angle$QYZ.
Triangles PXZ and QYZ are congruent since $\angle$XPZ $= \angle$YQZ, PZ = ZQ and $\angle$XZR $= \angle$YZQ (ASA). Hence XZ = YZ.
Triangles PRZ and QRZ are congruent since PR = RQ $\angle$RPZ $= \angle$RQZ and PZ = ZQ (SAS). Hence $\angle$RZX $= \angle$RZY.
Triangles RXZ and RYZ are congruent since $\angle$RXZ $= \angle$RYZ, XZ = YZ and $\angle$RZX $= \angle$RZY (ASA).
$\angle$QRZ $= 67°$ and thus $\angle$PRQ $= 67° + 67° = 134°$. Hence

$$\angle RPZ = \angle RQZ = \frac{180° - 134°}{2} = 23°$$

$\angle$RXZ $= 23° + 28° = 51°$ (external angle of a triangle equals the sum of the two interior opposite angles).

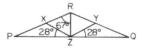

Fig. 14.16

Further problems on congruent triangles may be found in section 14.9, Problems 24 and 25, page 133.

14.6 Similar triangles

Two triangles are said to be **similar** if the angles of one triangle are equal to the angles of the other triangle. With reference to *Fig. 14.17*: Triangles ABC and PQR are similar and the corresponding sides are in proportion to each other, i.e.

$$\frac{p}{a} = \frac{q}{b} = \frac{r}{c}$$

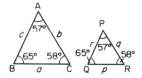

Fig. 14.17

Problem 22 In *Fig. 14.18*, find the length of side *a*.

In triangle ABC, $50° + 30° + \angle C = 180°$, from which $\angle C = 60°$.
In triangle DEF, $\angle E = 180° - 50° - 60° = 70°$. Hence triangles ABC and DEF are similar, since their angles are the same. Since corresponding sides are in proportion to each other then:

$$\frac{a}{d} = \frac{c}{f}, \text{ i.e. } \frac{a}{4.42} = \frac{12.0}{5.0}$$

Hence

$$a = \frac{12.0}{5.0}(4.42) = \textbf{10.61 cm}$$

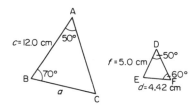

Fig. 14.18

Problem 23 In *Fig. 14.19*, find the dimensions marked *p*, *r* and *z*.

In triangle PQR, $\angle Q = 180° - 90° - 35° = 55°$.
In triangle XYZ, $\angle X = 180° - 90° - 55° = 35°$.

Hence triangles PQR and XYZ are similar since their angles are the same. The triangles may be redrawn as shown in *Fig. 14.20*.

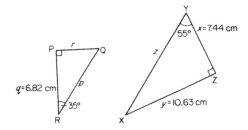

Fig. 14.19

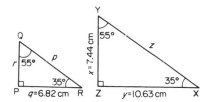

Fig. 14.20

By proportion: $\dfrac{p}{z} = \dfrac{r}{x} = \dfrac{q}{y}$

Hence $\dfrac{p}{z} = \dfrac{r}{7.44} = \dfrac{6.82}{10.63}$

from which, $r = 7.44\left(\dfrac{6.82}{10.63}\right) = \textbf{4.77 cm}$

Using Pythagoras' theorem on triangle XYZ gives:

$$z = \sqrt{[(7.44)^2 + (10.63)^2]} = \textbf{12.97 cm}$$

By proportion: $\dfrac{p}{z} = \dfrac{q}{y}$, i.e. $\dfrac{p}{12.97} = \dfrac{6.82}{10.63}$

Hence $p = 12.97\left(\dfrac{6.82}{10.63}\right) = \textbf{8.32 cm}$

Problem 24 In *Fig. 14.21*, show that triangles CBD and CAE are similar and hence find the length of CD and BD.

Since BD is parallel to AE then $\angle CBD = \angle CAE$ and $\angle CDB = \angle CEA$ (corresponding angles between parallel lines). Also $\angle C$ is common to triangles CBD and CAE.

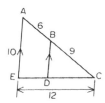

Fig. 14.21

Since the angles in triangle CBD are the same as in triangle CAE the triangles are similar. Hence, by proportion:

$$\frac{CB}{CA} = \frac{CD}{CE} \left(= \frac{BD}{AE} \right)$$

i.e. $\dfrac{9}{6+9} = \dfrac{CD}{12}$, from which $CD = 12\left(\dfrac{9}{15}\right) = \mathbf{7.2\ cm}$

Also, $\dfrac{9}{15} = \dfrac{BD}{10}$, from which $BD = 10\left(\dfrac{9}{15}\right) = \mathbf{6\ cm}$

Problem 25   A rectangular shed 2 m wide and 3 m high stands against a perpendicular building of height 5.5 m. A ladder is used to gain access to the roof of the building. Determine the minimum distance between the bottom of the ladder and the shed and also the minimum length of ladder required.

A side view is shown in *Fig. 14.22*, where AF is the minimum length of ladder. Since BD and CF are parallel, $\angle ADB = \angle DFE$ (corresponding angles between parallel lines). Hence triangles BAD and EDF are similar since their angles are the same. $AB = AC - BC = AC - DE = 5.5 - 3 = 2.5$ m.
By proportion:

$$\frac{AB}{DE} = \frac{BD}{EF}, \text{ i.e. } \frac{2.5}{3} = \frac{2}{EF}$$

Hence

$$EF = 2\left(\frac{3}{2.5}\right) = \mathbf{2.4\ m}$$

= minimum distance from bottom of ladder to the shed. Since $AC = 5.5$ m, $CF = BD + EF = 2 + 2.4 = 4.4$ m, then AF may be found using Pythagoras' theorem:

$$AF^2 = 5.5^2 + 4.4^2$$

Hence minimum length of ladder,

$$AF = \sqrt{(5.5^2 + 4.4^2)} = \mathbf{7.04\ m.}$$

Further problems on similar triangles may be found in section 14.9, Problems 26 to 30, page 133.

14.7 Construction of triangles

To construct any triangle the following drawing instruments are needed: (i) ruler and/or straight edge, (ii) compass, (iii) protractor, (iv) pencil. For actual constructions, see *Problems 26 to 29* which follow.

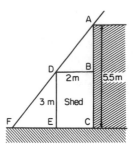

Fig. 14.22

Problem 26   Construct a triangle whose sides are 6 cm, 5 cm and 3 cm.

With reference to *Fig. 14.23*:

(i)   Draw a straight line of any length, and with a pair of compasses, mark out 6 cm length and label it AB.

(ii)   Set compass to 5 cm and with centre at A describe arc DE.

(iii)   Set compass to 3 cm and with centre at B describe arc FG.

(iv)   The intersection of the two curves at C is the vertex of the required triangle. Join AC and BC by straight lines.

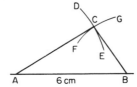

Fig. 14.23

It may be proved by measurement that the ratio of the angles of a triangle is not equal to the ratio of the sides (i.e. in this problem, the angle opposite the 3 cm side is not equal to half the angle opposite the 6 cm side).

Problem 27   Construct a triangle ABC such that $a = 6$ cm, $b = 3$ cm and $\angle C = 60°$.

With reference to *Fig. 14.24*:

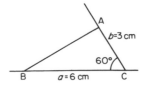

Fig. 14.24

(i)   Draw a line BC, 6 cm long.

(ii)   Using a protractor centred at C make an angle of 60° to BC.

(iii) From C measure a length of 3 cm and label A.

(iv) Join B to A by a straight line.

Problem 28 Construct a triangle PQR given that QR = 5 cm, ∠Q = 70° and ∠R = 44°.

With reference to *Fig. 14.25*:

(i) Draw a straight line 5 cm long and label it QR.

(ii) Use a protractor centred at Q and make an angle of 70°. Draw QQ'.

(iii) Use a protractor centred at R and make an angle of 44°. Draw RR'.

(iv) The intersection of QQ' and RR' forms the vertex P of the triangle. Join QP and RP by straight lines.

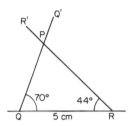

Fig. 14.25

Problem 29 Construct a triangle XYZ given that XY = 5 cm, the hypotenuse YZ = 6.5 cm and ∠X = 90°.

With reference to *Fig. 14.26*:

(i) Draw a straight line 5 cm long and label it XY.

(ii) Produce XY any distance to B. With compass centred at X make an arc at A and A'. (The length XA and XA' is arbitrary.) With compass centred at A draw the arc PQ. With the same compass setting and centred at A', draw the arc RS. Join the intersection of the arcs, C, to X, and a right angle to XY is produced at X. (Alternatively, a protractor can be used to construct a 90° angle.)

(iii) The hypotenuse is always opposite the right angle. Thus YZ is opposite ∠X. Using a compass centred at Y and set to 6.5 cm, describe the arc UV.

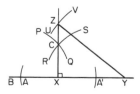

Fig. 14.26

(iv) The intersection of the arc UV with XC produced, forms the vertex Z of the required triangle. Join YZ by a straight line.

Further problems on constructions of triangles may be found in section 14.9, Problem 31, page 134.

14.8 The circle and its properties

A **circle** is a plain figure enclosed by a curved line, every point on which is equidistant from a point within, called the **centre**.

Properties of circles

(i) The distance from the centre to the curve is called the **radius**, *r*, of the circle (see OP in *Fig. 14.27*).

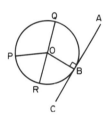

Fig. 14.27

(ii) The boundary of a circle is called the **circumference**, *c*.

(iii) Any straight line passing through the centre and touching the circumference at each end is called the **diameter**, *d* (see QR in *Fig. 14.27*). Thus $d = 2r$.

(iv) The ratio

$$\frac{\text{circumference}}{\text{diameter}} = \text{a constant for any circle}$$

This constant is denoted by the Greek letter π (pronounced 'pie'), where $\pi = 3.141\,59$, correct to 5 decimal places.

Hence $c/d = \pi$ or $c = \pi d$ or $c = 2\pi r$.

(v) A **semicircle** is one half of the whole circle.

(vi) A **quadrant** is one quarter of a whole circle.

(vii) A **tangent** to a circle is a straight line which meets the circle in one point only and does not cut the circle when produced. AC in *Fig. 14.27* is a tangent to the circle since it touches the curve at point B only. If radius OB is drawn, then angle ABO is a right angle.

(viii) A **sector** of a circle is the part of a circle between radii (for example, the portion OXY of *Fig. 14.28* is a sector). If a sector is less than a semicircle it is called a **minor sector**, if greater than a semicircle it is called a **major sector**.

(ix) A **chord** of a circle is any straight line which divides the circle into two parts and is terminated at each end by the circumference. ST, in *Fig. 14.28*, is a chord.

(x) A **segment** is the name given to the parts into which a circle is divided by a chord. If the segment is less than a semi-circle it is called a **minor segment** (see shaded area in *Fig. 14.28*). If the segment is greater than a semicircle it is called a **major segment** (see the unshaded area in *Fig. 14.28*).

(xi) An **arc** is a portion of the circumference of a circle. The distance SRT in *Fig. 14.28* is called a **minor arc** and the distance SXYT is called a **major arc**.

(xii) The angle at the centre of a circle, subtended by an arc, is double the angle at the circumference subtended by the same arc. With reference to *Fig. 14.29*, **Angle AOC = 2 × angle ABC**.

(xiii) The angle in a semicircle is a right angle (see angle BQP in *Fig. 14.29*).

One **radian** is defined as the angle subtended at the centre of a circle by an arc equal in length to the radius. With reference to *Fig. 14.30*, for arc length l, θ radians $= l/r$ or $l = r\theta$, where θ is in radians. When l = whole circumference $(= 2\pi r)$ then $\theta = l/r = 2\pi r/r = 2\pi$,

i.e. 2π radians $= 360°$ or $\boxed{\pi \text{ radians} = 180°.}$

Thus 1 rad $= 180°/\pi = 57.30°$, correct to 2 decimal places. Since π rad $= 180°$, then $\pi/2$ rad $= 90°$, $\pi/3$ rad $= 60°$, $\pi/4$ rad $= 45°$, and so on.

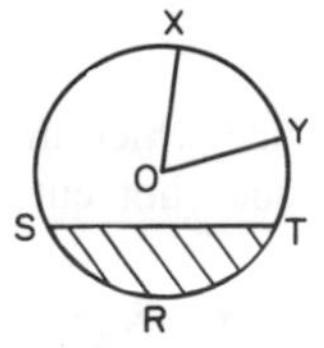

Fig. 14.28

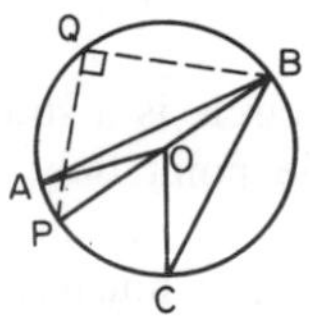

Fig. 14.29

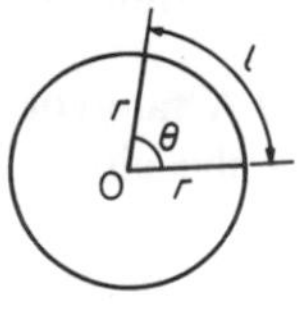

Fig. 14.30

Problem 30   Find the circumference of a circle of radius 12.0 cm.

Circumference, $c = 2 \times \pi \times$ radius $= 2\pi r = 2\pi(12.0) =$ **75.40 cm**

Problem 31   If the diameter of a circle is 75 mm, find its circumference.

Circumference, $c = \pi \times$ diameter $= \pi d = \pi(75) =$ **235.6 mm**

Problem 32   Determine the radius of a circle if its perimeter is 112 cm.

Perimeter $=$ circumference, $c = 2\pi r$.

Hence $r = \dfrac{c}{2\pi} = \dfrac{112}{2\pi} =$ **17.83 cm**

Problem 33   In *Fig. 14.31*, AB is a tangent to the circle at B. If the circle radius is 40 mm and AB = 150 mm, calculate the length AO.

A tangent to a circle is at right angles to a radius drawn from the point of contact, i.e. $\angle ABO = 90°$. Hence, using Pythagoras' theorem:

$$AO^2 = AB^2 + OB^2$$
$$AO = \sqrt{(AB^2 + OB^2)} = \sqrt{[(150)^2 + (40)^2]}$$
$$= \textbf{155.2 mm}$$

Fig. 14.31

Problem 34   Convert to radians: (a) 125° (b) 69°47′.

(a) Since $180° = \pi$ rad then $1° = \pi/180$ rad, therefore

$$125° = 125\left(\frac{\pi}{180}\right)^c = \textbf{2.182 radians}$$

(Note that ^{c} means 'circular measure' and indicates radian measure.)

(b) $69°47' = 69\dfrac{47°}{60} = 69.783°$

$69.783° = 69.783\left(\dfrac{\pi}{180}\right)^c = \textbf{1.218 radians}$

Problem 35 Convert to degrees and minutes:

(a) 0.749 radians (b) $3\pi/4$ radians.

(a) Since π rad $= 180°$ then 1 rad $= 180°/\pi$, therefore

0.749 rad $= 0.749\left(\dfrac{180}{\pi}\right)° = 42.915°$

$0.915° = (0.915 \times 60)' = 55'$, correct to the nearest minute, hence

0.749 radians = 42°55′

(b) Since

1 rad $= \left(\dfrac{180}{\pi}\right)°$ then

$\dfrac{3\pi}{4}$ rad $= \dfrac{3\pi}{4}\left(\dfrac{180}{\pi}\right)° = \dfrac{3}{4}(180)° = \textbf{135°}$

Problem 36 Express in radians, in terms of π:

(a) 45° (b) 60° (c) 90° (d) 150° (e) 270° (f) 37.5°

Since $180° = \pi$ rad then $1° = 180/\pi$ rad, hence

(a) $45° = 45\left(\dfrac{\pi}{180}\right)$ rad $= \dfrac{\pi}{4}$ **rad**

(b) $60° = 60\left(\dfrac{\pi}{180}\right)$ rad $= \dfrac{\pi}{3}$ **rad**

(c) $90° = 90\left(\dfrac{\pi}{180}\right)$ rad $= \dfrac{\pi}{2}$ **rad**

(d) $150° = 150\left(\dfrac{\pi}{180}\right)$ rad $= \dfrac{5\pi}{6}$ **rad**

(e) $270° = 270\left(\dfrac{\pi}{180}\right)$ rad $= \dfrac{3\pi}{2}$ **rad**

(f) $37.5° = 37.5\left(\dfrac{\pi}{180}\right)$ rad $= \dfrac{75\pi}{360}$ rad $= \dfrac{5\pi}{24}$ **rad**

Problem 37 Find the length of arc of a circle of radius 5.5 cm when the angle subtended at the centre is 1.20 radians.

Length of arc, $l = r\theta$, where θ is in radians, hence

$l = (5.5)(1.20) = \textbf{6.60 cm}$

Problem 38 Determine the diameter and circumference of a circle if an arc of length 4.75 cm subtends an angle of 0.91 radians.

Since $l = r\theta$ then

$r = \dfrac{l}{\theta} = \dfrac{4.75}{0.91} = 5.22$ cm

Diameter $= 2 \times$ radius $= 2 \times 5.22 = \textbf{10.44 cm}$.
Circumference, $c = \pi d = \pi(10.44) = \textbf{32.80 cm}$.

Problem 39 If an angle of 125° is subtended by an arc of a circle of radius 8.4 cm, find the length of (a) the minor arc, and (b) the major arc, correct to 3 significant figures.

Since $180° = \pi$ rad, then

$1° = \left(\dfrac{\pi}{180}\right)$ rad and $125° = 125\left(\dfrac{\pi}{180}\right)$ rad

Length of minor arc,

$l = r\theta = (8.4)(125)\left(\dfrac{\pi}{180}\right) = \textbf{18.3 cm}$, correct to 3 significant figures.

Length of major arc $=$ (circumference $-$ minor arc) $= 2\pi(8.4)$
$- 18.3 = \textbf{34.5 cm}$, correct to 3 significant figures.
(Alternatively, major arc $= r\theta = 8.4 (360 - 125)(\pi/180)$
$= 34.5$ cm.)

Problem 40 Determine the angle, in degrees and minutes, subtended at the centre of a circle of diameter 42 mm by an arc of length 36 mm.

Since length of arc, $l = r\theta$ then $\theta = l/r$.

Radius, $r = \dfrac{\text{diameter}}{2} = \dfrac{42}{2} = 21$ mm

hence $\theta = \dfrac{l}{r} = \dfrac{36}{21} = 1.714\ 3$ radians

$1.714\ 3$ rad $= 1.714\ 3 \times (180/\pi)° = 98.22° = \textbf{98°13′}$
$=$ angle subtended at centre of circle.

Problem 41 If an arc of length 11.48 cm subtends an angle of 168°27′ at the centre of a circle, find its radius correct to the nearest millimetre.

$168°27' = 168\dfrac{27°}{60} = 168.45° = 168.45(\pi/180)$ radians, hence

$\theta = 2.94$ radians. Since arc length $l = r\theta$ then

$$r = \frac{l}{\theta} = \frac{11.48}{2.94} = 3.905 \text{ cm}$$

$$= 39.05 \text{ mm}$$

$$= \mathbf{39\ mm} \text{ to the nearest millimetre}$$

Further problems on circles may be found in section 14.9, Problems 32 to 46, page 134.

14.9 Further problems on geometry

Angular measurement

1 Add together the following angles:
   (a) 32°19′ and 49°52′   (b) 29°42′, 56°37′ and 63°54′
   (c) 21°33′27″ and 78°42′36″
   (d) 48°11′19″, 31°41′27″ and 9°9′37″
   [(a) 82°11′ (b) 150°13′ (c) 100°16′3″ (d) 89°2′23″]
2 Determine
   (a) 17° − 9°49′   (b) 43°37′ − 15°49′
   (c) 78°29′41″ − 59°41′52″   (d) 114° − 47°52′37″
   [(a) 7°11′ (b) 27°48′ (c) 18°47′49″ (d) 66°7′23″]
3 Convert the following angles to degrees and decimals of a degree, correct to 3 decimal places:
   (a) 15°11′   (b) 29°53′   (c) 49°42′17″   (d) 135°7′19″
   [(a) 15.183° (b) 29.883° (c) 49.705° (d) 135.122°]
4 Convert the following angles into degrees, minutes and seconds:
   (a) 25.4°   (b) 36.48°   (c) 55.724°   (d) 231.025°
   [(a) 25°24′0″ (b) 36°28′48″ (c) 55°43′26″ (d) 231°1′30″]

Types and properties of angles

5 State the general name given to the following angles:
   (a) 63°   (b) 147°   (c) 250°
   [(a) acute (b) obtuse (c) reflex]
6 Determine the angles complementary to the following:
   (a) 69°   (b) 27°37′   (c) 41°3′43″
   [(a) 21° (b) 62°23′ (c) 48°56′17″]
7 Determine the angles supplementary to the following:
   (a) 78°   (b) 15°   (c) 169°41′11″
   [(a) 102° (b) 165° (c) 10°18′49″]
8 With reference to *Fig. 14.32*, what is the name given to the line XY.
   Give examples of each of the following:

(a) vertically opposite angles
(b) supplementary angles
(c) corresponding angles
(d) alternate angles.

> Transversal; (a) 1 & 3, 2 & 4, 5 & 7, 6 & 8
>   (b) 1 & 2, 2 & 3, 3 & 4, 4 & 1,
>   5 & 6, 6 & 7, 7 & 8, 8 & 5,
>   3 & 8, 1 & 6, 4 & 7 or 2 & 5
>   (c) 1 & 5, 2 & 6, 4 & 8, 3 & 7
>   (d) 3 & 5 or 2 & 8

9 In *Fig. 14.33*, find angle α.
   [59°20′]
10 In *Fig. 14.34*, find angles a, b and c.
   [$a = 69°$, $b = 21°$, $c = 82°$]
11 Find angle β in *Fig. 14.35*.
   [51°]

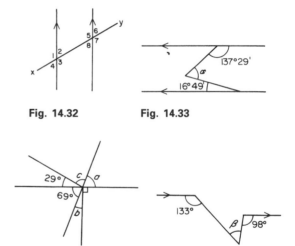

Fig. 14.32   **Fig. 14.33**

Fig. 14.34   **Fig. 14.35**

Properties of triangles

12 In *Fig. 14.36*, (*i*) and (*ii*), find angles w, x, y and z. What is the name given to the types of triangle shown in both (*i*) and (*ii*)?
   [40°, 70°, 70°, 125°, isosceles]

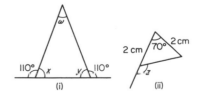

Fig. 14.36

13 Find the values of angles *a* to *g* in *Fig. 14.37 (i) and (ii)*.

$$\begin{bmatrix} a = 18°50', b = 71°10'; \\ c = 68°, d = 90°, e = 22°; \\ f = 49°, g = 41° \end{bmatrix}$$

14 Find the unknown angles *a* to *k* in *Fig. 14.38*.

$$\begin{bmatrix} a = 103°, b = 55°, c = 77°, \\ d = 125°, e = 55°, f = 22°, \\ g = 103°, h = 77°, i = 103°, \\ j = 77°, k = 81° \end{bmatrix}$$

15 Triangle ABC has a right angle at B and ∠BAC is 34°. BC is produced to D. If the bisectors of ∠ABC and ∠ACD meet at E, determine ∠BEC.
[17°]

16 If in *Fig. 14.39*, triangle BCD is equilateral, find the interior angles of triangle ABE.
[$A = 37°, B = 60°, E = 83°$]

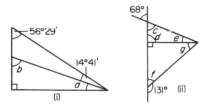

Fig. 14.37

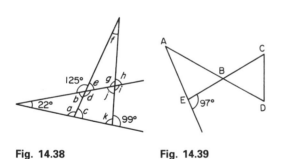

Fig. 14.38   **Fig. 14.39**

Theorem of Pythagoras

17 In a triangle ABC, ∠B is a right angle, AB = 6.92 cm and BC = 8.78 cm. Find the length of the hypotenuse.
[11.18 cm]

18 In a triangle CDE, ∠D = 90°, CD = 14.83 mm and CE = 28.31 mm. Determine the length of DE.
[24.11 mm]

19 Show that if a triangle has sides of 8, 15 and 17 cm it is right angled.

20 Triangle PQR is isosceles, ∠Q being a right angle. If the hypotenuse is 38.47 cm find (a) the lengths of sides PQ and QR, and (b) the value of ∠QPR.
[(a) 27.20 cm each (b) 45°]

21 A man cycles 24 km due south and then 20 km due east. Another man, starting at the same time as the first man, cycles 32 km due east and then 7 km due south. Find the distance between the two men.
[20.81 km]

22 A ladder 3.5 m long is placed against a perpendicular wall with its foot 2.0 m from the wall. How far up the wall (to the nearest centimetre) does the ladder reach? If the foot of the ladder is now moved 50 cm further away from the wall, how far does the top of the ladder fall?
[2.87 m, 42 cm]

23 Two ships leave a port at the same time. One travels due west at 18.4 km/h and the other due south at 27.6 km/h. Calculate how far apart the two ships are after 4 hours.
[132.7 km]

Congruent triangles

24 State which of the pairs of triangles in *Fig. 14.40* are congruent and name their sequence.

$$\begin{bmatrix} \text{(a) Congruent BAC, DAC (SAS)} \\ \text{(b) Congruent FGE, JHI (SSS)} \\ \text{(c) Not necessarily congruent} \\ \text{(d) Congruent QRT, SRT (RHS)} \\ \text{(e) Congruent UVW, XZY (ASA)} \end{bmatrix}$$

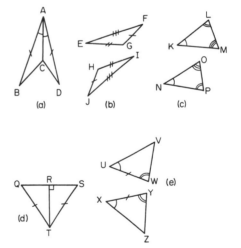

Fig. 14.40

25 In a triangle ABC, AB = BC and D and E are points on AB and BC, respectively, such that AD = CE. Show that triangles AEB and CDB are congruent.

Similar triangles

26 In *Fig. 14.41*, find the lengths *x* and *y*.
[$x = 16.54$ mm, $y = 4.18$ mm]

27 PQR is an equilateral triangle of side 4 cm. When PQ and PR are produced to S and T, respectively, ST is

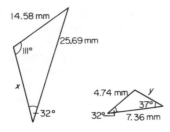

Fig. 14.41

found to be parallel with QR. If PS is 9 cm, find the length of ST. X is a point on ST between S and T such that the line PX is the bisector of ∠SPT. Find the length of PX.
[9 cm, 7.79 cm]

28 The triangle ABC is right angled at B. If AB = 4 cm and BC = 3 cm find the length of BD, which is the length of the perpendicular to the hypotenuse, and also the length AD and DC.
[BD = 2.4 cm, AD = 3.2 cm, DC = 1.8 cm]

29 In *Fig. 14.42*, find (a) the lengh of BC when AB = 6 cm, DE = 8 cm and DC = 3 cm, (b) the length of DE when EC = 2 cm, AC = 5 cm and AB = 10 cm.
[(a) 2.25 cm (b) 4 cm]

30 In *Fig. 14.43*, AF = 8 m, AB = 5 m and BC = 3 m. Find the length of BD.
[3 m]

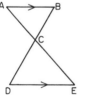
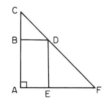

Fig. 14.42   **Fig. 14.43**

Construction of triangles

31 Construct triangles ABC, given:
(a) a = 8 cm, b = 6 cm and c = 5 cm
(b) a = 40 mm, b = 60 mm and C = 60°
(c) a = 6 cm, C = 45° and B = 75°
(d) c = 4 cm, A = 130° and C = 15°
(e) a = 90 mm, B = 90°, hypotenuse = 105 mm.

Circles

32 Calculate the length of the circumference of a circle of radius 7.2 cm.
[45.24 cm]

33 If the diameter of a circle is 82.6 mm, calculate the circumference of the circle.
[259.5 mm]

34 Determine the radius of a circle whose circumference is 16.52 cm.
[2.629 cm]

35 Find the diameter of a circle whose perimeter is 149.8 cm.
[47.68 cm]

36 Convert to radians in terms of π: (a) 30° (b) 75° (c) 225°.
$$\left[(a)\ \frac{\pi}{6}\ (b)\ \frac{5\pi}{12}\ (c)\ \frac{5\pi}{4}\right]$$

37 Convert to radians: (a) 48° (b) 84°51′ (c) 232°15′.
[(a) 0.838 (b) 1.481 (c) 4.054]

38 Convert to degrees: (a) $\dfrac{5\pi}{6}$ rad (b) $\dfrac{4\pi}{9}$ rad (c) $\dfrac{7\pi}{12}$ rad.
[(a) 150° (b) 80° (c) 105°]

39 Convert to degrees and minutes: (a) 0.012 5 rad (b) 2.69 rad (c) 7.241 rad.
[(a) 0°43′ (b) 154°8′ (c) 414°53′]

40 Find the length of an arc of a circle of radius 8.32 cm when the angle subtended at the centre is 2.14 radians.
[17.80 cm]

41 If the angle subtended at the centre of a circle of diameter 82 mm is 1.46 rad, find the lengths of the (a) minor arc, (b) major arc.
[(a) 59.86 mm (b) 197.8 mm]

42 A pendulum of length 1.5 m swings through an angle of 10° in a single swing. Find, in centimetres, the length of the arc traced by the pendulum bob.
[26.2 cm]

43 Determine the length of the radius and circumference of a circle if an arc length of 32.6 cm subtends an angle of 3.76 radians.
[8.67 cm, 54.48 cm]

44 An arc subtends an angle of 96° at the centre of a circle of radians 125 mm. Find the length of the arc.
[209.4 mm]

45 Determine the angle of lap, in degrees and minutes, if 180 mm of a belt drive are in contact with a pulley of diameter 250 mm.
[82°30]

46 Determine the number of complete revolutions a motorcycle wheel will make in travelling 2 km, if the wheel's diameter is 85.1 cm.
[748]

15

Areas and volumes

15.1 Mensuration

Mensuration is a branch of mathematics concerned with the determination of lengths, areas and volumes.

15.2 Properties of quadrilaterals

Polygon

A **polygon** is a closed plane figure bounded by straight lines. A polygon which has:

(i) 3 sides is called a **triangle**,
(ii) 4 sides is called a **quadrilateral**,
(iii) 5 sides is called a **pentagon**,
(iv) 6 sides is called a **hexagon**,
(v) 7 sides is called a **heptagon**,
(vi) 8 sides is called an **octagon**.

There are five types of **quadrilateral**, these being: (i) rectangle, (ii) square, (iii) parallelogram, (iv) rhombus, (v) trapezium. (The properties of these are given below.) If the opposite corners of any quadrilateral are joined by a straight line, two triangles are produced. Since the sum of the angles of a triangle is 180°, the sum of the angles of a quadrilateral is 360°.

In a **rectangle**, shown in *Fig. 15.1*:

(i) all four angles are right angles,
(ii) opposite sides are parallel and equal in length, and
(iii) diagonals AC and BD are equal in length and bisect one another.

In a **square**, shown in *Fig. 15.2*:

(i) all four angles are right angles,
(ii) opposite sides are parallel,

  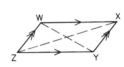

Fig. 15.1   **Fig. 15.2**   **Fig.15.3**

(iii) all four sides are equal in length, and
(iv) diagonals PR and QS are equal in length and bisect one another at right angles.

In a **parallelogram**, shown in *Fig. 15.3*:

(i) opposite angles are equal,
(ii) opposite sides are parallel and equal in length, and
(iii) diagonals WY and XZ bisect one another.

In a **rhombus**, shown in *Fig. 15.4*:

(i) opposite angles are equal,
(ii) opposite angles are bisected by a diagonal,
(iii) opposite sides are parallel,
(iv) all four sides are equal in length, and
(v) diagonals AC and BD bisect one another at right angles.

In a **trapezium**, shown in *Fig. 15.5*:

(i) only one pair of sides is parallel.

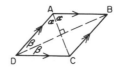

Fig. 15.4   **Fig. 15.5**

15.3 Areas of plane figures

Table 15.1

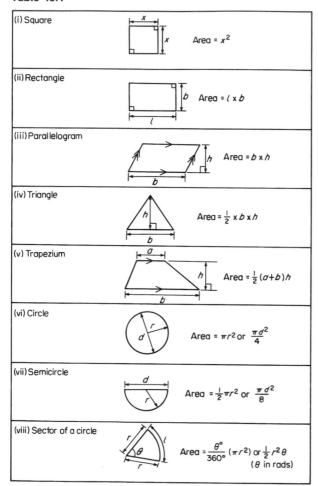

(i) Square	Area = x^2
(ii) Rectangle	Area = $l \times b$
(iii) Parallelogram	Area = $b \times h$
(iv) Triangle	Area = $\frac{1}{2} \times b \times h$
(v) Trapezium	Area = $\frac{1}{2}(a+b)h$
(vi) Circle	Area = πr^2 or $\frac{\pi d^2}{4}$
(vii) Semicircle	Area = $\frac{1}{2}\pi r^2$ or $\frac{\pi d^2}{8}$
(viii) Sector of a circle	Area = $\frac{\theta°}{360°}(\pi r^2)$ or $\frac{1}{2}r^2\theta$ (θ in rads)

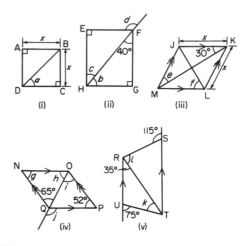

(i)   (ii)   (iii)

(iv)   (v)

Fig. 15.6

> **Problem 1** State the types of quadrilateral shown in Fig. 15.6 and determine the angles marked *a* to *l*.

(i) *ABCD is a square*
 The diagonals of a square bisect each of the right angles, hence

$$a = \frac{90°}{2} = \mathbf{45°}$$

(ii) *EFGH is a rectangle*
 In triangle FGH, $40° + 90° + b = 180°$ (angles in a triangle add up to $180°$) from which, $\boldsymbol{b = 50°}$. Also $c = \mathbf{40°}$ (alternate angles between parallel lines EF and HG).
 (Alternatively, *b* and *c* are complementary, i.e. add up to $90°$.)

$d = 90° + c$ (external angle of a triangle equals the sum of the interior opposite angles), hence

$$d = 90° + 40° = \mathbf{130°}$$

(iii) *JKLM is a rhombus*
 The diagonals of a rhombus bisect the interior angles and opposite internal angles are equal. Thus $\angle JKM = \angle MKL = \angle JMK = \angle LMK = 30°$, hence

$$e = \mathbf{30°}$$

 In triangle KLM, $30° + \angle KLM + 30° = 180°$ (angles in a triangle add up to $180°$), hence $\angle KLM = 120°$.

 The diagonal JK bisects $\angle KLM$, hence $f = \dfrac{120°}{2} = \mathbf{60°}$

(iv) *NOPQ is a parallelogram*
 $g = \mathbf{52°}$ (since opposite interior angles of a parallelogram are equal). In triangle NOQ, $g + h + 65° = 180°$ (angles in a triangle add up to $180°$), from which,

$$h = 180° - 65° - 52° = \mathbf{63°}$$

 $i = \mathbf{65°}$ (alternate angles between parallel lines NQ and OP)
 $j = 52° + i = 52° + 65° = \mathbf{117°}$ (external angle of a triangle equals the sum of the interior opposite angles).

(v) *RSTU is a trapezium*
 $35° + k = 75°$ (external angle of a triangle equals the sum of the interior opposite angles), hence

$$k = \mathbf{40°}$$

 $\angle STR = 35°$ (alternate angles between parallel lines RU and ST). $l + 35° = 115°$ (external angle of a triangle equals the sum of the interior opposite angles), hence

$$l = 115° - 35° = \mathbf{80°}$$

Problem 2 A rectangular tray is 820 mm long and 400 mm wide. Find its area in (a) mm², (b) cm², (c) m².

(a) Area = length × width = 820 × 400 = **328 000 mm²**.

(b) 1 cm² = 100 mm². Hence

$$328\,000 \text{ mm}^2 = \frac{328\,000}{100} \text{ cm}^2 = \textbf{3280 cm}^2$$

(c) 1 m² = 10 000 cm². Hence

$$3280 \text{ cm}^2 = \frac{3280}{10\,000} \text{ m}^2 = \textbf{0.328 0 m}^2$$

Problem 3 Find (a) the cross-sectional area of the girder shown in *Fig. 15.7(a)*, and (b) the area of the path shown in *Fig. 15.7(b)*.

(a) The girder may be divided into three separate rectangles as shown.
Area of rectangle A = 50 × 5 = 250 mm².
Area of rectangle B = (75 − 8 − 5) × 6 = 62 × 6 = 372 mm².
Area of rectangle C = 70 × 8 = 560 mm².
Total area of girder = 250 + 372 + 560 = **1182 mm²** or **11.82 cm²**.
(b) Area of path = area of large rectangle − area of small rectangle

= (25 × 20) − (21 × 16)
= 500 − 336 = **164 m²**

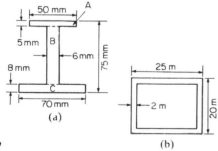

Fig. 15.7

Problem 4 Find the area of the parallelogram shown in *Fig. 15.8* (dimensions are in mm).

Area of parallelogram = base × perpendicular height. The perpendicular height h is found using Pythagoras' theorem.

i.e.
$$BC^2 = CE^2 + h^2$$
$$15^2 = (34 - 25)^2 + h^2$$
$$h^2 = 15^2 - 9^2 = 225 - 81 = 144$$

Hence, $h = \sqrt{144} = 12$ mm (−12 can be neglected)
Hence, area of ABCD = 25 × 12 = **300 mm²**

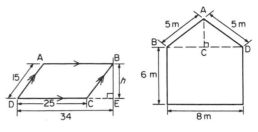

Fig. 15.8   Fig. 15.9

Problem 5 *Fig. 15.9* shows the gable end of a building. Determine the area of brickwork in the gable end.

The shape is that of a rectangle and a triangle. Area of rectangle = 6 × 8 = 48 m². Area of triangle = $\frac{1}{2}$ × base × height. CD = 4 m, AD = 5 m, hence AC = 3 m (since it is a 3,4,5 triangle). Hence, area of triangle

$$ABD = \frac{1}{2} \times 8 \times 3 = 12 \text{ m}^2$$

Total area of brickwork

= 48 + 12 = **60 m²**

Problem 6 Determine the area of the shape shown in *Fig. 15.10*.

The shape shown is a trapezium.
Area of trapezium = $\frac{1}{2}$(sum of parallel sides)(perpendicular distance between them)

$$= \frac{1}{2}(27.4 + 8.6)(5.5)$$

$$= \frac{1}{2} \times 36 \times 5.5 = \textbf{99 mm}^2$$

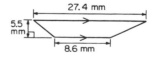

Fig. 15.10

Problem 7 Find the areas of the circles having (a) a radius of 5 cm, (b) a diameter of 15 mm, (c) a circumference of 70 mm.

Area of a circle $= \pi r^2$ or $\pi d^2/4$.
(a) Area $= \pi r^2 = \pi(5)^2 = 25\pi = 78.54$ cm^2.
(b) Area $= \pi d^2/4 = \pi(15)^2/4 = 225\pi/4 = 176.7$ mm^2.
(c) Circumference, $c = 2\pi r$ (from Chapter 14), hence

$$r = \frac{c}{2\pi} = \frac{70}{2\pi} = \frac{35}{\pi} \text{ mm}$$

Area of circle $= \pi r^2$

$$= \pi\left(\frac{35}{\pi}\right)^2 = \frac{35^2}{\pi} = 389.9 \text{ mm}^2 \text{ or } 3.899 \text{ cm}^2$$

Problem 8   Calculate the areas of the following sectors of circles:

(a) having radius 6 cm with angle subtended at centre 50°,
(b) having diameter 80 mm with angle subtended at centre 107°42′,
(c) having radius 8 cm with angle subtended at centre 1.15 radians.

Area of sector of a circle

$$= \frac{\theta°}{360}(\pi r^2) \text{ or } \frac{1}{2}r^2\theta \ (\theta \text{ in radians})$$

(a) Area of sector

$$= \frac{50}{360}(\pi 6^2) = \frac{50 \times \pi \times 36}{360} = 5\pi = 15.71 \text{ cm}^2$$

(b) If diameter $= 80$ mm, then radius, $r = 40$ mm, and area of sector

$$= \frac{107°42′}{360}(\pi 40^2) = \frac{107\frac{42}{60}}{360}(\pi 40^2) = \frac{107.7}{360}(\pi 40^2)$$

$$= 1504 \text{ mm}^2 \text{ or } 15.04 \text{ cm}^2$$

(c) Area of sector

$$= \frac{1}{2}r^2\theta = \frac{1}{2} \times 8^2 \times 1.15 = 36.8 \text{ cm}^2$$

Problem 9   A hollow shaft has an outside diameter of 5.45 cm and an inside diameter of 2.25 cm. Calculate the cross-sectional area of the shaft.

The cross-sectional area of the shaft is shown by the shaded part in *Fig. 15.11* (often called an **annulus**).
Area of shaded part = area of large circle − area of small circle

$$= \frac{\pi D^2}{4} - \frac{\pi d^2}{4} = \frac{\pi}{4}(D^2 - d^2) = \frac{\pi}{4}(5.45^2 - 2.25^2)$$

$$= 19.35 \text{ cm}^2$$

Problem 10   Calculate the area of a regular octagon, if each side is 5 cm and the width across the flats is 12 cm.

An octagon is an 8-sided polygon. If radii are drawn from the centre of the polygon to the vertices then 8 equal triangles are produced (see *Fig. 15.12*).

$$\text{Area of one triangle} = \frac{1}{2} \times \text{base} \times \text{height}$$

$$= \frac{1}{2} \times 5 \times \frac{12}{2} = 15 \text{ cm}^2$$

$$\text{Area of octagon} \quad = 8 \times 15 = 120 \text{ cm}^2$$

Problem 11   Determine the area of a regular hexagon which has sides 8 cm long.

A hexagon is a 6-sided polygon which may be divided into 6 equal triangles as shown in *Fig. 15.13*. The angle subtended at the centre of each triangle is 360°/6 = 60°. The other two angles in the triangle add up to 120° and are equal to each other. Hence each of the triangles is equilateral with each angle 60° and each side 8 cm.
Area of one triangle

$$= \frac{1}{2} \times \text{base} \times \text{height} = \frac{1}{2} \times 8 \times h$$

h is calculated using Pythagoras' theorem:

$$8^2 = h^2 + 4^2$$

from which

$$h = \sqrt{(8^2 - 4^2)}$$

$$= 6.928 \text{ cm}$$

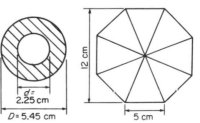

Fig. 15.11   **Fig. 15.12**   **Fig. 15.13**

Hence area of one triangle

$$= \frac{1}{2} \times 8 \times 6.928 = 27.71 \text{ cm}^2$$

Area of hexagon = $6 \times 27.71 = \textbf{166.3 cm}^2$

Problem 12 *Fig. 15.14* shows a plan of a floor of a building which is to be carpeted. Calculate the area of the floor in square metres. Calculate the cost, correct to the nearest pound, of carpeting the floor with carpet costing £16.80 per m², assuming 30% extra carpet is required due to wastage in fitting.

Area of floor plan

$$= \text{area of triangle ABC} + \text{area of semicircle}$$
$$+ \text{area of rectangle CGLM} + \text{area of rectangle CDEF}$$
$$- \text{area of trapezium HIJK}$$

Triangle ABC is equilateral since AB = BC = 3 m and hence angle B'CB = 60°. sin B'CB = BB'/3, i.e.

$$\text{BB}' = 3 \sin 60° = 2.598 \text{ m}$$

Area of triangle ABC = $\frac{1}{2}$(AC)(BB')

$$= \frac{1}{2}(3)(2.598) = 3.897 \text{ m}^2$$

Area of semicircle

$$= \frac{1}{2}\pi r^2 = \frac{1}{2}\pi(2.5)^2 = 9.817 \text{ m}^2$$

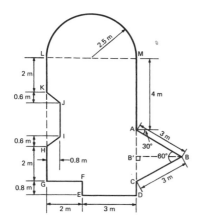

Fig. 15.14

Area of CGLM = $5 \times 7 = 35 \text{ m}^2$. Area of CDEF = $0.8 \times 3 = 2.4 \text{ m}^2$.

Area of HIJK = $\frac{1}{2}$(KH + IJ)(0.8). Since MC = 7 m then LG = 7 m, hence

$$\text{JI} = 7 - 5.2 = 1.8 \text{ m}$$

Hence area of HIJK = $\frac{1}{2}$(3 + 1.8)(0.8)

$$= 1.92 \text{ m}^2$$

Total floor area = $3.897 + 9.817 + 35 + 2.4 - 1.92$
= 49.194 m^2

To allow for 30% wastage, amount of carpet required = $1.3 \times 49.194 = 63.95 \text{ m}^2$.

Cost of carpet at £16.80 per m² = $63.95 \times 16.80 = \textbf{£1074}$, correct to the nearest pound.

Further problems on areas of plane figures may be found in section 15.7, Problems 1 to 15.

15.4 Volumes and surface areas of regular solids

For a summary of volumes and surface areas of regular solids, see Table 15.2 on page 140.

Problem 13 A water tank is the shape of a rectangular prism having length 2 m, breadth 75 cm and height 50 cm. Determine the capacity of the tank in (a) m³ (b) cm³ (c) litres.

Volume of rectangular prism = $l \times b \times h$ (see Table 15.2).

(a) Volume of tank = $2 \times 0.75 \times 0.5 = \textbf{0.75 m}^3$
(b) 1 m³ = 10^6 cm³. Hence

$$0.75 \text{ m}^3 = 0.75 \times 10^6 \text{ cm}^3 = \textbf{750 000 cm}^3$$

(c) 1 litre = 1000 cm³. Hence

$$750\,000 \text{ cm}^3 = \frac{750\,000}{1000} \text{ litres} = \textbf{750 litres}$$

Problem 14 Find the volume and total surface area of a cylinder of length 15 cm and diameter 8 cm.

Volume of cylinder = $\pi r^2 h$ (see Table 15.2). Since diameter = 8 cm, then radius, $r = 4$ cm. Hence

$$\text{volume} = \pi \times 4^2 \times 15 = \textbf{754 cm}^3$$

Table 15.2

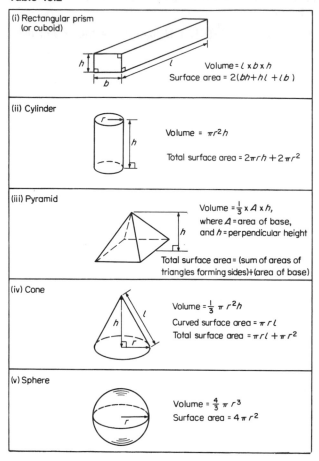

(i) Rectangular prism (or cuboid)

Volume = $l \times b \times h$
Surface area = $2(bh + hl + lb)$

(ii) Cylinder

Volume = $\pi r^2 h$

Total surface area = $2\pi rh + 2\pi r^2$

(iii) Pyramid

Volume = $\frac{1}{3} \times A \times h$,
where A = area of base,
and h = perpendicular height

Total surface area = (sum of areas of triangles forming sides) + (area of base)

(iv) Cone

Volume = $\frac{1}{3}\pi r^2 h$
Curved surface area = $\pi r l$
Total surface area = $\pi r l + \pi r^2$

(v) Sphere

Volume = $\frac{4}{3}\pi r^3$
Surface area = $4\pi r^2$

Total surface area (i.e. including the two ends)

$$= 2\pi rh + 2\pi r^2$$
$$= (2 \times \pi \times 4 \times 15) + (2 \times \pi \times 4^2)$$
$$= \textbf{477.5 cm}^2$$

Problem 15 Determine the volume (in cm³) of the shape shown in *Fig. 15.15*.

The solid shown in *Fig. 15.15* is a triangular prism. The volume V of any prism is given by: $V = Ah$, where A is the cross-sectional area and h is the perpendicular height. Hence

$$\text{volume} = \left(\frac{1}{2} \times 16 \times 12\right) \times 40$$

$$= 3840 \text{ mm}^3$$

$$= \textbf{3.840 cm}^3$$

(since 1 cm³ = 1000 mm³)

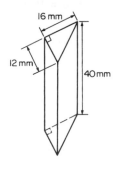

Fig. 15.15

Problem 16 Calculate the volume and total surface area of the solid prism shown in *Fig. 15.16*.

The solid shown in *Fig 15.16* is a trapezoidal prism.

Volume

$$= \text{cross-sectional area} \times \text{height}$$

$$= \left[\frac{1}{2}(11 + 5)4\right] \times 15 = 32 \times 15 = \textbf{480 cm}^3$$

Surface area = sum of two trapeziums + 4 rectangles
$$= (2 \times 32) + (5 \times 15) + (11 \times 15)$$
$$\quad + 2(5 \times 15)$$
$$= 64 + 75 + 165 + 150 = \textbf{454 cm}^2$$

Problem 17 Determine the volume and the total surface area of the square pyramid shown in *Fig. 15.17* if its perpendicular height is 12 cm.

Volume of pyramid

$$= \frac{1}{3}(\text{area of base}) \times \text{perpendicular height}$$

$$= \frac{1}{3}(5 \times 5) \times 12 = \textbf{100 cm}^3$$

The total surface area consists of a square base and 4 equal triangles. Area of triangle $ADE = \frac{1}{2} \times$ base $\times$ perpendicular height $= \frac{1}{2} \times 5 \times AC$.

The length AC may be calculated using Pythagoras' theorem on triangle ABC, where AB = 12 cm, BC = $\frac{1}{2} \times 5$ = 2.5 cm.

$$AC = \sqrt{(AB^2 + BC^2)} = \sqrt{(12^2 + 2.5^2)} = 12.26 \text{ cm}$$

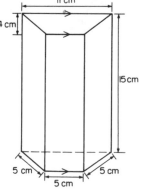

Fig. 15.16   **Fig. 15.17**

Hence area of triangle ADE

$$= \frac{1}{2} \times 5 \times 12.26 = 30.65 \text{ cm}^2$$

Total surface area of pyramid

$$= (5 \times 5) + 4(30.65)$$
$$= \textbf{147.6 cm}^2$$

Problem 18 Determine the volume and total surface area of a cone of radius 5 cm and perpendicular height 12 cm.

The cone is shown in *Fig. 15.18*. Volume of cone

$$= \frac{1}{3}\pi r^2 h = \frac{1}{3} \times \pi \times 5^2 \times 12 = \textbf{314.2 cm}^3$$

Total surface area

$$= \text{curved surface area} + \text{area of base}$$
$$= \pi r l + \pi r^2$$

From *Fig. 15.18*, slant height l may be calculated using Pythagoras' theorem

$$l = \sqrt{(12^2 + 5^2)} = 13 \text{ cm}$$

Hence total surface area

$$= (\pi \times 5 \times 13) + (\pi \times 5^2) = \textbf{282.7 cm}^2$$

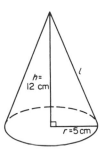

Fig. 15.18

Problem 19 Find the volume and surface area of a sphere of diameter 8 cm.

Since diameter = 8 cm, then radius, r = 4 cm. Volume of sphere

$$= \frac{4}{3}\pi r^3 = \frac{4}{3} \times \pi \times 4^3 = \textbf{268.1 cm}^3$$

Surface area of sphere

$$= 4\pi r^2 = 4 \times \pi \times 4^2 = \textbf{201.1 cm}^2$$

Problem 20 A wooden section is shown is *Fig. 15.19*. Find (a) its volume (in m³), and (b) its total surface area.

The section of wood is a prism whose end comprises a rectangle and a semicircle. Since the radius of the semicircle is 8 cm, the diameter is 16 cm. Hence the rectangle has dimensions 12 cm by 16 cm. Area of end

$$= (12 \times 16) + \frac{1}{2}\pi 8^2$$
$$= 292.5 \text{ cm}^2$$

Volume of wooden section

$$= \text{area of end} \times \text{perpendicular height}$$
$$= 292.5 \times 300 = 87\,750 \text{ cm}^3 = \frac{87\,750 \text{ m}^3}{10^6}$$
$$= \textbf{0.087 75 cm}^3$$

The total surface area comprises the two ends (each of area 292.5 cm²), 3 rectangles and a curved surface (which is half a cylinder), hence total surface area

$$= (2 \times 292.5) + 2(12 \times 300) + (16 \times 300)$$
$$+ \frac{1}{2}(2\pi \times 8 \times 300)$$
$$= 585 + 7200 + 4800 + 2400\pi$$
$$= \textbf{20 125 cm}^2 \text{ or } \textbf{2.012 5 m}^2$$

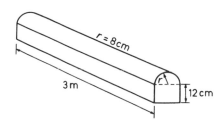

Fig. 15.19

Problem 21 A pyramid has a rectangular base 3.60 cm by 5.40 cm. Determine the volume and total surface area of the pyramid if each of its sloping edges is 15.0 cm.

The pyramid is shown in *Fig. 15.20*. To calculate the volume of the pyramid the perpendicular height EF is required. Diagonal BD is calculated using Pythagoras' theorem: i.e.

$$BD = \sqrt{\{(3.60)^2 + (5.40)^2\}} = 6.490 \text{ cm}$$

Hence

$$EB = \frac{1}{2}BD = \frac{6.490}{2} = 3.245 \text{ cm}$$

Using Pythagoras' theorem on triangle BEF gives $BF^2 = EB^2 + EF^2$, from which, $EF = \sqrt{(BF^2 - EB^2)} = \sqrt{\{(15.0)^2 - (3.245)^2\}} = 14.64$ cm. Volume of pyramid

$$= \frac{1}{3}(\text{area of base})(\text{perpendicular height})$$

$$= \frac{1}{3}(3.60 \times 5.40)(14.64) = \textbf{94.87 cm}^3$$

Area of triangle ADF (which equals triangle BCF) = $\frac{1}{2}(AD)(FG)$, where G is the midpoint of AD. Using Pythagoras' theorem on triangle FGA gives

$$FG = \sqrt{\{(15.0)^2 - (1.80)^2\}}$$

$$= 14.89 \text{ cm}$$

Hence area of triangle

$$ADF = \frac{1}{2}(3.60)(14.89) = 26.80 \text{ cm}^2$$

Similarly, if H is the mid-point of AB, then FH = $\sqrt{\{(15.0)^2 - (2.70)^2\}} = 14.75$ cm, hence area of triangle ABF (which equals triangle CDF)

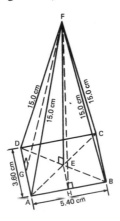

Fig. 15.20

$$= \frac{1}{2}(5.40)(14.75) = 39.83 \text{ cm}^2$$

Total surface area of pyramid

$$= 2(26.80) + 2(39.83) + (3.60)(5.40)$$
$$= 53.60 \quad + 79.66 \quad + 19.44 \qquad = \textbf{152.7 cm}^2$$

Problem 22   Calculate the volume and total surface area of a hemisphere of diameter 5.0 cm.

Volume of hemisphere

$$= \frac{1}{2}(\text{volume of sphere})$$

$$= \frac{2}{3}\pi r^3 = \frac{2}{3}\pi\left(\frac{5.0}{2}\right)^3 = \textbf{32.7 cm}^3$$

Total surface area

$$= \text{curved surface area} + \text{area of circle}$$

$$= \frac{1}{2}(\text{surface area of sphere}) + \pi r^2$$

$$= 2\pi r^2 + \pi r^2 = 3\pi r^2 = 3\pi\left(\frac{5.0}{2}\right)^2 = \textbf{58.9 cm}^2$$

Problem 23   A rectangular piece of metal having dimensions 4 cm by 3 cm by 12 cm is melted down and recast into a pyramid having a rectangular base measuring 2.5 cm by 5 cm. Calculate the perpendicular height of the pyramid.

Volume of rectangular prism of metal = $4 \times 3 \times 12$ = 144 cm³. Volume of pyramid = $\frac{1}{3}$(area of base)(perpendicular height). Assuming no waste of metal, $144 = \frac{1}{3}(2.5 \times 5)(\text{height})$

i.e. perpendicular height = $\dfrac{144 \times 3}{2.5 \times 5}$ = **34.56 cm**

Problem 24   A rivet consists of a cylindrical head, of diameter 1 cm and depth 2 mm, and a shaft of diameter 2 mm and length 1.5 cm. Determine the volume of metal in 2000 such rivets.

Radius of cylindrical head = $\dfrac{1}{2}$ cm = 0.5 cm; height of cylindrical head = 2 mm = 0.2 cm. Hence, volume of cylindrical head $= \pi r^2 h = \pi (0.5)^2 (0.2) = 0.157\ 1\ \text{cm}^3$.

Volume of cylindrical shaft $= \pi r^2 h = \pi \left(\dfrac{1}{10}\right)^2 (1.5) = 0.047\ 1\ \text{cm}^3$

Total volume of 1 rivet $= 0.157\ 1 + 0.047\ 1 = 0.204\ 2\ \text{cm}^3$. Volume of metal in 2000 such rivets $= 2000 \times 0.204\ 2 = \textbf{408.4 cm}^3$.

Problem 25 A solid metal cylinder of radius 6 cm and height 15 cm is melted down and recast into a shape comprising a hemisphere surmounted by a cone. Assuming that 8% of the metal is wasted in the process, determine the height of the conical portion, if its diameter is to be 12 cm.

Volume of cylinder $= \pi r^2 h = \pi \times 6^2 \times 15 = 540\pi\ \text{cm}^3$. If 8% of metal is lost then 92% of 540π gives the volume of the new shape (shown in *Fig. 15.21*). Hence volume of (hemisphere + cone) $= 0.92 \times 540\pi\ \text{cm}^3$, i.e.

$$\frac{1}{2}\left(\frac{4}{3}\pi r^3\right) + \frac{1}{3}\pi r^2 h = 0.92 \times 540\pi$$

Dividing throughout by π gives:

$$\frac{2}{3}r^3 + \frac{1}{3}r^2 h = 0.92 \times 540$$

Since the diameter of the new shape is to be 12 cm, then radius, $r = 6$ cm, hence

$$\frac{2}{3}(6)^3 + \frac{1}{3}(6)^2 h = 0.92 \times 540$$

$$144 + 12h = 496.8$$

i.e. height of conical portion,

$$h = \frac{496.8 - 144}{12} = \textbf{29.4 cm}$$

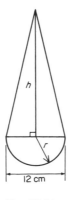

Fig. 15.21

Problem 26 A block of copper having a mass of 50 kg is drawn out to make 500 m of wire of uniform cross-section. Given that the density of copper is 8.91 g/cm³, calculate (a) the volume of copper, (b) the cross-sectional area of the wire, and (c) the diameter of the cross-section of the wire.

(a) A density of 8.91 g/cm³ means that 8.91 g of copper has a volume of 1 cm³, or 1 g of copper has a volume of (1/8.91) cm³.

Hence 50 kg, i.e. 50 000 g, has a volume

$$\frac{50\ 000}{8.91}\ \text{cm}^3 = \textbf{5 612 cm}^3$$

(b) Volume of wire = area of circular cross-section × length of wire. Hence 5612 cm³ = area × (500 × 100 cm),

from which, $\quad \text{area} = \dfrac{5612}{500 \times 100}\ \text{cm}^2 = \textbf{0.112 2 cm}^2$

(c) Area of circle πr^2 or $\pi d^2/4$, hence $0.112\ 2 = \dfrac{\pi d^2}{4}$, from

which $\quad d = \sqrt{\left(\dfrac{4 \times 0.112\ 2}{\pi}\right)} = 0.378\ 0\ \text{cm}$

i.e. diameter of cross-section is 3.780 mm.

Problem 27 A boiler consists of a cylindrical section of length 8 m and diameter 6 m, on one end of which is surmounted a hemispherical section of diameter 6 m, and on the other end a conical section of height 4 m. Calculate the volume of the boiler and the total surface area.

The boiler is shown in *Fig. 15.22*

Volume of hemisphere, $\text{P} = \dfrac{2}{3}\pi r^3 = \dfrac{2}{3} \times \pi \times 3^3 = 18\pi\ \text{m}^3$

Volume of cylinder, $\quad \text{Q} = \pi r^2 h = \pi \times 3^2 \times 8 = 72\pi\ \text{m}^3$

Volume of cone, $\quad\quad \text{R} = \dfrac{1}{3}\pi r^2 h = \dfrac{1}{3} \times \pi \times 3^2 \times 4$

$$= 12\pi\ \text{m}^3$$

Total volume of boiler $= 18\pi + 72\pi + 12\pi = 102\pi$

$$= \textbf{320.4 m}^3$$

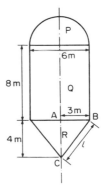

Fig. 15.22

Surface area of hemisphere,

$$P = \frac{1}{2}(4\pi r^2) = 2 \times \pi \times 3^2 = 18\pi \, \text{m}^2$$

Curved surface area of cylinder,

$$Q = 2\pi rh = 2 \times \pi \times 3 \times 8 = 48\pi \, \text{m}^2$$

The slant height of the cone, l, is obtained by Pythagoras' theorem on triangle ABC.

$$l = \sqrt{(4^2 + 3^2)} = 5$$

Curved surface area of cone,

$$R = \pi rl = \pi \times 3 \times 5 = 15\pi \, \text{m}^2$$

Total surface area of boiler $= 18\pi + 48\pi + 15\pi = 81\pi$
= 254.5 m².

Further problems on volumes and surface areas of regular solids may be found in section 15.7, Problems 16 to 37, page 148.

15.5   Volumes and surface areas of frusta of pyramids and cones

The **frustum** of a pyramid or cone is the portion remaining when a part containing the vertex is cut off by a plane parallel to the base.

The **volume of a frustum of a pyramid or cone** is given by the volume of the whole pyramid or cone minus the volume of the small pyramid or cone cut off.

The **surface area of the sides of a frustum of a pyramid or cone** is given by the surface area of the whole pyramid or cone minus the surface area of the small pyramid or cone cut off. This gives the lateral surface area of the frustum. If the total surface area of the frustum is required then the surface area of the two parallel ends are added to the lateral surface area.

There is an alternative method for finding the volume and surface area of a **frustum of a cone**. With reference to *Fig. 15.23*:

Fig. 15.23

$$\text{Volume} = \frac{1}{3}\pi h(R^2 + Rr + r^2)$$

Curved surface area $= \pi l(R + r)$
Total surface area $\quad = \pi l(R + r) + \pi r^2 + \pi R^2$

> **Problem 28**   Determine the volume of a frustum of a cone if the diameter of the ends are 6.0 cm and 4.0 cm and its perpendicular height is 3.6 cm.

Method 1
A section through the vertex of a complete cone is shown in *Fig. 15.24*.

Using similar triangles $\dfrac{AP}{DP} = \dfrac{DR}{BR}$

Hence $\dfrac{AP}{2.0} = \dfrac{3.6}{1.0}$, from which $AP = \dfrac{(2.0)(3.6)}{1.0} = 7.2 \, \text{cm}$

The height of the large cone $= 3.6 + 7.2 = 10.8$ cm.
Volume of frustum of cone

$$= \text{volume of large cone} - \text{volume of small cone cut off}$$

$$= \frac{1}{3}\pi(3.0)^2(10.8) - \frac{1}{3}\pi(2.0)^2(7.2)$$

$$= 101.79 - 30.16 = \textbf{71.6 cm}^3$$

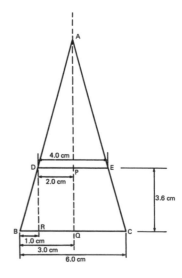

Fig. 15.24

Method 2
From above, volume of the frustum of a cone

$$= \frac{1}{3}\pi h(R^2 + Rr + r^2), \text{ where } R = 3.0 \text{ cm}, r = 2.0 \text{ cm and}$$

$h = 3.6$ cm. Hence volume of frustum

$$= \frac{1}{3}\pi(3.6)[(3.0)^2 + (3.0)(2.0) + (2.0)^2]$$

$$= \frac{1}{3}\pi(3.6)(19.0) = \textbf{71.6 cm}^3$$

Problem 29 Find the total surface area of the frustum of the cone in *Problem 28*.

Method 1

Curved surface area of frustum = curved surface area of large cone − curved surface area of small cone cut off. From *Fig. 15.24*, using Pythagoras' theorem:

$AB^2 = AQ^2 + BQ^2$, from which

$AB = \sqrt{\{(10.8)^2 + (3.0)^2\}} = 11.21$ cm

and

$AD^2 = AP^2 + DP^2$, from which

$AD = \sqrt{\{(7.2)^2 + (2.0)^2\}} = 7.47$ cm

Curved surface area of large cone = $\pi rl = \pi(BQ)(AB) = \pi(3.0)(11.21) = 105.7$ cm^2, and curved surface area of small cone = $\pi(DP)(AD) = \pi(2.0)(7.47) = 46.94$ cm^2. Hence curved surface area of frustum = $105.7 - 46.94 = 58.76$ cm^2. Total surface area of frustum

= curved surface area + area of two circular ends
= $58.76 + \pi(2.0)^2 + \pi(3.0)^2$
= $58.76 + 12.57 + 28.27 = \textbf{99.6 cm}^2$

Method 2

From page 144, total surface area of frustum = $\pi l(R + r) + \pi r^2 + \pi R^2$, where $l = BD = 11.21 - 7.47 = 3.74$ cm, $R = 3.0$ cm and $r = 2.0$ cm. Hence total surface area of frustum

= $\pi(3.74)(3.0 + 2.0) + \pi(2.0)^2 + \pi(3.0)^2$
= $\textbf{99.6 cm}^2$

Problem 30 A storage hopper is in the shape of a frustum of a pyramid. Determine its volume if the ends of the frustum are squares of sides 8.0 m and 4.6 m, respectively, and the perpendicular height between its ends is 3.6 m.

The frustum is shown shaded in *Fig. 15.25(b)* as part of a complete pyramid. A section perpendicular to the base through the vertex is shown in *Fig. 15.25(a)*.

By similar triangles: $\dfrac{CG}{BG} = \dfrac{BH}{AH}$

Hence $CG = BG\left(\dfrac{BH}{AH}\right) = \dfrac{(2.3)(3.6)}{(1.7)} = 4.87$ m

Height of complete pyramid

= $3.6 + 4.87 = 8.47$ m.

Volume of large pyramid

$$= \frac{1}{3}(8.0)^2(8.47) = 180.7 \text{ m}^3$$

Volume of small pyramid cut off

$$= \frac{1}{3}(4.6)^2(4.87) = 34.35 \text{ m}^3$$

Hence volume of storage hopper

$$= 180.7 - 34.35 = \textbf{146.4 m}^3$$

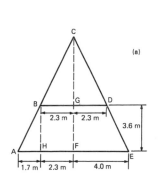
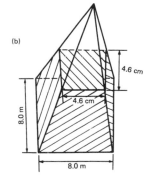

Fig. 15.25

Problem 31 Determine the lateral surface area of the storage hopper in *Problem 30*.

The lateral surface area of the storage hopper consists of four equal trapeziums. From *Fig. 15.26*, area of trapezium PRSU

$$= \frac{1}{2}(PR + SU)(QT)$$

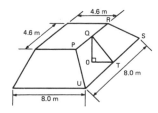

Fig. 15.26

OT = 1.7 m (same as AH in *Fig. 15.25(b)* and OQ = 3.6 m. By Pythagoras' theorem,

$$QT = \sqrt{(OQ^2 + OT^2)} = \sqrt{\{(3.6^2 + 1.7)^2\}} = 3.98 \text{ m}$$

Area of trapezium PRSU

$$= \frac{1}{2}(4.6 + 8.0)(3.98) = 25.07 \text{ m}^2$$

Lateral surface area of hopper = 4(25.07) = **100.3 m²**.

Problem 32   A lampshade is in the shape of a frustum of a cone. The vertical height of the shade is 25.0 cm and the diameters of the ends are 20.0 cm and 10.0 cm, respectively. Determine the area of the material needed to form the lampshade, correct to 3 significant figures.

The curved surface area of a frustum of a cone = $\pi l(R + r)$, from page 144. Since the diameters of the ends of the frustum are 20.0 cm and 10.0 cm, then, from *Fig. 15.27*, r = 5.0 cm, R = 10.0 cm and $l = \sqrt{\{(25.0)^2 + (5.0)^2\}}$ = 25.50 cm, by Pythagoras' theorem. Hence curved surface area = $\pi(25.50)$ (10.0 + 5.0) = 1201.7 cm², i.e. the area of material needed to form the lampshade is **1200 cm²**, correct to 3 significant figures.

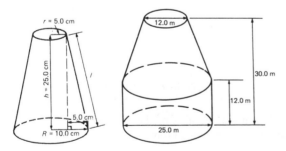

Fig. 15.27   **Fig. 15.28**

Problem 33   A cooling tower is in the form of a cylinder surmounted by a frustum of a cone as shown in *Fig. 15.28*. Determine the volume of air space in the tower if 40% of the space is used for pipes and other structures.

Volume of cylindrical portion = $\pi r^2 h = \pi(25.0/2)^2(12.0)$ = 5890 m³. Volume of frustum of cone = $\frac{1}{3}\pi h(R^2 + Rr + r^2)$,

where h = 30.0 − 12.0 = 18.0 m, R = 25.0/2 = 12.5 m and r = 12.0/2 = 6.0 m. Hence volume of frustum of cone

$$= \frac{1}{3}\pi(18.0)[(12.5)^2 + (12.5)(6.0) + (6.0)^2]$$

$$= 5038 \text{ m}^3$$

Total volume of cooling tower

$$= 5890 + 5038 = 10\,928 \text{ m}^3$$

If 40% of space is occupied then volume of air space = 0.6 × 10 928 = **6557 m³**.

Further problems on volumes and surface areas of frustra of pyramids and cones may be found in section 15.7, Problems 38 to 44, page 149.

15.6 Areas and volumes of similar shapes

The areas of similar shapes are proportional to the squares of corresponding linear dimensions. For example, *Fig. 15.29* shows two squares, one of which has sides three times as long as the other.

Area of Fig. 15.29(a) = $(x)(x)$ $= x^2$
Area of Fig. 15.29(b) = $(3x)(3x) = 9x^2$

Hence *Fig. 15.29(b)* has an area $(3)^2$, i.e. 9 times the area of Fig. 15.29(a).

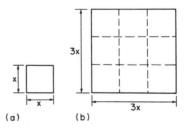

(a)   (b)

Fig 15.29

The volumes of similar bodies are proportional to the cubes of corresponding linear dimensions. For example, *Fig. 15.30* shows two cubes, one of which has sides three times as long as those of the other.

Volume of Fig. 30(a) = $(x)(x)(x)$ $= x^3$
Volume of Fig. 30(b) = $(3x)(3x)(3x) = 27x^3$

Hence *Fig. 15.30(b)* has a volume $(3)^3$, i.e. 27 times the volume of *Fig. 15.30(a)*.

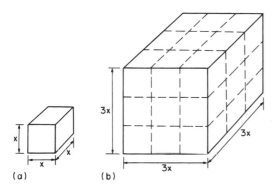

(a) (b)

Fig. 15.30

Problem 34 A rectangular garage is shown on a building plan having dimensions 10 mm by 20 mm. If the plan is drawn to a scale of 1 to 250, determine the true area of the garage in square metres.

Area of garage on the plan = 10 mm × 20 mm = 200 mm². Since the areas of similar shapes are proportional to the squares of corresponding dimensions then:

$$\text{True area of garage} = 200 \times (250)^2$$
$$= 12.5 \times 10^6 \text{ mm}^2$$
$$= \frac{12.5 \times 10^6}{10^6} \text{ m}^2 = \mathbf{12.5 \ m^2}$$

Problem 35 A car has a mass of 1000 kg. A model of the car is made to a scale of 1 to 50. Determine the mass of the model if the car and its model are made of the same material.

$\dfrac{\text{Volume of model}}{\text{Volume of car}} = \left(\dfrac{1}{50}\right)^3$ since the volume of similar bodies are proportional to the cube of corresponding dimensions. Mass = density × volume, and since both car and model are made of the same material then:

$$\frac{\text{Mass of model}}{\text{Mass of car}} = \left(\frac{1}{50}\right)^3$$

Hence mass of model = (mass of car) $\left(\dfrac{1}{50}\right)^3 = \dfrac{1000}{50^3}$

$$= \mathbf{0.008 \ kg \ or \ 8 \ g}$$

Further problems on areas and volumes of similar shapes may be found in section 15.7, Problems 45 to 47, page 150.

15.7 Further problems on areas and volumes

Areas of plane figures

1 A rectangular plate is 85 mm long and 42 mm wide. Find its area in square centimetres.
[35.7 cm²]
2 A rectangular field has an area of 1.2 hectares and a length of 150 m.
Find (a) its width and (b) the length of a diagonal (1 hectare = 10 000 m²).
[(a) 80 m (b) 170 m]
3 Determine the area of each of the angle iron sections shown in *Fig. 15.31*.
[(a) 29 cm² (b) 650 mm²]

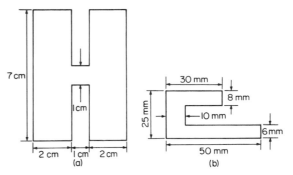

Fig. 15.31

4 A rectangular garden measures 40 m by 15 m. A 1 m flower border is made round the two shorter sides and one long side. A circular swimming pool of diameter 8 m is constructed in the middle of the garden. Find, correct to the nearest square metre, the area remaining.
[482 m²]
5 The area of a trapezium is 13.5 cm² and the perpendicular distance between its parallel sides is 3 cm. If the length of one of the parallel sides is 5.6 cm, find the length of the other parallel side.
[3.4 cm]
6 Find the angles p, q, r, s and t in *Figs 15.32(a)* to (c).
[p = 105°, q = 35°, r = 142°, s = 95°, t = 146°]

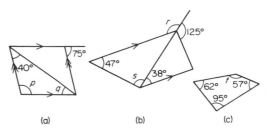

(a) (b) (c)

Fig. 15.32

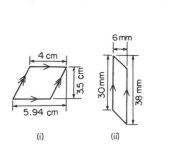

Fig. 15.33

7 Name the types of quadrilateral shown in *Figs 15.33(i)* to *(iv)*, and determine (a) the area, and (b) the perimeter of each.

$$\begin{bmatrix} \text{(i)   rhombus (a) 14 cm}^2 \text{ (b) 16 cm} \\ \text{(ii) parallelogram (a) 180 mm}^2 \text{ (b) 80 mm} \\ \text{(iii) rectangle (a) 3600 mm}^2 \text{ (b) 300 mm} \\ \text{(iv) trapezium (a) 190 cm}^2 \text{ (b) 62.91 cm} \end{bmatrix}$$

8 Determine the area of circles having (a) a radius of 4 cm (b) a diameter of 30 mm (c) a circumference of 200 mm.
[(a) 50.27 cm^2 (b) 706.9 mm^2 (c) 3183 mm^2]

9 An annulus has an outside diameter of 60 mm and an inside diameter of 20 mm. Determine its area.
[2513 mm^2]

10 If the area of a circle is 320 mm^2, find (a) its diameter, and (b) its circumference.
[(a) 20.19 mm (b) 63.41 mm]

11 Calculate the areas of the following sectors of circles: (a) radius 9 cm, angle subtended at centre 75° (b) diameter 35 mm, angle subtended at centre 48°37′ (c) diameter 5 cm, angle subtended at centre 2.19 radians.
[(a) 53.01 cm^2 (b) 129.9 mm^2 (c) 6.84 cm^2]

12 Calculate the area of a regular octagon if each side is 20 mm and the width across the flats is 48.3 mm.
[1932 mm^2]

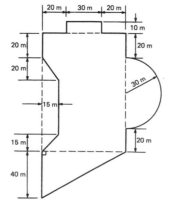

Fig. 15.34

13 Determine the area of a regular hexagon which has sides 25 mm.
[1624 mm^2]

14 Find the area of triangle ABC if $\angle B = 90°$, $a = 4.87$ cm and $b = 7.54$ cm.
[14.02 cm^2]

15 A plot of land is in the shape shown in *Fig. 15.34*. Determine (a) its area in hectares (1 ha = 10^4 m^2), and (b) the length of fencing required, to the nearest metre, to completely enclose the plot of land.
[(a) 0.918 ha (b) 456 m]

Volumes and surface areas of regular solids

16 A rectangular block of metal has dimensions of 40 mm by 25 mm by 15 mm. Determine its volume. Find also its mass if the metal has a density of 9 g/cm^3.
[15 cm^3, 135 g]

17 Determine the maximum capacity, in litres, of a fish tank measuring 50 cm by 40 cm by 2.5 m (1 litre = 1000 cm^3).
[500 l]

18 Determine how many cubic metres of concrete are required for a 120 m long path, 150 mm wide and 80 mm deep.
[1.44 m^3]

19 Calculate the volume of a metal tube whose outside diameter is 8 cm and whose inside diameter is 6 cm, if the length of the tube is 4 m.
[8796 cm^3]

20 The volume of a cylinder is 400 cm^3. If its radius is 5.20 cm, find its height. Determine also its curved surface area.
[4.709 cm, 153.9 cm^2]

21 If a cone has a diameter of 80 mm and a perpendicular height of 120 mm calculate its volume in cm^3 and its curved surface area.
[201.1 cm^3, 159.0 cm^2]

22 A cylinder is cast from a rectangular piece of alloy 5 cm by 7 cm by 12 cm. If the length of the cylinder is to be 60 cm, find its diameter.
[2.99 cm]

23 Find the volume and the total surface area of a regular hexagonal bar of metal of length 3 m if each side of the hexagon is 6 cm.
[28.060 cm^3, 1.099 m^2]

24 A square pyramid has a perpendicular height of 4 cm. If a side of the base is 2.4 cm long find the volume and total surface area of the pyramid.
[7.68 cm^3, 25.81 cm^2]

25 A sphere has a diameter of 6 cm. Determine its volume and surface area.
[113.1 cm^3, 113.1 cm^2]

26 Find the total surface area of a hemisphere of diameter 50 mm.
[5890 mm^2 or 58.90 cm^2]

27 Determine the mass of a hemispherical copper container whose external and internal radii are 12 cm and 10 cm. Assume that 1 cm^3 of copper weighs 8.9 g.
[13.57 kg]

28 If the volume of a sphere is 566 cm^3, find its radius.
[5.131 cm]

29 A metal plumb bob comprises a hemisphere surmounted by a cone. If the diameter of the hemisphere and cone are each 4 cm and the total length is 5 cm, find its total volume.
[29.32 cm^3]

30 A marquee is in the form of a cylinder surmounted by a cone. The total height is 6 m and the cylindrical portion has a height of 3.5 m, with a diameter of 15 m. Calculate the surface area of material needed to make the marquee assuming 12% of the material is wasted in the process.
[393.4 m^2]

31 Determine (a) the volume and (b) the total surface area of the following solids:
  (i) a cone of radius 8.0 cm and perpendicular height 10 cm,
      [(a) 670 cm^3 (b) 523 cm^2]
  (ii) a sphere of diameter 7.0 cm,
      [(a) 180 cm^3 (b) 154 cm^2]
  (iii) a hemisphere of radius 3.0 cm,
      [(a) 56.5 cm^3 (b) 84.8 cm^2]
  (iv) a 2.5 cm by 2.5 cm square pyramid of perpendicular height 5.0 cm,
      [(a) 10.4 cm^3 (b) 32.0 cm^2]
  (v) a 4.0 cm by 6.0 cm rectangular pyramid of perpendicular height 12.0 cm,
      [(a) 96.0 cm^3 (b) 146 cm^2]
  (vi) a 4.2 cm by 4.2 cm square pyramid whose sloping edges are each 15.0 cm,
      [(a) 86.5 cm^3 (b) 142 cm^2]
  (vii) a pyramid having an octagonal base of side 5.0 cm and perpendicular height 20 cm.
      [(a) 805 cm^3 (b) 539 cm^2]

32 The volume of a sphere is 325 cm^3. Determine its diameter.
[8.53 cm]

33 A metal sphere weighing 24 kg is melted down and recast into a solid cone of base radius 8.0 cm. If the density of the metal is 8000 kg/m^3 determine (a) the diameter of the metal sphere and (b) the perpendicular height of the cone, assuming that 15% of the metal is lost in the process.
[(a) 17.9 cm (b) 38.0 cm]

34 Find the volume of a regular hexagonal pyramid if the perpendicular height is 16.0 cm and the side of base is 3.0 cm.
[125 cm^3]

35 A buoy consists of a hemisphere surmounted by a cone. The diameter of the cone and hemisphere is 2.5 m and the slant height of the cone is 4.0 m. Determine the volume and surface area of the buoy.
[10.3 m^3, 25.5 m^2]

36 A petrol container is in the form of a central cylindrical portion 5.0 m long with a hemispherical section surmounted on each end. If the diameters of the hemisphere and cylinder are both 1.2 m determine the capacity of the tank in litres (1 litre = 1000 cm^3).
[6560 l]

37 *Fig. 15.35* shows a metal rod section. Determine its volume and total surface area.
[657.1 cm^3, 1027 cm^2]

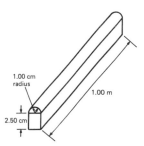

1.00 cm radius

1.00 m

2.50 cm

Fig. 15.35

Volumes and surface areas of frustra of pyramids and cones

38 The radii of the faces of a frustum of a cone are 2.0 cm and 4.0 cm and the thickness of the frustum is 5.0 cm. Determine its volume and total surface area.
[147 cm^3, 164 cm^2]

39 A frustum of a pyramid has square ends, the squares having sides 9.0 cm and 5.0 cm, respectively. Calculate the volume and total surface area of the frustum if the perpendicular distance between its ends is 8.0 cm.
[403 cm^3, 337 cm^2]

40 A cooling tower is in the form of a frustum of a cone. The base has a diameter of 32.0 m, the top has a diameter of 14.0 m and the vertical height is 24.0 m. Calculate the volume of the tower and the curved surface area.
[10 480 m^3, 1852 m^2]

41 A loudspeaker diaphragm is in the form of a frustum of a cone. If the end diameters are 28.0 cm and 6.00 cm and the vertical distance between the ends is 30.0 cm, find the area of material needed to cover the curved surface of the speaker.
[1707 cm^2]

42 A rectangular prism of metal having dimensions 4.3 cm by 7.2 cm by 12.4 cm is melted down and recast into a frustum of a square pyramid, 10% of the metal being lost in the process. If the ends of the frustum are squares

of size 3 cm and 8 cm, respectively, find the thickness of the frustum.
[10.69 cm]

43 Determine the volume and total surface area of a bucket consisting of an inverted frustum of a cone, of slant height 36.0 cm and end diameters 55.0 cm and 35.0 cm.
[55 910 cm^3, 6051 cm^2]

44 A cylindrical tank of diameter 2.0 m and perpendicular height 3.0 m is to be replaced by a tank of the same capacity but in the form of a frustum of a cone. If the diameters of the ends of the frustum are 1.0 m and 2.0 m, respectively, determine the vertical height required.
[5.14 m]

Areas and volumes of similar shapes

45 The area of a park on a map is 500 mm^2. If the scale of the map is 1 to 40 000 determine the true area of the park in hectares (1 hectare = 10^4 m^2).
[80 ha]

46 The diameter of two spherical bearings are in the ratio 2 : 5. What is the ratio of their volumes?
[8 : 125]

47 An engineering component has a mass of 400 g. If each of its dimensions are reduced by 30% determine its new mass.
[137.2 g]

16

Irregular areas and volumes and mean values of waveforms

16.1 Areas of irregular figures

Areas of irregular plane surfaces may be approximately determined by using (a) a planimeter, (b) the trapezoidal rule, (c) the mid-ordinate rule, and (d) Simpson's rule. Such methods may be used, for example, by engineers estimating areas of indicator diagrams of steam engines, surveyors estimating areas of plots of land or naval architects estimating areas of water planes or transverse sections of ships.

(a) **A planimeter** is an instrument for directly measuring small areas bounded by an irregular curve.

(b) **Trapezoidal rule**
    To determine the area PQRS in *Fig. 16.1*:
    (i) Divide base PS into any number of equal intervals, each of width d (the greater the number of intervals, the greater the accuracy).
    (ii) Accurately measure ordinates y_1, y_2, y_3, etc.
    (iii) Area PQRS = $d\left[\dfrac{y_1 + y_7}{2} + y_2 + y_3 + y_4 + y_5 + y_6\right]$

In general, the trapezoidal rule states:

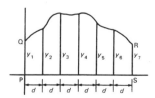

Fig. 16.1

$$\text{Area} = (\text{width of interval}) \left[\frac{1}{2}(\text{first} + \text{last ordinate}) + \text{sum of remaining ordinates}\right]$$

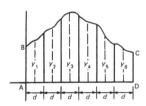

Fig. 16.2

(c) **Mid-ordinate rule**
    To determine the area ABCD of *Fig. 16.2*:
    (i)   Divide base AD into any number of equal intervals, each of width d (the greater the number of intervals, the greater the accuracy).
    (ii) Erect ordinates in the middle of each interval (shown by broken lines in *Fig. 16.2*).
    (iii) Accurately measure ordinates y_1, y_2, y_3, etc.
    (iv) Area ABCD = $d(y_1 + y_2 + y_3 + y_4 + y_5 + y_6 + y_7)$.

In general, the mid-ordinate rule states:

$$\text{Area} = (\text{width of interval})(\text{sum of mid-ordinates})$$

(d) **Simpson's rule**
    To determine the area PQRS of *Fig. 16.1*:

(i) Divide base PS into an **even** number of intervals, each of width d (the greater the number of intervals, the greater the accuracy).

(ii) Accurately measure ordinates y_1, y_2, y_3, etc.

(iii) Area

$$PQRS = \frac{d}{3}[(y_1 + y_7) + 4(y_2 + y_4 + y_6) + 2(y_3 + y_5)]$$

In general, Simpson's rule states:

$$\text{Area} = \frac{1}{3}\left(\begin{array}{c}\textbf{width of}\\\textbf{interval}\end{array}\right)\left[\left(\begin{array}{c}\textbf{first + last}\\\textbf{ordinate}\end{array}\right) + 4\left(\begin{array}{c}\textbf{sum of even}\\\textbf{ordinate}\end{array}\right) + 2\left(\begin{array}{c}\textbf{sum of remaining}\\\textbf{odd ordinates}\end{array}\right)\right]$$

Problem 1 A car starts from rest and its speed is measured every second for 6 s.

Time t (s)	0	1	2	3	4	5	6
Speed v (m/s)	0	2.5	5.5	8.75	12.5	17.5	24.0

Determine the distance travelled in 6 seconds (i.e. the area under the v/t graph) by (a) the trapezoidal rule, (b) the mid-ordinate rule, and (c) Simpson's rule.

A graph of speed/time is shown in *Fig. 16.3*.

(a) **Trapezoidal rule** (see para. (b) above)
The time base is divided into 6 strips each of width 1 s, and the length of the ordinates measured. Thus

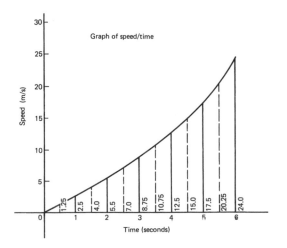

Fig. 16.3

area

$$= (1)\left[\left(\frac{0 + 24.0}{2}\right) + 2.5 + 5.5 + 8.75 + 12.5 + 17.5\right]$$

$$= \textbf{58.75 m}$$

(b) **Mid-ordinate rule** (see para. (c) above)
The time base is divided into 6 strips each of width 1 second. Mid-ordinates are erected as shown in *Fig. 16.3* by the broken lines. The length of each mid-ordinate is measured. Thus

$$\text{area} = (1)[1.25 + 4.0 + 7.0 + 10.75 + 15.0 + 20.25]$$

$$= \textbf{58.25 m}$$

(c) **Simpson's rule** (see para. (d) above)
The time base is divided into 6 strips each of width 1 s, and the length of the ordinates measured. Thus

$$\text{area} = \frac{1}{3}(1)[(0 + 24.0) + 4(2.5 + 8.75 + 17.5) + 2(5.5 + 12.5)]$$

$$= \textbf{58.33 m}$$

Problem 2 A river is 15 m wide. Soundings of the depth are made at equal intervals of 3 m across the river and are as shown below.

Depth (m)	0	2.2	3.3	4.5	4.2	2.4	0

Calculate the cross-sectional area of the flow of water at this point using Simpson's rule.

From para. (d) above,

$$\text{Area} = \frac{1}{3}(3)[(0 + 0) + 4(2.2 + 4.5 + 2.4) + 2(3.3 + 4.2)]$$

$$= (1)[0 + 36.4 + 15] = \textbf{51.4 m}^2$$

Further problems on areas of irregular figures may be found in section, 16.4, Problems 1 to 5, page 156.

16.2 Volumes of irregular solids

If the cross-sectional areas A_1, A_2, A_3, etc., of an irregular solid bounded by two parallel planes are known at equal intervals of width d (as shown in *Fig. 16.4*), then by Simpson's rule:

$$\text{Volume, } V = \frac{d}{3}[(A_1 + A_7) + 4(A_2 + A_4 + A_6) + 2(A_3 + A_5)]$$

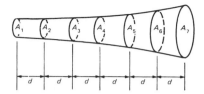

Fig. 16.4

Problem 3   A tree trunk is 12 m in length and has a varying circular cross-section. The cross-sectional areas at intervals of 2 m measured from one end are: 0.52, 0.55, 0.59, 0.63, 0.72, 0.84, 0.97 m². Estimate the volume of the tree trunk.

A sketch of the tree trunk is similar to that shown in *Fig. 16.4* above, where $d = 2$ m, $A_1 = 0.52$ m², $A_2 = 0.55$ m², and so on. Using Simpson's rule for volumes gives:

$$\text{Volume} = \frac{2}{3}[(0.52 + 0.97) + 4(0.55 + 0.63 + 0.84)$$
$$+ \; 2(0.59 + 0.72)]$$

$$= \frac{2}{3}[1.49 + 8.08 + 2.62]$$

$$= \textbf{8.13 m}^3$$

Problem 4   The areas of 7 horizontal cross-sections of a water reservoir at intervals of 10 m are: 210, 250, 320, 350, 290, 230, 170 m². Calculate the capacity of the reservoir in litres.

Using Simpson's rule for volumes gives:

$$\text{Volume} = \frac{10}{3}[(210 + 170) + 4(250 + 350 + 230)$$
$$+ \; 2(320 + 290)]$$

$$= \frac{10}{3}[380 + 3320 + 1220]$$

$$= \textbf{16 400 m}^3$$

$16\,400$ m³ $= 16\,400 \times 10^6$ cm³

Since 1 litre = 1000 cm³, capacity of reservoir

$$= \frac{16\,400 \times 10^6}{1000} \text{ litres}$$

$$= 16\,400\,000 = \textbf{1.64} \times \textbf{10}^7 \textbf{ litres}$$

Further poblems on the volume of irregular solids may be found in section 16.4, Problems 6 to 8, page 156.

16.3   The mean or average value of a waveform

The mean or average value, y, of the waveform shown in *Fig. 16.5* is given by:

$$y = \frac{\textbf{area under curve}}{\textbf{length of base, } b}$$

If the mid-ordinate rule is used to find the area under the curve, then:

$$y = \frac{\textbf{sum of mid-ordinates}}{\textbf{number of mid-ordinates}}$$

$$\left(= \frac{y_1 + y_2 + y_3 + y_4 + y_5 + y_6 + y_7}{7} \text{ for } Fig.\ 16.5 \right)$$

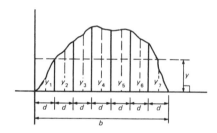

Fig. 16.5

For a **sine wave**, the mean or average value:

(i)   over one complete cycle is zero (see *Fig. 16.6(a)*),
(ii) over half a cycle is **0.637 × maximum value**, or **2/π × maximum value**,
(iii) of a full-wave rectified waveform (see *Fig. 16.6(b)*) is **0.637 × maximum value**,
(iv) of a half-wave rectified waveform (see *Fig. 16.6(c)*) is

0.318 × maximum value, or **1/π × maximum value.**

Problem 5   Determine the average values over half a cycle of the periodic waveforms shown in *Fig. 16.7*.

(a) Area under triangular waveform (a) for a half cycle is given by:

$$\text{Area} = \frac{1}{2}(\text{base})(\text{perpendicular height})$$

$$= \frac{1}{2}(2 \times 10^{-3})(20) = 20 \times 10^{-3} \text{ V s}$$

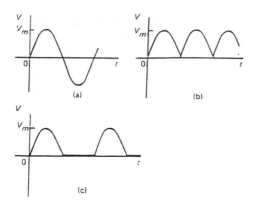

(a)

(b)

(c)

Fig. 16.6

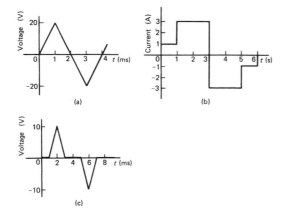

(a)

(b)

(c)

Fig. 16.7

Average value of waveform

$$= \frac{\text{area under curve}}{\text{length of base}} = \frac{20 \times 10^{-3} \text{ V s}}{2 \times 10^{-3} \text{ s}} = \textbf{10 V}$$

(b) Area under waveform (b) for a half cycle = (1×1) + (3×2) = 7 A s.
Average value of waveform

$$= \frac{\text{area under curve}}{\text{length of base}} = \frac{7 \text{ A s}}{3 \text{ s}} = \textbf{2.33 A}$$

(c) A half cycle of the voltage waveform (c) is completed in 4 ms.

$$\text{Area under curve} = \frac{1}{2}\{(3-1)10^{-3}\} (10) = 10 \times 10^{-3} \text{ V s}.$$

Average value of waveform

$$= \frac{\text{area under curve}}{\text{length of base}} = \frac{10 \times 10^{-3} \text{ V s}}{4 \times 10^{-3} \text{ s}} = \textbf{2.5 V}$$

Problem 6 Determine the mean value of current over one complete cycle of the periodic waveforms shown in *Fig. 16.8*.

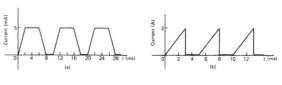

(a)

(b)

Fig. 16.8

(a) One cycle of the trapezoidal waveform (a) is completed in 10 ms (i.e. the periodic time is 10 ms).

Area under curve
= area of trapezium

$$= \frac{1}{2}(\text{sum of parallel sides}) \text{ (perpendicular distance between parallel sides)}$$

$$= \frac{1}{2}\{(4+8) \times 10^{-3}\}(5 \times 10^{-3})$$

$$= 30 \times 10^{-6} \text{ A s}$$

Mean value over one cycle

$$= \frac{\text{area under curve}}{\text{length of base}} = \frac{30 \times 10^{-6} \text{ A s}}{10 \times 10^{-3} \text{ s}} = \textbf{3 mA}$$

(b) One cycle of the sawtooth waveform (b) is completed in 5 ms.

$$\text{Area under curve} = \frac{1}{2}(3 \times 10^{-3})(2) = 3 \times 10^{-3} \text{ A s}.$$

Mean value over one cycle

$$= \frac{\text{area under curve}}{\text{length of base}} = \frac{3 \times 10^{-3} \text{ A s}}{5 \times 10^{-3} \text{ s}} = \textbf{0.6 A}$$

Problem 7 The power used in a manufacturing process during a 6 hour period is recorded at intervals of 1 hour as shown below.

Time (h)	0	1	2	3	4	5	6
Power (kW)	0	14	29	51	45	23	0

Plot a graph of power against time and, by using the mid-ordinate rule, determine (a) the area under the curve and (b) the average value of the power.

The graph of power/time is shown in *Fig. 16.9*.

(a) The time base is divided into 6 equal intervals, each of width 1 hour. Mid-ordinates are erected (shown b

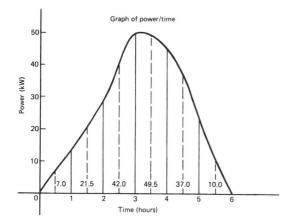

Fig. 16.9

broken lines in *Fig. 16.9*) and measured. The values are shown in *Fig. 16.9*.

    Area under curve
= (width of interval)(sum of mid-ordinates)
= (1)(7.0 + 21.5 + 42.0 + 49.5 + 37.0 + 10.0)
= **167 kW h** (i.e. a measure of electrical energy)

(b) Average value of waveform

$$= \frac{\text{area under curve}}{\text{length of base}} = \frac{167 \text{ kW h}}{6 \text{ h}} = \textbf{27.83 kW}$$

Alternatively, average value

$$= \frac{\text{sum of mid-ordinates}}{\text{number of mid-ordinates}}$$

Problem 8   *Fig. 16.10* shows a sinusoidal output voltage of a full-wave rectifier. Determine, using the mid-ordinate rule with 6 intervals, the mean output voltage.

One cycle of the output voltage is completed in π radians or 180°. The base is divided into 6 intervals, each of width 30°. The mid-ordinate of each interval will lie at 15°, 45°, 75°, etc.

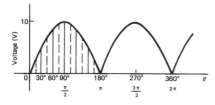

Fig. 16.10

At 15° the height of the mid-ordinate is 10 sin 15° = 2.588 V. At 45° the height of the mid-ordinate is 10 sin 45° = 7.071 V, and so on.
The results are tabulated below:

Mid-ordinate	Height of mid-ordinate
15°	10 sin 15° = 2.588 V
45°	10 sin 45° = 7.071 V
75°	10 sin 75° = 9.659 V
105°	10 sin 105° = 9.659 V
135°	10 sin 135° = 7.071 V
165°	10 sin 165° = 2.588 V
	Sum of mid-ordinates = 38.636 V

Mean or average value of output voltage

$$= \frac{\text{sum of mid-ordinates}}{\text{number of mid-ordinates}} = \frac{38.636}{6} = \textbf{6.439 V}$$

(With a larger number of intervals a more accurate answer may be obtained).
For a sine wave, the actual mean value is 0.637 × maximum value, which in this problem gives 6.37 V.

Problem 9   An indicator diagram for a steam engine is shown in *Fig. 16.11*. The base line has been divided into 6 equally spaced intervals and the lengths of the 7 ordinates measured with the results shown in centimetres. Determine (a) the area of the indicator diagram using Simpson's rule, and (b) the mean pressure in the cylinder given that 1 cm represents 100 kPa.

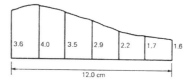

Fig. 16.11

(a) The width of each interval is 12.0/6 cm. Using Simpson's rule,

$$\text{area} = \frac{1}{3}(2.0)[(3.6 + 1.6) + 4(4.0 + 2.9 + 1.7) + 2(3.5 + 2.2)]$$

$$= \frac{2}{3}[5.2 + 34.4 + 11.4] = \textbf{34 cm}^2$$

(b) Mean height of ordinates

$$= \frac{\text{area of diagram}}{\text{length of base}} = \frac{34}{12} = 2.83 \text{ cm}$$

Since 1 cm represents 100 kPa, the mean pressure in the cylinder

$$= 2.83 \text{ cm} \times 100 \text{ kPa/cm} = \textbf{283 kPa}$$

Further problems on mean or average values of waveforms may be found in section 16.4, Problems 9 to 13.

16.4 Further problems on irregular areas and volumes and mean values of waveforms

Areas of irregular figures

1 Plot a graph of $y = 3x - x^2$ by compiling a table of values of y from $x = 0$ to $x = 3$. Determine the area enclosed by the curve, the x-axis and ordinate $x = 0$ and $x = 3$ by (a) the trapezoidal rule, (b) the mid-ordinate rule and (c) by Simpson's rule.

$$\left[4\frac{1}{2} \text{ square units}\right]$$

2 Plot the graph of $y = 2x^2 + 3$ between $x = 0$ and $x = 4$. Estimate the area enclosed by the curve, the ordinates $x = 0$ and $x = 4$, and the x-axis by an approximate method. [54.7 square units]

3 The velocity of a car at one second intervals is given in the following table:

time t (s)	0	1	2	3	4	5	6
velocity v (m/s)	0	2.0	4.5	8.0	14.0	21.0	29.0

Determine the distance travelled in 6 seconds (i.e. the area under the v/t graph) using an approximate method. [63 m]

4 The shape of a piece of land is shown in *Fig. 16.12*. To estimate the area of the land, a surveyor takes measurements at intervals of 50 m, perpendicular to the straight portion with the results shown (the dimensions being in metres). Estimate the area of the land in hectares (1 ha = 10^4 m²) [4.70 ha]

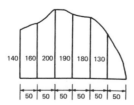

Fig. 16.12

5 The deck of a ship is 35 m long. At equal intervals of 5 m the width is given by the following table:

Width (m)	0	2.8	5.2	6.5	5.8	4.1	3.0	2.3

Estimate the area of the deck.
[143 m²]

Volumes of irregular solids

6 The areas of equidistantly spaced sections of the underwater form of a small boat are as follows:

1.76, 2.78, 3.10, 3.12, 2.61, 1.24, 0.85 m²

Determine the underwater volume if the sections are 3 m apart.
[42.59 m³]

7 To estimate the amount of earth to be removed when constructing a cutting the cross-sectional area at intervals of 8 m were estimated as follows:

0, 2.8, 3.7, 4.5, 4.1, 2.6, 0 m³

Estimate the volume of earth to be excavated.
[147 m³]

8 The circumference of a 12 m long log of timber of varying circular cross-section is measured at intervals of 2 m along its length and the results are:

Distance from one end (m)	0	2	4	6	8	10	12
Circumference (m)	2.80	3.25	3.94	4.32	5.16	5.82	6.36

Estimate the volume of the timber in cubic metres.
[20.42 m³]

Mean or average values of waveforms

9 Determine the mean value of the periodic waveforms shown in *Fig. 16.13* over a half cycle.
[(a) 2 A (b) 50 V (c) 2.5 A]

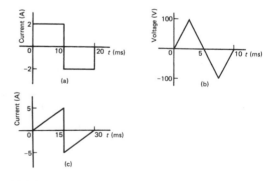

Fig. 16.13

10 Find the average value of the periodic waveforms shown in *Fig. 16.14* over one complete cycle.
[(a) 2.5 V (b) 3 A]

11 An alternating current has the following values at equal intervals of 5 ms.

Time (ms)   0   5   10   15   20   25   30
Current (A)   0   0.9   2.6   4.9   5.8   3.5   0

Plot a graph of current against time and estimate the area under the curve over the 30 ms period using the mid-ordinate rule and determine its mean value.
[0 093 A s, 3.1 A]

(a)

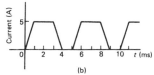

(b)

Fig. 16.14

12 Determine, using an approximate method, the average value of a sine wave of maximum value 50 V for (a) a half cycle and (b) a complete cycle.
    [(a) 31.83 V (b) 0]

13 An indicator diagram of a steam engine is 12 cm long. Seven evenly spaced ordinates, including the end ordinates, are measured as follows:

5.90,   5.52,   4.22,   3.63,   3.32,   3.24,   3.16 cm

Determine the area of the diagram and the mean pressure in the cylinder if 1 cm represents 90 kPa.
[49.13 cm², 368.5 kPa]

17

Centroids of simple shapes

17.1 Introduction to centroids

(a) A **lamina** is a thin, flat sheet having uniform thickness. The **centre of gravity** of a lamina is the point where it balances perfectly, i.e. the lamina's **centre of mass**. When dealing with a shape or area (a lamina of negligible thickness and mass) the term **centre of area** or **centroid** is used for the point where the centre of gravity of a lamina of that shape would lie.

(b) (i) The centroid C of a **rectangle** lies on the intersection of the diagonals (see *Fig. 17.1(a)*).

(ii) The centroid C of a **triangle** lies on the intersection of its medians, a median being a line which joins the vertices of a triangle with the mid-point of the opposite side. It may be shown that the centroid lies at one-third of the perpendicular height above any side as base (see *Fig. 17.1(b)*).

(iii) The centroid C of a **circle** lies at its centre (see *Fig. 17.1(c)*).

(iv) The centroid C of a **semicircle** of radius r lies on the centre line at a distance $4r/3\pi$ from the diameter (see *Fig. 17.1(d)*).

(c) The **first moment of area** is defined as the product of the area and the perpendicular distance of its centroid from a given axis in the plane of the area. In *Fig. 17.2*, the first moment of area A about axis XX is given by (Ay) cubic units.

(d) A **composite area** consists of two or more areas having different shapes joined together. The centroid of a com-

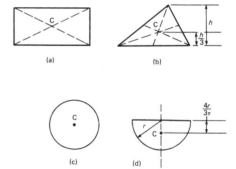

Fig. 17.1

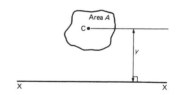

Fig. 17.2

posite area is found by dividing the whole area into parts, the centroids of which are known, and then taking moments (i.e. finding the first moment of area) about two orthogonal axes (i.e. two axes lying in the same place and at right angles to each other). For the composite area shown in *Fig. 17.3*:

Sum of moments about YY, $\Sigma ax = a_1x_1 + a_2x_2 + a_3x_3$
Sum of moments about XX, $\Sigma ay = a_1y_1 + a_2y_2 + a_3y_3$

If $A = a_1 + a_2 + a_3$ and $\bar{x}$ and $\bar{y}$ are the distances of the centroid of the composite area about axes YY and XX respectively, then:

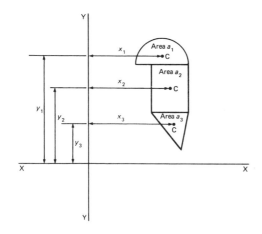

Fig. 17.3

$A\bar{x} = \Sigma ax$, from which

$$\bar{x} = \frac{\Sigma ax}{A} = \frac{\text{first moment of area about YY}}{\text{total area}}$$

and

$A\bar{y} = \Sigma ay$, from which

$$\bar{y} = \frac{\Sigma ay}{A} = \frac{\text{first moment of area about XX}}{\text{total area}}$$

17.2 Worked problems on centroids of simple shapes

> *Problem 1* Determine the first moment of area of each of the shapes shown in *Fig. 17.4* about axes XX and YY.

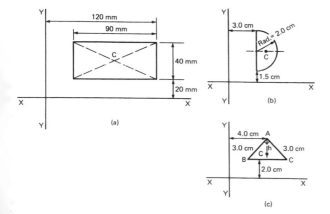

Fig. 17.4

(a) Area of rectangle = $90 \times 40 = 3600 \text{ mm}^2$. Distance of centroid C of rectangle from XX = $20 + 40/2$ = 40 mm. First moment of area of rectangle about XX

= area × perpendicular distance to centroid
= $(3600)(40) = \mathbf{144\ 000\ mm^3}$

Distance of centroid C of rectangle from YY = $(120 - 90) + 90/2 = 75$ mm. First moment of area of rectangle about

YY = $(3600)(75) = \mathbf{270\ 000\ mm^3}$

(b) Area of semicircle = $\pi r^2/2 = \pi(2)^2/2 = 2\pi \text{ cm}^2$. Distance of centroid C of semicircle from XX = $1.5 + 2.0 = 3.5$ cm. First moment of area of semicircle about XX

= $(2\pi)(3.5) = 7\pi = \mathbf{22.0\ cm^3}$

Distance of centroid C of semicircle from YY

= $3.0 + \dfrac{4r}{3\pi} = 3.0 + \dfrac{4(2.0)}{3\pi} = 3.849$ cm

First moment of area of semicircle about YY

= $(2\pi)(3.849) = \mathbf{24.2\ cm^3}$

(c) Area of triangle ABC = $\dfrac{1}{2}(3.0)(3.0) = 4.5 \text{ cm}^2$. Since the triangle ABC is isosceles $\angle B = \angle C = 45°$. Hence $\sin 45° = h/3.0$, from which $h = 3.0 \sin 45° = 2.121$ cm. Distance of centroid C from XX

= $2.0 + \dfrac{h}{3.0} = 2.0 + \dfrac{2.121}{3.0} = 2.707$ cm

First moment of area of triangle about XX

= $(4.5)(2.707) = \mathbf{12.2\ cm^3}$

Distance of centroid C from YY = 4.0 cm. First moment of area about YY = $(4.5)(4.0) = \mathbf{18.0\ cm^3}$

> *Problem 2* Determine the position of the centroid of the symmetrical T-section shown in *Fig. 17.5*.

Since the T-section is symmetrical the centroid will lie on axis YY. The T-section is divided into rectangles A and B with centroids at C_A and C_B, respectively. Let the centroid, C, of the whole area be at a distance $\bar{y}$ from axis XX.

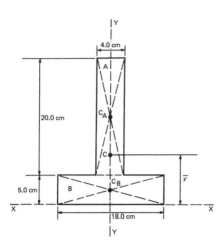

Fig. 17.5

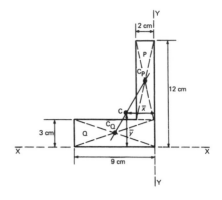

Fig. 17.6

Area of rectangle A = (4.0)(20.0) = 80.0 cm²
Area of rectangle B = (5.0)(18.0) = 90.0 cm²
Total area A of T-section = 80.0 + 90.0 = 170.0 cm²

First moment of area of rectangle A about XX

= (area) × (perpendicular distance to centroid
of A from XX)

$$= (80.0)\left(5.0 + \frac{20.0}{2}\right) = 1200 \text{ cm}^3$$

First moment of area of rectangle B about XX

$$= (90.0)\left(\frac{5.0}{2}\right) = 225 \text{ cm}^3$$

Hence, from para. (d)

$$A\bar{y} = \Sigma ax = 1200 + 225 = 1425 \text{ cm}^3$$

i.e. $\bar{y} = \dfrac{1425}{A} = \dfrac{1425}{170.0} = 8.38$ cm

Hence the centroid of the T-section lies on the axis of symmetry 8.38 cm from the base.

Problem 3 Calculate the position of the centroid of the angle-iron section shown in *Fig. 17.6* about axes XX and YY.

The L-section is divided into two rectangles P and Q with centroids at C_P and C_Q respectively. Let the centroid, C, of the whole area be at a perpendicular distance $\bar{x}$ from axis YY and at a perpendicular distance $\bar{y}$ from axis XX.

Area of rectangle P = (12 − 3)(2) = 18 cm²
Area of rectangle Q = (9)(3) = 27 cm²
Total area, A, of L-section = 18 + 27 = 45 cm²

The first moment of area of rectangle P about YY

= (area)(perpendicular distance to centroid of P
from YY)

= (18)(1) = 18 cm³

The first moment of area of rectangle Q about YY = (27)(4.5)
= 121.5 cm³.
From para. (d), page 159,

$$A\bar{x} = \Sigma ax = 18 + 121.5 = 139.5 \text{ cm}^3$$

Hence $\bar{x} = \dfrac{139.5}{A} = \dfrac{139.5}{45} = 3.1$ cm

The first moment of area of rectangle P about XX =
(18)(3 + 4.5) = 135 cm³.
The first moment of area of rectangle Q about XX = (27)(1.5)
= 40.5 cm³.
From para. (d), page 159,

$$A\bar{y} = \Sigma ay = 135 + 40.5 = 175.5 \text{ cm}^3$$

Hence $\bar{y} = \dfrac{175.5}{A} = \dfrac{175.5}{45} = 3.9$ cm

Thus the centroid of the L-section lies at a point outside of the cross-section 3.1 cm from YY and 3.9 cm from XX.
(When there are two parts forming a composite area, the centroid always lies on a straight line joining the centroids of the two separate parts – a fact which can be used as a check. In this problem, C_P, C and C_Q lie on the same straight line, as shown in *Fig. 17.6*.)

Problem 4 A metal gusset plate is as shown in *Fig. 17.7*. Determine the position of its centroid.

Table 17.1

Part	Area a mm²	Distance of centroid from YY (i.e. *x* mm)	First moment of area about YY (i.e. *ax* mm³)	Distance of centroid from XX (i.e. *y* mm)	First moment of area about XX (i.e. *ay* mm³)
P	90 × 10 = 900	5	(900)(5) = 4500	$15 + \frac{1}{2}(90) = 60$	(900)(60) = 54 000
Q	$\frac{1}{2}(72)(90) = 3240$	$10 + \frac{1}{3}(72) = 34$	(3240)(34) = 110 160	$15 + \frac{1}{3}(90) = 45$	(3240)(45) = 145 800
R	82 × 15 = 1230	41	(1230)(41) = 50 430	7.5	(1230)(7.5) = 9225
$\Sigma a = A = 900 + 3240$ $+ 1230 = 5370$		$\Sigma ax = 4500 + 110\ 160 + 50\ 430 = 165\ 090$		$\Sigma ay = 54\ 000 + 145\ 800 + 9225 = 209\ 025$	

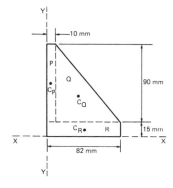

Fig. 17.7

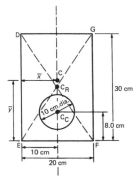

Fig. 17.8

The gusset plate is divided into rectangle P, triangle Q and rectangle R, with their centroids C_P, C_Q and C_R, respectively. When determining positions of centroids it is often more convenient to use a tabular approach as shown in *Table 17.1*. Let $\bar{x}$ and $\bar{y}$ be the distances of the centroid from YY and XX, respectively. From para. (d), page 159, $A\bar{x} = \Sigma ax$, from which

$$\bar{x} = \frac{\Sigma ax}{A} = \frac{165\ 090}{5370} = 30.7 \text{ mm}$$

$A\bar{y} = \Sigma ay$, from which

$$\bar{y} = \frac{\Sigma ay}{A} = \frac{209\ 025}{5370} = 38.9 \text{ mm}$$

Hence the centroid of the gusset plate lies at a point 30.7 mm from YY and 38.9 mm from XX.

Problem 5 A rectangular template has dimensions of 30 cm by 20 cm. A 10 cm diameter hole is removed from the plate in the position shown in *Fig. 17.8*. Determine the position of the centroid of the template.

The centroids of the rectangle and the circle to be removed are denoted by C_R and C_C, respectively. In *Table 17.2*, the area, and thus the first moment of area, of the circle is shown as negative since the circle is removed. Although the position of $\bar{x}$ can be seen by inspection, it is calculated in *Table 17.2* to illustrate the general method.

If $\bar{x}$ and $\bar{y}$ are the distances of the centroid from DE and EF, respectively, then, from para. (d), page 159,

$$A\bar{x} = \Sigma ax, \text{ from which}$$

$$\bar{x} = \frac{\Sigma ax}{A} = \frac{5214.6}{521.46} = 10.0 \text{ cm}$$

and

$$A\bar{y} = \Sigma ay, \text{ from which}$$

$$\bar{y} = \frac{\Sigma ay}{A} = \frac{8371.68}{521.46} = 16.1 \text{ cm}$$

Hence the centroid of the template lies at a point 10.0 cm from its left-hand edge and 16.1 cm from its bottom edge. (Note that C_R, C_C and C lie on the same straight line.)

Problem 6 Determine the position of the centroid of the area shown in *Fig. 17.9*.

Table 17.2

Part	Area *a* cm²	Distance of centroid from DE (i.e. *x* cm)	First moment of area about DE (i.e. *ax* cm³)	Distance of centroid from EF (i.e. *y* cm)	First moment of area about EF (i.e. *ay* cm³)
Rectangle	30 × 20 = 600	10	(600)(10) = 6000	15	(600)(15) = 9000
Circle	$-\pi r^2$ $= -\pi e\left(\dfrac{10}{2}\right)^2$ $= -78.54$	10	(−78.54)(10) = −785.4	8.0	(−78.54)(8.0) = −628.32
$\Sigma a = A = 600 - 78.54$ $= 521.46$		$\Sigma ax = 6000 - 785.4 = 5214.6$		$\Sigma ay = 9000 - 628.32 = 8371.68$	

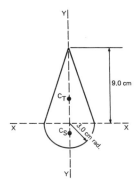

Fig. 17.9

Table 17.3

Part	Area *a* cm²	Distance of centroid from XX (i.e. *y* cm)	First moment of area about XX (i.e. *ay* cm³)
Triangle	$\dfrac{1}{2}(6.0)(9.0)$ $= 27.0$	$\dfrac{1}{3}(9.0)$ $= +3.0$	(27.0)(3.0) $= +81.0$
Semicircle	$\dfrac{1}{2}\pi(3.0)^2$ $= 14.14$	$\dfrac{-4(3.0)}{3\pi}$ $= -1.27$	(14.14)(−1.27) $= -17.96$
$\Sigma a = A = 41.14$		$\Sigma ay = +81.0 - 17.96 = +63.04$	

Let C_T and C_S be the centroids of the triangle and semi-circle, respectively. Since the area is symmetrical about its axis of symmetry YY, the centroid of the whole area will lie on YY. The horizontal axis XX has areas above and below it. In such cases, distances of centroids above XX are considered positive whilst distances of centroids below are considered negative, as shown in *Table 17.3*.

If $\bar{y}$ is the distance of the centroid from XX then, from para. (d), page 159,

$A\bar{y} = \Sigma ay$, from which

$$\bar{y} = \frac{\Sigma ay}{A} = \frac{+63.04}{41.14} = +1.53 \text{ cm}$$

Hence the centroid of the area lies on the axis of symmetry YY, 1.53 cm above the axis XX.

Problem 7 Find the position of the centroid of the metal template shown in *Fig. 17.10* about the right-hand edge and the diameter of the semicircle. The circular area is removed.

Table 17.4 on page 163 shows the analysis.

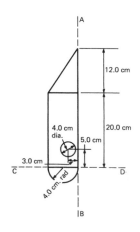

Fig. 17.10

If $\bar{x}$ and $\bar{y}$ are the distances of the centroid from AB and CD respectively then, from para. (d), page 159,

$A\bar{x} = \Sigma ax$, from which

$$\bar{x} = \frac{\Sigma ax}{A} = \frac{830.8}{220.567} = 3.77 \text{ cm}$$

Table 17.4

Part	Area a cm^3	Distance of centroid from AB (i.e. x cm)	First moment of area about AB (i.e. ax cm^3)	Distance of centroid from CD (i.e. y cm)	First moment of area about CD (i.e. ay cm^3)
Triangle	$\frac{1}{2}(8.0)(12.0)$ $= 48.0$	$\frac{1}{3}(8.0) = 2.667$	$(48.0)(2.667) = 128.0$	$20.0 + \frac{1}{3}(12.0)$ $= 24.0$	$(48.0)(24.0) = 1152.0$
Rectangle	$(20.0)(8.0)$ $= 160.0$	$\frac{1}{2}(8.0) = 4.0$	$(160.0)(4.0) = 640.0$	$\frac{1}{2}(20.0) = 10.0$	$(160.0)(10.0) = 1600.0$
Circle	$-\pi(2.0)^2$ $= -12.566$ (minus since circle is removed)	3.0	$(-12.566)(3.0)$ $= -37.70$	5.0	$(-12.566)(5.0)$ $= -62.83$
Semicircle	$\frac{1}{2}\pi(4.0)^2$ $= 25.133$	4.0	$(25.133)(4.0) = 100.5$	$\frac{-4(4.0)}{3\pi} = -1.698$ (minus since below CD)	$(25.133)(-1.698)$ $= -42.68$
$\Sigma a = A = 220.567$		$\Sigma ax = 830.8$		$\Sigma ay = 2646.49$	

and

$A\bar{y} = \Sigma ay$, from which

$$\bar{y} = \frac{\Sigma ay}{A} = \frac{2646.49}{220.567} = 12.0 \text{ cm}$$

Hence the centroid lies at a point 3.77 cm to the left of AB and 12.0 cm above CD.

Further problems on centroids of simple shapes may be found in section 17.3, Problems 1 to 8.

17.3 Further problems on centroids of simple shapes

1 Determine the first moment of area of each of the shapes shown in *Fig. 17.11* about axes XX and YY.

$\left[\begin{array}{l}\text{(a) } 427.5 \text{ cm}^3, 247.5 \text{ cm}^3 \text{ (b) } 37\,700 \text{ mm}^3, 43\,980 \text{ mm}^3 \\ \text{(c) } 252 \text{cm}^3, 252 \text{ cm}^3 \text{ (d) } 3808 \text{ mm}^3, 3927 \text{ mm}^3\end{array}\right]$

2 Determine the distances from axes XX and YY of the centroids for the shapes shown in *Fig. 17.12*.

$\left[\begin{array}{l}\text{(a) } 16.3 \text{ cm below XX on YY} \\ \text{(b) } 14.2 \text{ mm below XX and 28.0 mm to the right of YY} \\ \text{(c) } 6.66 \text{ cm above XX on YY}\end{array}\right]$

3 Find the positions of the centroids for the shapes shown in *Fig. 17.13*.

$\left[\begin{array}{l}\text{(a) } 3.30 \text{ cm to the right of AB, 1.60 cm above BC} \\ \text{(b) } 78.9 \text{ mm to the right of DE, 70.4 mm above EF} \\ \text{(d) } 10.1 \text{ cm above GH on IJ}\end{array}\right]$

4 Determine the positions of the centroids for the templates shown in *Fig. 17.14*. (In Fig. *17.14(a)*, the circular area is removed.)

$\left[\begin{array}{l}\text{(a) } 56.0 \text{ mm from bottom edge, 44.3 mm} \\ \text{from the left-hand edge (b) } 4.56 \text{ cm below} \\ \text{the top edge, 3.56 cm from the right-hand edge.}\end{array}\right]$

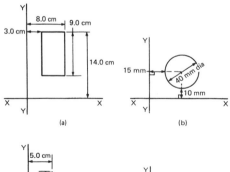

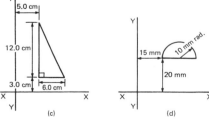

Fig. 17.11

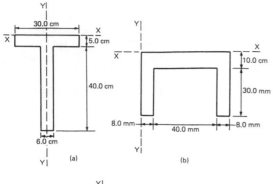

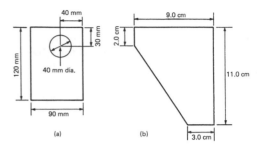

Fig. 17.14

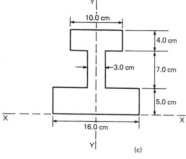

Fig. 17.12

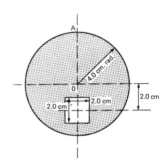

Fig. 17.15

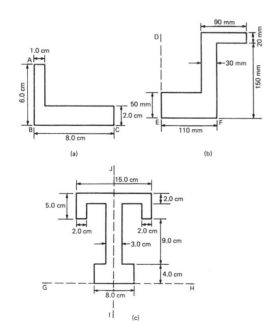

Fig. 17.13

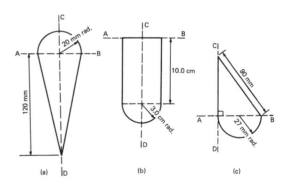

Fig. 17.16

6 Find the positions of the centroids of the shapes shown in *Fig. 17.16* about axes AB and CD.

[(a) 29.9 mm below AB on CD
(b) 6.20 cm below AB on CD
(c) 10.9 mm above AB, 21.3 mm to the right of CD]

7 Determine the positions of the centroids for the letters shown in *Fig. 17.17*.

[(a) 11.8 cm from bottom, 2.86 cm from left-hand edge
(b) 11.8 cm from bottom, 4.00 cm from left-hand edge]

5 Determine the centroid of the shaded area in *Fig. 17.15*.
[On OA, 0.173 cm from 0]

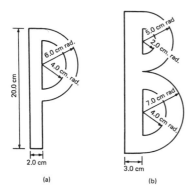

(a)

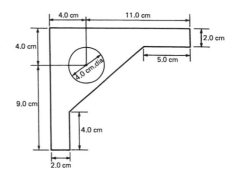

(b)

Fig. 17.17

Fig. 17.18

8 In the template shown in *Fig. 17.18*, the circular area is removed.

Determine the position of the centroid of the template.

$\begin{bmatrix} \text{3.95 cm from the top edge, 4.86 cm from the} \\ \text{left-hand edge} \end{bmatrix}$

18

Further areas and volumes

18.1 The ellipse

An **ellipse** is the name given to the regular oval shape PRQS shown in *Fig. 18.1*. PQ is called the **major axis** and a is the semi-major axis. RS is the **minor axis** and b the semi-minor axis.

> **Area of ellipse PRQS** $= \pi ab$

> **Perimeter of ellipse PRQS** $= \pi(a + b) =$
> $\dfrac{\pi}{2}$ **(sum of major and minor axes)**

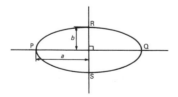

Fig. 18.1

> *Problem 1*   The major axis of an ellipse is 12.0 cm and the minor axis is 7.0 cm. Determine the perimeter and the area of the ellipse.

From above, the perimeter of an ellipse $= \pi(a + b)$, where a and b are the semi-major and semi-minor axes, respectively. Hence

$$a = \frac{12.0}{2} = 6.0 \text{ cm and } b = \frac{7.0}{2} = 3.5 \text{ cm.}$$

Perimeter of ellipse

$$= \pi(a + b) = \pi(6.0 + 3.5) = \mathbf{29.85 \text{ cm}}$$

Area of ellipse

$$= \pi ab = \pi(6.0)(3.5) = \mathbf{65.97 \text{ cm}^2}$$

> *Problem 2*   Determine the total surface area and the volume of a regular elliptical cylinder, the diameters being 20.0 cm (major axis) and 15.0 cm (minor axis) respectively and the height 12.0 cm.

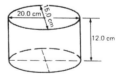

Fig. 18.2

The cylinder is shown in *Fig. 18.2*. Since the major axis is 20.0 cm the semi-major axis is $20.0/2 = 10.0$ cm, and since the minor axis is 15.0 cm the semi-minor axis is $15.0/2 = 7.5$ cm.

From above, area of elliptical end $= \pi ab = \pi(10.0)(7.5)$ $= 75\pi \text{ cm}^2$ and the perimeter of elliptical end $= \pi(a + b)$ $= \pi(10.0 + 7.5) = 17.5\pi$ cm. Total surface area of cylinder

$$= \text{curved surface area} + \text{area of two ends}$$
$$= (17.5\pi)(12.0) + 2(75\pi)$$
$$= 210\pi + 150\pi = 360\pi = \mathbf{1131 \text{ cm}^2}$$

Volume of cylinder

= area of base × perpendicular height
= $(75\pi)(12.0) = \mathbf{2827\ cm^3}$

Problem 3 A rectangular metal plate measures 16 cm by 10 cm. If the maximum possible sized ellipse is cut from the plate, determine the amount of metal wasted.

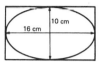

Fig. 18.3

The plate is shown in *Fig. 18.3*. Area of metal plate = 16 × 10 = 160 cm².
Area of ellipse, having semi-major axis, $a = 16/2 = 8$ cm and semi-minor axis, $b = 10/2 = 5$ cm, is given by:

$$\text{Area} = \pi ab = \pi(8)(5) = 40\pi = 125.7\ cm^2$$

Hence metal wasted = 160 − 125.7 = **34.3 cm²**.

Problem 4 An elliptical plot of land has an area of 150 m² and a circumference of 50 m. Determine the maximum length and maximum width of the plot.

Let the semi-major and semi-minor axes be a and b, respectively (as in *Fig. 18.1*).

$$\text{Area of ellipse} = \pi ab = 150 \qquad (1)$$

Circumference (or perimeter) of ellipse
$$= \pi(a + b) = 50 \qquad (2)$$

From equation (2), $a + b = 50/\pi$, from which $a = (50/\pi) - b$. Substituting, $a = 50/\pi - b$ into equation (1) gives:

$$\pi\left(\frac{50}{\pi} - b\right)b = 150, \text{ i.e. } 50b - \pi b^2 = 150$$

from which, $\pi b^2 - 50b + 150 = 0$.
Using the quadratic formula,

$$b = \frac{50 \pm \sqrt{[(-50)^2 - 4(\pi)(150)]}}{2\pi}$$

$$= \frac{50 \pm 24.8}{2\pi} = 11.90\ \text{m or } 4.01\ \text{m}$$

Substituting in equation (1) gives: $\pi a(11.90) = 150$, i.e.

$$a = \frac{150}{\pi(11.90)} = 4.01\ \text{m or } \pi a(4.01) = 150$$

i.e. $a = \dfrac{150}{\pi(4.01)} = 11.91\ \text{m}$

Hence the semi-major and semi-minor axes are 11.91 m and 4.01 m. The maximum length of the plot = 2 × 11.91 = **23.82 m** and the maximum width of the plot = 2 × 4.01 = **8.02 m**.

Further problems on the ellipse may be found in section 17.5, Problems 1 to 6, page 173.

18.2 The frustum and zone of a sphere

Volume of sphere = $\dfrac{4}{3}\pi r^3$.

Surface area of sphere = $4\pi r^2$.

A **frustum of a sphere** is the portion contained between two parallel planes. In *Fig. 18.4*, PQRS is a frustum of the sphere. A **zone of a sphere** is the curved surface of a frustum. With reference to *Fig. 18.4*:

Surface area of a zone of a sphere = $2\pi rh$

Volume of frustum of sphere

$$= \frac{\pi h}{6}(h^2 + 3r_1^2 + 3r_2^2)$$

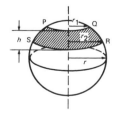

Fig. 18.4

Problem 5   Determine the volume of a frustum of a sphere of diameter 49.74 cm if the diameter of the ends of the frustum are 24.0 cm and 40.0 cm, and the height of the frustum is 7.00 cm.

From above, volume of frustum of a sphere

$$= \frac{\pi h}{6}(h^2 + 3r_1^2 + 3r_2^2)$$

where $h = 7.00$ cm, $r_1 = 24.0/2 = 12.0$ cm and $r_2 = 40.0/2 = 20.0$ cm. Hence volume of frustum

$$= \frac{\pi(7.00)}{6}[(7.00)^2 + 3(12.0)^2 + 3(20.0)^2]$$

$$= \textbf{6161 cm}^3$$

Problem 6   Determine for the frustum of *Problem 5* the curved surface area of the frustum.

The curved surface area of the frustum = surface area of zone = $2\pi rh$ (from above), where r = radius of sphere = 49.74/2 = 24.87 cm and $h = 7.00$ cm. Hence, surface area of zone = $2\pi(24.87)(7.00) = \textbf{1094 cm}^2$.

Problem 7   The diameters of the ends of the frustum of a sphere are 14.0 cm and 26.0 cm, respectively, and the thickness of the frustum is 5.0 cm. Determine, correct to 3 significant figures (a) the volume of the frustum of the sphere, (b) the radius of the sphere and (c) the area of the zone formed.

The frustum is shown shaded in the cross-section of *Fig. 18.5*.

(a) Volume of frustum of sphere

$$= \frac{\pi h}{6}(h^2 + 3r_1^2 + 3r_2^2)$$

from above, where $h = 5.0$ cm, $r_1 = 14.0/2 = 7.0$ cm and $r_2 = 26.0/2 = 13.0$ cm.
Hence volume of frustum of sphere

$$= \frac{\pi(5.0)}{6}[(5.0)^2 + 3(7.0)^2 + 3(13.0)^2]$$

$$= \frac{\pi(5.0)}{6}[25.0 + 147.0 + 507.0]$$

$$= \textbf{1780 cm}^3, \text{ correct to 3 significant figures}$$

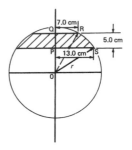

Fig. 18.5

(b) The radius, r, of the sphere may be calculated using *Fig. 18.5*.
Using Pythagoras' theorem

$$OS^2 = PS^2 + OP^2, \text{ i.e. } r^2 = (13.0)^2 + OP^2$$
$$OR^2 = QR^2 + OQ^2, \text{ i.e. } r^2 = (7.0)^2 + OQ^2 \quad (1)$$

However $OQ = QP + OP = 5.0 + OP$, therefore

$$r^2 = (7.0)^2 + (5.0 + OP)^2. \quad (2)$$

Equating equations (1) and (2) gives:

$$(13.0)^2 + OP^2 = (7.0)^2 + (5.0 + OP)^2$$
$$169.0 + OP^2 = 49.0 + 25.0 + 10.0(OP) + OP^2$$
$$169.0 = 74.0 + 10.0(OP)$$

Hence $OP = \dfrac{169.0 - 74.0}{10.0} = 9.50$ cm

Substituting $OP = 9.50$ cm into equation (1) gives

$$r^2 = (13.0)^2 + (9.50)^2$$

from which $r = \sqrt{[(13.0)^2 + (9.50)^2]}$

i.e. **radius of sphere, $r = 16.1$ cm.**

(c) Area of zone of sphere = $2\pi rh$

$$= 2\pi(16.1)(5.0)$$
$$= \textbf{506 cm}^2, \text{ correct to 3 significant figures}$$

Problem 8   A frustum of a sphere of diameter 12.0 cm is formed by two parallel planes, one through the diameter and the other distance h from the diameter. The curved surface area of the frustum is required to be $\dfrac{1}{4}$ of the total surface area of the sphere. Determine (a) the volume and surface area of the sphere, (b) the thickness h of the frustum, (c) the volume of the frustum and (d) the volume of the frustum expressed as a percentage of the sphere.

(a) Volume of sphere

$$V = \frac{4}{3}\pi r^3 = \frac{4}{3}\pi\left(\frac{12.0}{2}\right)^3 = \textbf{904.8 cm}^3$$

Surface area of sphere $= 4\pi r^2$

$$= 4\pi\left(\frac{12.0}{2}\right)^2 = \textbf{452.4 cm}^2$$

(b) Curved surface area of frustum

$$= \frac{1}{4} \times \text{surface area of sphere}$$

$$= \frac{1}{4} \times 452.4 = 113.1 \text{ cm}^2$$

From above, $113.1 = 2\pi r h = 2\pi\left(\frac{12.0}{2}\right)h$. Hence

thickness of frustum,

$$h = \frac{113.1}{2\pi (6.0)} = \textbf{3.0 cm}$$

(c) Volume of frustum, $V = \dfrac{\pi h}{6}(h^2 + 3r_1^2 + 3r_2^2)$

where $h = 3.0$ cm, $r_2 = 6.0$ cm and $r_1 = \sqrt{(OQ^2 - OP^2)}$, from *Fig. 18.6*, i.e. $r_1 = \sqrt{[(6.0)^2 - (3.0)^2]} = 5.196$ cm.

Hence volume of frustum

$$= \frac{\pi(3.0)}{6}[(3.0)^2 + 3(5.196)^2 + 3(6.0)^2]$$

$$= \frac{\pi}{2}[9.0 + 81 + 108.0] = \textbf{311.0 cm}^3$$

(d) $\dfrac{\text{Volume of frustum}}{\text{Volume of sphere}} = \dfrac{311.0}{904.8} \times 100\%$

$$= \textbf{34.37\%}$$

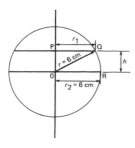

Fig. 18.6

Problem 9 A spherical storage tank is filled with liquid to a depth of 20 cm. If the internal diameter of the vessel is 30 cm, determine the number of litres of liquid in the container (1 litre = 1000 cm³).

The liquid is represented by the shaded area in the section shown in *Fig. 18.7*. The volume of liquid comprises a hemisphere and a frustum of thickness 5 cm. Hence volume of liquid

$$= \frac{2}{3}\pi r^3 + \frac{\pi h}{6}[h^2 + 3r_1^2 + 3r_2^2]$$

where $r_2 = 30/2 = 15$ cm and $r_1 = \sqrt{[(15)^2 - (5)^2]} = 14.14$ cm.
Volume of liquid

$$= \frac{2}{3}\pi(15)^3 + \frac{\pi(5)}{6}[5^2 + 3(14.14)^2 + 3(15)^2]$$

$$= 7069 + 3403$$

$$= 10\,470 \text{ cm}^3$$

Since 1 litre = 1000 cm³, the number of litres of liquid = $10\,470/1000 = $ **10.47 litres**.

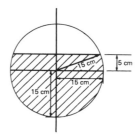

Fig. 18.7

Further problems on frustums and zones of spheres may be found in section 18.5, Problems 7 to 11, page 173.

18.3 Prismoidal rule for finding volumes

The prismoidal rule applies to a solid of length x divided by only three equidistant plane areas, A_1, A_2 and A_3 as shown in *Fig. 18.8* and is merely an extension of Simpson's rule for volumes (see Chapter 16, section 2).

With reference to *Fig. 18.8*:

$$\text{Volume, } V = \frac{x}{6}[A_1 + 4A_2 + A_3]$$

The prismoidal rule gives precise values of volume for regular solids such as pyramids, cones, spheres and prismoids.

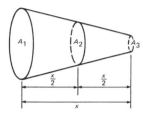

Fig. 18.8

Problem 10   A container is in the shape of a frustum of a cone. Its diameter at the bottom is 18 cm and at the top 30 cm. If the depth is 24 cm determine the capacity of the container, correct to the nearest litre, by the prismoidal rule (1 litre = 1000 cm³).

The container is shown in *Fig. 18.9*. At the midpoint, i.e. at a distance of 12 cm from one end the radius r_2 is $(9 + 15)/2 = 12$ cm, since the sloping sides change uniformly.

Volume of container by the prismoidal rule = $(x/6)[A_1 + 4A_2 + A_3]$, from above, where $x = 24$ cm, $A_1 = \pi(15)^2$ cm², $A_2 = \pi(12)^2$ cm² and $A_3 = \pi(9)^2$ cm².

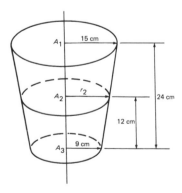

Fig. 18.9

Hence volume of container

$$-\frac{24}{6}[\pi(15)^2 + 4\pi(12)^2 + \pi(9)^2]$$

$$= 4[706.86 + 1809.56 + 254.47]$$

$$= 11\ 080 \text{ cm}^3 = \frac{11\ 080}{1000} \text{ litres}$$

$$= \textbf{11 litres, correct to the nearest litre}$$

(*Check*: Volume of frustum of cone

$$= \frac{1}{3}\pi h[R^2 + Rr + r^2] \text{ from Chapter 15}$$

$$= \frac{1}{3}\pi(24)[(15)^2 + (15)(9) + (9)^2]$$

$$= 11\ 080 \text{ cm}^3 \text{ as shown above})$$

Problem 11   A frustum of a sphere of radius 13 cm is formed by two parallel planes on opposite sides of the centre, each at distances of 5 cm from the centre. Determine the volume of the frustum (a) by using the prismoidal rule, and (b) by using the formula for the volume of a frustum of a sphere.

The frustum of the sphere is shown by the section in *Fig. 18.10*. Radius $r_1 = r_2 = PQ = \sqrt{(13^2 - 5^2)} = 12$ cm, by Pythagoras' theorem.

(a) Using the prismoidal rule, volume of frustum,

$$V = \frac{x}{6}[A_1 + 4A_2 + A_3]$$

$$= \frac{10}{6}[\pi(12)^2 + 4\pi(13)^2 + \pi(12)^2]$$

$$= \frac{10\pi}{6}[144 + 676 + 144]$$

$$= \textbf{5047 cm}^3$$

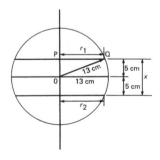

Fig. 18.10

(b) Using the formula for the volume of a frustum of a sphere:

Volume $V = \dfrac{\pi h}{6}(h^2 + 3r_1^2 + 3r_2^2)$

$= \dfrac{\pi(10)}{6}[10^2 + 3(12)^2 + 3(12)^2]$

$= \dfrac{10\pi}{6}(100 + 432 + 432) = \textbf{5047 cm}^3$

Problem 12 A hole is to be excavated in the form of a prismoid. The bottom is to be a rectangle 16 m long by 12 m wide; the top is also a rectangle, 26 m long by 20 m wide. Find the volume of earth to be removed, correct to 3 significant figures, if the depth of the hole is 6.0 m.

The hole is shown in *Fig. 18.11*. Let A_1 represent the area of the top of the hole, i.e. $A_1 = 20 \times 26 = 520$ m². Let A_3 represent the area of the bottom of the hole, i.e. $A_3 = 16 \times 12 = 192$ m². Let A_2 represent the rectangular area through the middle of the hole parallel to areas A_1 and A_3. The length of this rectangle is $(26 + 16)/2 = 21$ m and the width is $(20 + 12)/2 = 16$ m, assuming the sloping edges are uniform. Thus area $A_2 = 21 \times 16 = 336$ m².

Using the prismoidal rule, volume of hole

$= \dfrac{x}{6}[A_1 + 4A_2 + A_3] = \dfrac{6}{6}[520 + 4(336) + 192]$

$= 2056$ m³ $= \textbf{2060 m}^3$, correct to 3 significant figures

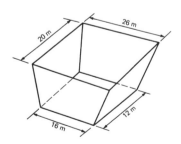

Fig. 18.11

Problem 13 The roof of a building is in the form of a frustum of a pyramid with a square base of side 5.0 m. The flat top is a square of side 1.0 m and all the sloping sides are pitched at the same angle. The vertical height of the flat top above the level of the eaves is 4.0 m. Calculate, using the prismoidal rule, the volume enclosed by the roof.

Let area of top of frustum be $A_1 = (1.0)^2 = 1.0$ m². Let area of bottom of frustum be $A_3 = (5.0)^2 = 25.0$ m². Let area of section through the middle of the frustum parallel to A_1 and A_3 be A_2. The length of the side of the square forming A_2 is the average of the sides forming A_1 and A_3, i.e. $(1.0 + 5.0)/2 = 3.0$ m. Hence $A_2 = (3.0)^2 = 9.0$ m². Using the prismoidal rule, volume of frustum

$= \dfrac{x}{6}[A_1 + 4A_2 + A_3]$

$= \dfrac{4.0}{6}[1.0 + 4(9.0) + 25.0]$

Hence volume enclosed by roof = 41.3 m³.

Further problems on the prismoidal rule may be found in section 18.5, Problems 12 to 15, page 173.

18.4 Theorem of Pappus

A theorem of Pappus states:

> If a plane area is rotated about an axis in its own plane but not intersecting it, the volume of the solid formed is given by the product of the area and the distance moved by the centroid of the area.

i.e. volume generated = area × distance moved through by the centroid

In *Fig. 18.12*, let C be the centroid of area A, and let $\bar{y}$ be the perpendicular distance of C from axis OX, then the distance moved by the centroid when area A makes one revolution about OX is $2\pi\bar{y}$ (i.e. the circumference of a circle). Hence by Pappus' theorem:
Volume generated = area × $2\pi\bar{y}$, i.e.

$$V = 2\pi A\bar{y} \text{ cubic units}$$

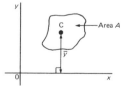

Fig. 18.12

Problem 14 Use the theorem of Pappus to calculate the position of the centroid of a semicircle of radius r.

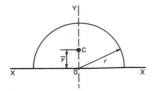

Fig. 18.13

If the semicircular area shown in *Fig. 18.13* is rotated about axis XX, then the volume of the solid generated is that of a sphere of volume $\frac{4}{3}\pi r^3$. The area of the semicircle is $\pi r^2/2$. Centroid C is shown in *Fig. 18.13* on the axis of symmetry OY. By the theorem of Pappus, volume generated = area × distance moved through by centroid, i.e.

$$\frac{4}{3}\pi r^3 = \left(\frac{\pi r^2}{2}\right)(2\pi \bar{y})$$

Hence $\bar{y} = \dfrac{\frac{4}{3}\pi r^3}{\left(\dfrac{\pi r^2}{2}\right)(2\pi)} = \dfrac{4r}{3\pi}$ **units**

i.e. the centroid of a semicircle lies on the axis of symmetry at a distance of $4r/3\pi$ from the diameter (see Chapter 17).

Problem 15   Determine the volume of an anchor ring formed by rotating a circle of radius 6.0 cm about an axis at a distance of 25 cm from its centre. Give the answer in cubic metres, correct to 3 significant figures.

The 6.0 cm radius circle is rotated about axis XX as shown by the side elevation in *Fig. 18.14*.
Area of circle = $\pi(6.0)^2 = 36.0\pi\ \text{cm}^2$. Distance moved by centroid C in one complete revolution = $2\pi(25) = 50\pi\ \text{cm}$ (i.e. circumference of a circle of radius 25 cm).
By the theorem of Pappus, volume generated = area × distance moved by centroid, i.e. volume of anchor ring

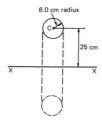

6.0 cm radius

25 cm

Fig. 18.14

$\doteq (36.0\pi)(50\pi) = 17\,770\ \text{cm}^3$

$= \dfrac{17\,770}{10^6}\ \text{m}^3$

$= \mathbf{0.017\,8\ m^3}$ correct to 3 significant figures

Problem 16   A metal disc has a radius of 5.0 cm and is of thickness 2.0 cm. A semicircular groove of diameter 2.0 cm is machined centrally around the rim to form a pulley. Determine, using Pappus' theorem, the volume of metal removed and the volume of the pulley.

A side view of the rim of the disc is shown in *Fig. 18.15*. When area PQRS is rotated about axis XX the volume generated is that of the pulley. The centroid of the semicircular area removed is at a distance $4r/3\pi$ from its diameter (see *Problem 14*), i.e. $4(1.0)/3\pi$, i.e. 0.424 cm from PQ. Thus the distance of the centroid from XX is $5.0 - 0.424$, i.e. 4.576 cm. The distance moved through in one revolution by the centroid is $2\pi(4.576)$ cm.

Area of semicircle $= \dfrac{\pi r^2}{2} = \dfrac{\pi(1.0)^2}{2} = \dfrac{\pi}{2}\ \text{cm}^2$

By the theorem of Pappus,

volume generated = area × distance moved by centroid

$$= \left(\frac{\pi}{2}\right)(2\pi 4.576)$$

i.e. **volume of metal removed = 45.16 cm³. Volume of pulley**

= volume of cylindrical disc − volume of metal removed

$= \pi(5.0)^2(2.0) - 45.16$

$= \mathbf{111.9\ cm^3}$

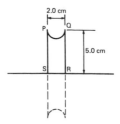

Fig. 18.15

Further problems on the theorem of Pappus may be found in section 18.5, Problems 16 to 21, page 173.

18.5 Further problems on areas and volumes

The ellipse

1 The major axis of an ellipse is 200 mm and the minor axis is 125 mm. Determine the perimeter and the area of the ellipse.
[510.5 mm, 19 630 mm²]

2 A regular elliptical closed cylinder has diameters of 12.0 cm (major axis) and 8.0 cm (minor axis) and the perpendicular height is 15.0 cm. Determine the total surface area and the volume of the cylinder.
[622 cm², 1131 cm³]

3 An elliptical fish pond has an area of 120 m². If its greatest length is 25.0 m, find the perimeter of the pond.
[48.87 m]

4 Determine the cost, correct to the nearest pound, of enclosing an elliptical plot of land, having major and minor diameter lengths of 80.2 m and 30.0 m, with fencing costing £8 per metre length.
[£1385]

5 An ellipse has an area of 125 cm² and a perimeter of 64 cm. Determine the maximum length and maximum breadth of the ellipse.
[36.36 cm, 4.38 cm]

6 A greyhound track is in the form of an ellipse, the axes being 160 m and 100 m, respectively, for the inner boundary and 175 m and 115 m for the outer boundary. Calculate the area of the track.
[3240 m²]

Frustums and zones of spheres

7 Determine the volume and surface area of a frustum of a sphere of diameter 47.85 cm if the radii of the ends of the frustum are 14.0 cm and 22.0 cm and the height of the frustum is 10.0 cm.
[11 210 cm³, 1503 cm²]

8 Determine the volume (in cm³) and the surface area (in cm²) of a frustum of a sphere if the diameter of the ends are 80.0 mm and 120.0 mm and the thickness is 30.0 mm.
[259.2 cm³, 118.3 cm²]

9 A sphere has a radius of 6.50 cm. Determine its volume and surface area. A frustum of the sphere is formed by two parallel planes, one through the diameter and the other at a distance h from the diameter. If the curved surface area of the frustum is to be $\frac{1}{5}$ of the surface area of the sphere, find the height h and the volume of the frustum.
[1150 cm³, 531 cm², 2.60 cm, 326.7 cm³]

10 A sphere has a diameter of 32.0 mm. Calculate the volume (in cm³) of the frustum of the sphere contained between two parallel planes distances 12.0 mm and 10.0 mm from the centre and on opposite sides of it.
[14.84 cm³]

11 A spherical storage tank is filled with liquid to a depth of 30.0 cm. If the inner diameter of the vessel is 45.0 cm determine the number of litres of liquid in the container (1 litre = 1000 cm³).
[35.34 l]

The prismoidal rule

12 Use the prismoidal rule to find the volume of a frustum of a sphere contained between two parallel planes on opposite sides of the centre each of radius 7.0 cm and each 4.0 cm from the centre.
[1500 cm³]

13 Determine the volume of a cone of perpendicular height 16.0 cm and base diameter 10.0 cm by using the prismoidal rule.
[418.9 cm³]

14 A bucket is in the form of a frustum of a cone. The diameter of the base is 28.0 cm and the diameter of the top is 42.0 cm. If the depth is 32.0 cm, determine the capacity of the bucket (in litres) using the prismoidal rule (1 litre = 1000 cm³).
[31.20 litre]

15 Determine the capacity of a water reservoir, in litres, the top being a 30.0 m by 12.0 m rectangle, the bottom being a 20.0 m by 8.0 m rectangle and the depth being 5.0 m (1 litre = 1000 cm³).
[1.267 × 10⁶ litre]

Theorem of Pappus

16 A right-angled isosceles triangle having a hypotenuse of 8 cm is revolved one revolution about one of its equal sides as axis. Determine the volume of the solid generated using Pappus' theorem.
[189.6 cm³]

17 A rectangle measuring 10.0 cm by 6.0 cm rotates one revolution about one of its longest sides as axis. Determine the volume of the resulting cylinder by using the theorem of Pappus.
[1131 cm²]

18 Determine the volume of a metal ring formed by rotating a circle of radius 9.0 cm about an axis at a perpendicular distance of 20.0 cm from its centre.
[31 980 cm³]

19 A steel ring is formed by rotating an ellipse having major and minor axes of 10.0 cm and 7.0 cm about an axis at a perpendicular distance of 30.0 cm from its centre. Determine the volume of the ring using the theorem of Pappus.
[10 360 cm³]

20 A metal disc has a radius of 7.0 cm and is of thickness

2.5 cm. A semicircular groove of diameter 2.0 cm is machined centrally around the rim to form a pulley. Determine the volume of metal removed using Pappus' theorem and express this as a percentage of the original volume of the disc.

[64.90 cm³, 16.86%]

21 One edge of an equilateral triangle of side 10.0 cm is parallel to an axis XX lying at a distance of 12.0 cm from the edge and the apex is on the opposite side to the axis. If the triangle is rotated 360° about axis XX, determine the volume of the solid generated.

[4050 cm³]

19

An introduction to trigonometry

19.1 Trigonometry

Trigonometry is the branch of mathematics which deals with the measurement of sides and angles of triangles, and their relationships with each other.

19.2 Trigonometric ratios of acute angles

(a) With reference to the right-angled triangle shown in *Fig. 19.1*:

(i) sine $\theta = \dfrac{\text{opposite side}}{\text{hypotenuse}}$, i.e. $\sin \theta = \dfrac{b}{c}$

(ii) cosine $\theta = \dfrac{\text{adjacent side}}{\text{hypotenuse}}$, i.e. $\cos \theta = \dfrac{a}{c}$

(iii) tangent $\theta = \dfrac{\text{opposite side}}{\text{adjacent side}}$, i.e. $\tan \theta = \dfrac{b}{a}$

(iv) secant $\theta = \dfrac{\text{hypotenuse}}{\text{adjacent side}}$, i.e. $\sec \theta = \dfrac{c}{a}$

(v) cosecant $\theta = \dfrac{\text{hypotenuse}}{\text{opposite side}}$, i.e. $\text{cosec } \theta = \dfrac{c}{b}$

(vi) cotangent $\theta = \dfrac{\text{adjacent side}}{\text{opposite side}}$, i.e. $\cot \theta = \dfrac{a}{b}$

(b) From above,

(i) $\dfrac{\sin \theta}{\cos \theta} = \dfrac{\dfrac{b}{c}}{\dfrac{a}{c}} = \dfrac{b}{a} = \tan \theta$, i.e. $\tan \theta = \dfrac{\sin \theta}{\cos \theta}$

Fig. 19.1   **Fig. 19.2**

(ii) $\dfrac{\cos \theta}{\sin \theta} = \dfrac{\dfrac{a}{c}}{\dfrac{b}{c}} = \dfrac{a}{b} = \cot \theta$, i.e. $\cot \theta = \dfrac{\cos \theta}{\sin \theta}$

(iii) $\sec \theta = \dfrac{1}{\cos \theta}$

(iv) $\text{cosec } \theta = \dfrac{1}{\sin \theta}$ (Note: 's' and 'c' go together)

(v) $\cot \theta = \dfrac{1}{\tan \theta}$

Secants, cosecants and cotangents are called the **reciprocal ratios**.

> *Problem 1*   From *Fig. 19.2*, find sin *D*, cos *D* and tan *F*.

By Pythagoras' theorem, $17^2 = 8^2 + \text{EF}^2$, from which,

$\text{EF} = \sqrt{(17^2 - 8^2)} = 15$

$$\sin D = \frac{\text{EF}}{\text{DF}} = \frac{15}{17} \text{ or } \mathbf{0.882\ 4}$$

$$\cos D = \frac{\text{DE}}{\text{DF}} = \frac{8}{17} \text{ or } \mathbf{0.470\ 6}$$

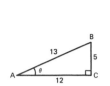

Not to scale

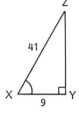

Fig. 19.3　　　　　　　　**Fig. 19.4**

$$\tan F = \frac{DE}{EF} = \frac{8}{15} \text{ or } \mathbf{0.533\ 3}$$

Problem 2　Determine the values of the six trigonometric ratios for angle θ shown in the right-angled triangle ABC shown in *Fig. 19.3*.

By definition:

$$\sin \theta = \frac{\text{opposite side}}{\text{hypotenuse}} = \frac{5}{13} = \mathbf{0.384\ 6}$$

Hence

$$\operatorname{cosec} \theta = \frac{13}{5} = \mathbf{2.600\ 0}$$

$$\cos \theta = \frac{\text{adjacent side}}{\text{hypotenuse}} = \frac{12}{13} = \mathbf{0.923\ 1}$$

Hence

$$\sec \theta = \frac{13}{12} = \mathbf{1.083\ 3}$$

$$\tan \theta = \frac{\text{opposite side}}{\text{adjacent side}} = \frac{5}{12} = \mathbf{0.416\ 7}$$

Hence

$$\cot \theta = \frac{12}{5} = \mathbf{2.400\ 0}$$

Problem 3　If $\cos X = \dfrac{9}{41}$ determine the values of the other five trigonometric ratios.

Figure 19.4 shows a right-angled triangle. Since $\cos X = \dfrac{9}{41}$, then XY = 9 units and XZ = 41 units.

Using Pythagoras' theorem: $41^2 = 9^2 + YZ^2$ from which

$YZ = \sqrt{(41^2 - 9^2)} = 40$ units.

Thus

$$\sin X = \frac{40}{41}, \ \tan X = \frac{40}{9} = 4\frac{4}{9}$$

$$\operatorname{cosec} X = \frac{41}{40} = 1\frac{1}{40}, \ \sec X = \frac{41}{9} = 4\frac{5}{9}$$

and

$$\cot X = \frac{9}{40}$$

Problem 4　If $\sin \theta = 0.625$ and $\cos \theta = 0.500$ determine, without using trigonometrical tables or calculators, the values of $\operatorname{cosec} \theta$, $\sec \theta$, $\tan \theta$ and $\cot \theta$.

$$\operatorname{cosec} \theta = \frac{1}{\sin \theta} = \frac{1}{0.625} = \mathbf{1.60}$$

$$\sec \theta = \frac{1}{\cos \theta} = \frac{1}{0.500} = \mathbf{2.00}$$

$$\tan \theta = \frac{\sin \theta}{\cos \theta} = \frac{0.625}{0.500} = \mathbf{1.25}$$

$$\cot \theta = \frac{\cos \theta}{\sin \theta} = \frac{0.500}{0.625} = \mathbf{0.80}$$

Further problems on trigonometric ratios of acute angles may be found in section 19.10, Problems 1 to 7, page 185.

19.3　The production of a sine and cosine wave

In *Fig. 19.5*, let OR be a vector 1 unit long and free to rotate anticlockwise about 0. In one revolution a circle is produced and is shown with 15° sectors. Each radius arm has a vertical and a horizontal component. For example, at 30°, the vertical component is TS and the horizontal component is OS.

From trigonometric ratios,

$$\sin 30° = \frac{TS}{OT} = \frac{TS}{1}, \text{ i.e. TS} = \sin 30° \text{ and}$$

$$\cos 30° = \frac{OS}{OT} = \frac{OS}{1}, \text{ i.e. OS} = \cos 30°$$

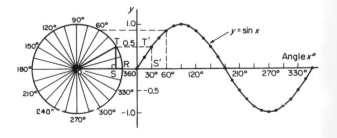

Fig. 19.5

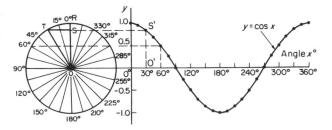

Fig. 19.6

The vertical component TS may be projected across to T'S', which is the corresponding value of 30° on the graph of y against angle x°. If all such vertical components as TS are projected on to the graph, then a **sine wave** is produced as shown in *Fig. 19.5*. If all horizontal components such as OS are projected on to a graph of y against angle x°, then a **cosine wave** is produced. It is easier to visualize these projections by redrawing the circle with the radius arm OR initially in a vertical position as shown in *Fig. 19.6*. From *Figs 19.5* and *19.6* it is seen that a cosine curve is of the same form as the sine curve but is displaced by 90° (or $\pi/2$ radians).

19.4   Evaluating trigonometric ratios

Four-figure tables are available which give sines, cosines, tangents, secants, cosecants and cotangents for angles between 0° and 90°. However, the easiest method of evaluating trigonometric functions of any angle is by using a **calculator**.

The following values, correct to 4 decimal places, may be checked:

sine 18° = 0.309 0, cosine 56° = 0.559 2
tangent 29° = 0.554 3, sine 172° = 0.139 2
cosine 115° = −0.422 6, tangent 178° = −0.034 9
sine 241.63° = −0.879 9, cosine 331.78° = 0.881 1
tangent 296.42° = −2.012 7

Most calculators contain only sine, cosine and tangent functions. Thus to evaluate secants, cosecants and cotangents, reciprocals need to be used.

The following values, correct to 4 decimal places, may be checked:

$$\text{secant } 32° = \frac{1}{\cosine 32°} = 1.179\ 2$$

$$\text{secant } 215.12° = \frac{1}{\cosine 215.12°} = -1.222\ 6$$

$$\text{cosecant } 75° = \frac{1}{\sine 75°} = 1.035\ 3$$

$$\text{cosecant } 321.62° = \frac{1}{\sine 321.62°} = -1.610\ 6$$

$$\text{cotangent } 41° = \frac{1}{\tangent 41°} = 1.150\ 4$$

$$\text{cotangent } 263.59° = \frac{1}{\tangent 263.59°} = 0.112\ 3$$

To evaluate, say, sine 42°23′ using a calculator means finding sine $42\dfrac{23°}{60}$ since there are 60 minutes in 1 degree.

$\dfrac{23}{60} = 0.383\ \dot{3}$, thus $42°23′ \equiv 42.383\ \dot{3}°$

Thus sine 42°23′ = sine 42.383 $\dot{3}$° = 0.674 1, correct to 4 decimal places.

Similarly, cosine 72°38′ = cosine $72\dfrac{38°}{60} = 0.298\ 5$, correct to 4 decimal places.

Problem 5   Evaluate, correct to 4 decimal places:

(a) sine 11°   (b) sine 121.68°   (c) sine 259°10′

(a) sine 11° = **0.190 8**
(b) sine 121.68° = **0.851 0**

(c) sine 259°10′ = sine $259\dfrac{10°}{60}$ = **−0.982 2**

Problem 6   Evaluate, correct to 4 decimal places:

(a) cosine 23°   (b) cosine 159.32°
(c) cosine 321°41′

(a) cosine 23° = **0.920 5**
(b) cosine 159.32° = **−0.935 6**

(c) cosine 321°41′ = cosine $321\dfrac{41°}{60}$ = **0.784 6**

Problem 7   Evaluate, correct to 4 significant figures:

(a) tangent 276°   (b) tangent 131.29°
(c) tangent 76°58′

(a) tan 276° = **−9.514**
(b) tan 131.29° = **−1.139**

(c) tan 76°58′ = tan $76\dfrac{58°}{60}$ = **4.320**

Problem 8   Evaluate, correct to 4 decimal places:

(a) secant 161°   (b) secant 22.45°   (c) secant 302°29′

(a)   sec 161° = 1/cos 161° = **−1.057 6**

(b)   sec 22.45° = 1/cos 22.45° = **1.082 0**

(c)   sec 302°29′ = 1/cos 302°29′ = 1/cos $302\dfrac{29°}{60}$ = **1.862 0**

Problem 9   Evaluate, correct to 4 significant figures:

(a) cosecant 97°   (b) cosecant 279.16°
(c) cosecant 49°7′

(a)   cosec 97° = 1/sin 97° = **1.008**

(b)   cosec 279.16° = 1/sin 279.16° = **−1.013**

(c)   cosec 49°7′ = 1/sin 49°7′ = 1/sin $49\dfrac{7}{60}°$ = **1.323**

Problem 10   Evaluate, correct to 4 decimal places:

(a) cotangent 341°   (b) cotangent 17.49°
(c) cotangent 163°52′

(a)   cot 341° = 1/tan 341° = **−2.904 2**

(b)   cot 17.49° = 1/tan 17.49° = **3.173 5**

(c)   cot 163°52′ = 1/tan 163°52′ = 1/tan $163\dfrac{52°}{60}$ = **−3.457 0**

Problem 11   Evaluate, correct to 4 significant figures:

(a) sin 2.162   (b) cos $(3\pi/8)$ (c) tan 1.16

(a) sin 2.162 means the sine of 2.162 radians. Hence a calculator needs to be on the radian function. Hence sin 2.162 = **0.830 3**

(b) cos $(3\pi/8)$ = cos 1.178 097 . . . = **0.382 7**

(c) tan 1.16 = **2.296**

Problem 12   Evaluate, correct to 4 decimal places:

(a) secant 5.37   (b) cosecant $\pi/4$
(c) cotangent $=\pi/24$

(a) Again, with no degrees sign, it is assumed that 5.37 means 5.37 radians. Hence sec 5.37 = 1/cos 5.37 = **1.004 4**

(b)   cosec $(\pi/4)$ = 1/sin $(\pi/4)$ = 1/sin 0.785 398 . . . = **1.273 2**

(c)   cot $(5\pi/24)$ = 1/tan $(5\pi/24)$ = 1/tan 0.654 498 . . . = **1.527 9**

Problem 13   Determine the acute angle:

(a) arcsin 0.732 1   (b) arccos 0.417 4
(c) arctan 1.469 5

(a) Note that 'arcsin θ' is an abbreviation for 'the angle whose sine is equal to θ'. 0.732 1 is entered into a calculator and then the inverse sine (or sin^{-1}) key is pressed. Hence arcsin 0.732 1 = 47.062 73 . . . ° Subtracting 47 leaves 0.062 73 . . . ° Multiplying by 60 gives 4′ to the nearest minute.
Hence arcsin 0.732 1 = **47.06°** or **47°4′**.
Alternatively, in radians, arcsin 0.732 1 = **0.821 radians**.

(b)   arccos 0.417 4 = **65.33°** or **65°20′** or **1.140 radians**.

(c)   arctan 1.469 5 = **55.76°** or **55°46′** or **0.973 radians**.

Problem 14   Determine the acute angles:

(a) arcsec 2.316 4   (b) arccosec 1.178 4
(c) arccot 2.127 3

(a)   arcsec 2.316 4 = arccos $\left(\dfrac{1}{2.316\ 4}\right)$

= arccos 0.431 7 . . .
= **64.42°** or **64°25′** or **1.124 radians**

(b)   arccosec 1.178 4 = arcsin $\left(\dfrac{1}{1.178\ 4}\right)$

= arcsin 0.848 6 . . .
= **58.06°** or **58°4′** or **1.013 radians**

(c)   arccot 2.127 3 = arctan $\left(\dfrac{1}{2.127\ 3}\right)$

= arctan 0.470 0 . . .
= **25.18°** or **25°11′** or **0.439 radians**

Problem 15   Evaluate $\dfrac{4.2 \tan 49°26′ - 3.7 \sin 66°1′}{7.1 \cos 29°34′}$

correct to 3 significant figures.

By calculator, tan 49°26′ = tan $\left(49\dfrac{26}{60}\right)°$ = 1.168 1, sin 66°1′
= 0.913 6 and cos 29°34′ = 0.869 8. Hence

$$\frac{4.2 \tan 49°26' - 3.7 \sin 66°1'}{7.1 \cos 29°34'}$$

$$= \frac{(4.2 \times 1.168\ 1) - (3.7 \times 0.913\ 6)}{(7.1 \times 0.869\ 8)}$$

$$= \frac{4.906\ 0 - 3.380\ 3}{6.175\ 6} = \frac{1.525\ 7}{6.175\ 6}$$

$= 0.247\ 1 = \mathbf{0.247}$, correct to 3 significant figures

---

Problem 16 Evaluate the following expression, correct to 4 significant figures:

$$\frac{4 \sec 32°10' - 2 \cot 15°19'}{3 \cosec 63°8' \tan 14°57'}$$

By calculator:

$\sec 32°10' = 1.181\ 4$, $\cot 15°19' = 3.651\ 3$
$\cosec 63°8' = 1.121\ 0$, $\tan 14°57' = 0.267\ 0$

Hence

$$\frac{4 \sec 32°10' - 2 \cot 15°19'}{3 \cosec 63°8' \tan 14°57'}$$

$$= \frac{4(1.181\ 4) - 2(3.651\ 3)}{3(1.121\ 0)(0.267\ 0)}$$

$$= \frac{4.725\ 6 - 7.302\ 6}{0.897\ 9} = \frac{-2.577\ 0}{0.897\ 9}$$

$= -2.870$, correct to 4 significant figures

Problem 17 Evaluate correct to 4 decimal places:
(a) sine (−112°) (b) tangent (−217.29°)
(c) secant (−93°16′)

(a) Positive angles are shown anticlockwise and negative angles are shown clockwise. From *Fig. 19.7*, −112° is actually the same as +248° (i.e. 360° − 112°). Hence by calculator sine (−112°) = sine 248° = **−0.927 2**.
(b) tangent (−217.29°) = **−0.761 5** (which is the same as tan (360° − 217.29°), i.e. tan 142.71°).
(c) secant (−93°16′) = $\dfrac{1}{\text{cosine } (-93\frac{16°}{60})}$ = **−17.549 0**

Further problems on evaluating trigonometric ratios may be found in section 19.10, Problems 8 to 23, page 186.

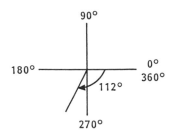

Fig. 19.7

19.5 Graphs of trigonometric functions

By drawing up tables of values from 0° to 360°, graphs of $y = \sin A$, $y = \cos A$ and $y = \tan A$ may be plotted. Values obtained with a calculator (correct to 3 decimal places – which is more than sufficient for plotting graphs), using 30° intervals, are shown below, with the respective graphs shown in *Fig. 19.8*.

(a) $y = \sin A$

A	0	30°	60°	90°	120°	150°	180°
sin A	0	0.500	0.866	1.000	0.866	0.500	0

A		210°	240°	270°	300°	330°	360°
sin A	0	−0.500	−0.866	−1.000	−0.866	−0.500	0

(b) $y = \cos A$

A	0	30°	60°	90°	120°	150°	180°
cos A	1.000	0.866	0.500	0	−0.500	−0.866	−1.000

A		210°	240°	270°	300°	330°	360°
cos A		−0.866	−0.500	0	0.50	0.866	1.000

(c) $y = \tan A$

A	0	30°	60°	90°	120°	150°	180°
tan A	0	0.577	1.732	∞	−1.732	−0.577	0

A		210°	240°	270°	300°	330°	360°
tan A		0.577	1.732	∞	−1.732	−0.577	0

From *Fig. 19.8* it is seen that:

(i) Sine and cosine graphs oscillate between peak values of ±1.

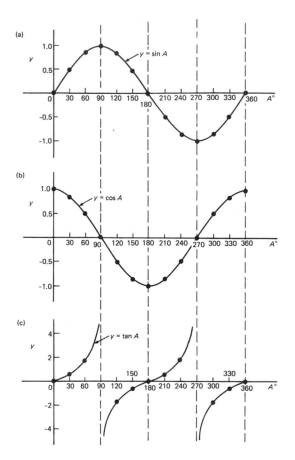

(a)

$y = \sin A$

(b)

$y = \cos A$

(c)

$y = \tan A$

Fig. 19.8

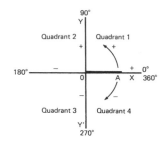

Fig. 19.9

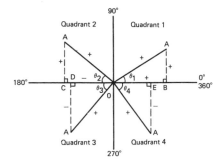

Fig. 19.10

(ii) The cosine curve is the same shape as the sine curve but displaced by 90°.
(iii) The sine and cosine curves are continuous and they repeat at intervals of 360°; the tangent curve appears to be discontinuous and repeats at intervals of 180°.

19.6 Angles of any magnitude

(i) *Fig. 19.9* shows rectangular axes XX′ and YY′ intersecting at origin 0. As with graphical work, measurements made to the right and above 0 are positive while those to the left and downwards are negative. Let OA be free to rotate about 0. By convention, when OA moves anticlockwise angular measurement is considered positive, and vice-versa.

(ii) Let OA be rotated anticlockwise so that θ_1 is any angle in the first quadrant and let perpendicular AB be constructed to form the right-angled triangle OAB (see *Fig. 19.10*). Since all three sides of the triangle are positive, all six trigonometric ratios are positive in the first quadrant. (*Note*: OA is always positive since it is the radius of a circle.)

(iii) Let OA be further rotated so that θ_2 is any angle in the second quadrant and let AC be constructed to form the right-angled triangle OAC. Then:

$$\sin \theta_2 = \frac{+}{+} = +, \cos \theta_2 = \frac{-}{+} = -, \tan \theta_2 = \frac{+}{-} = -$$

$$\operatorname{cosec} \theta_2 = \frac{+}{+} = +, \sec \theta_2 = \frac{+}{-} = -, \cot \theta_2 = \frac{-}{+} = -$$

(iv) Let OA be further rotated so that θ_3 is any angle in the third quadrant and let AD be constructed to form the right-angled triangle OAD. Then:

$$\sin \theta_3 = \frac{-}{+} = - \quad \text{(and hence cosec } \theta_3 \text{ is } -)$$

$$\cos \theta_3 = \frac{-}{+} = - \quad \text{(and hence sec } \theta_3 \text{ is } -)$$

$$\tan \theta_3 = \frac{-}{-} = + \quad \text{(and hence cot } \theta_3 \text{ is } +)$$

(v) Let OA be further rotated so that θ_4 is any angle in the fourth quadrant and let AE be constructed to form the right-angled triangle OAE. Then:

$$\sin \theta_4 = \frac{-}{+} = - \quad \text{(and hence cosec } \theta_4 \text{ is } -)$$

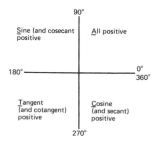

Fig. 19.11

$$\cos \theta_4 = \frac{+}{+} = + \ \text{(and hence sec } \theta_4 \text{ is +)}$$

$$\tan \theta_4 = \frac{-}{+} = - \ \text{(and hence cot } \theta_4 \text{ is -)}$$

(vi) The results obtained in (ii) to (v) are summarized in *Fig. 19.11*. The letters underlined spell the word CAST when starting in the fourth quadrant and moving in an anticlockwise direction.

(vii) In the first quadrant of *Fig. 19.8* all the curves have positive values; in the second only sine is positive; in the third only tangent is positive; in the fourth only cosine is positive (exactly as summarized in *Fig. 19.11*).

A knowledge of angles of any magnitude is needed when finding, for example, all the angles between 0° and 360° whose sine is, say, 0.326 1. If 0.326 1 is entered into a calculator and then the inverse sine key pressed (or sin⁻¹ key) the answer 19.03° appears. However there is a second angle between 0° and 360° which the calculator does not give. Sine is also positive in the second quadrant (either from CAST or from *Fig. 19.8 (a)*). The other angle is shown in *Fig. 19.12* as angle θ where $\theta = 180° - 19.03° = 160.97°$. Thus 19.03° **and** 160.97° are the angles between 0° and 360° whose sine is 0.326 1 (check that sin 160.97° = 0.326 1 on your calculator).

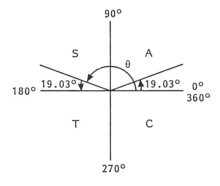

Fig. 19.12

Be careful! Your calculator only gives you one of these answers. The second answer needs to be deduced from a knowledge of angles of any magnitude, as shown in the following problems.

Problem 18 Determine all the angles between 0° and 360° (a) whose sine is −0.463 8 and (b) whose tangent is 1.762 9.

(a) The angles whose sine is −0.463 8 occurs in the third and fourth quadrants since sine is negative in these quadrants (see *Fig. 19.13(a)*). From *Fig. 19.13(b)*, $\theta =$ arcsin 0.463 8 = 27°38′.
Measured from 0°, the two angles between 0° and 360° whose sine is −0.463 8 are 180° + 27°38′, i.e. **207°38′** and 360° − 27°38′, i.e. **332°22′**. (Note that a calculator generally only gives one answer, i.e. − 27.632 588°.)

(b) A tangent is positive in the first and third quadrants (see *Fig. 19.13(c)*). From *Fig. 19.13(d)*, $\theta =$ arctan 1.762 9 = 60°26′. Measured from 0°, the two angles between 0° and 360° whose tangent is 1.762 9 are **60°26′** and 180° + 60°26′, i.e. **240°26′**.

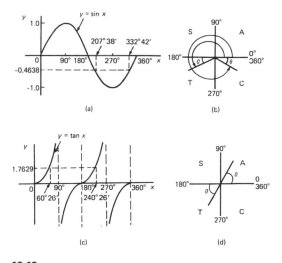

Fig. 19.13

Problem 19 Solve for angles of α between 0° and 360°: (a) arcsec (−2.149 9) = α (b) arccot 1.311 1 = α.

(a) Secant is negative in the second and third quadrants (i.e. the same as for cosine). From *Fig. 19.14(a)*, $\theta =$ arcsec 2.149 9 = 62°17′. Measured from 0°, the two angles between 0° and 360° whose secant is −2.149 9 are

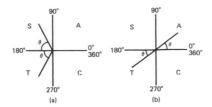

Fig. 19.14

$$\alpha = 180° - 62°17' = \mathbf{117°43'}$$
and $\quad \alpha = 180° + 62°17' = \mathbf{242°17'}$

(b) Cotangent is positive in the first and third quadrants (i.e. same as for tangent). From *Fig. 19.14(b)*, θ = arccot 1.311 1 = 37°20' . Hence

$$\alpha = \mathbf{37°20'} \text{ and } 180° + 37°20' = \mathbf{217°20'}$$

Further problems on evaluating trigonometric ratios of any magnitude may be found in section 19.10, Problems 24 and 25, page 186.

19.7 Fractional and surd forms of trigonometric ratios

In *Fig. 19.15*, ABC is an equilateral triangle of side 2 units. AD bisects angle A and bisects the side BC. Using Pythagoras' theorem on triangle ABD gives: AD = $\sqrt{(2^2 - 1^2)} = \sqrt{3}$. Hence

$$\sin 30° = \frac{BD}{AB} = \frac{1}{2} \quad \cos 30° = \frac{AD}{AB} = \frac{\sqrt{3}}{2}$$

$$\tan 30° = \frac{BD}{AD} = \frac{1}{\sqrt{3}}$$

$$\sin 60° = \frac{AD}{AB} = \frac{\sqrt{3}}{2} \quad \cos 60° = \frac{BD}{AB} = \frac{1}{2}$$

$$\tan 60° = \frac{AD}{BD} = \sqrt{3}$$

In *Fig. 19.16*, PQR is an isosceles triangle with PQ = QR = 1 unit. By Pythagoras' theorem, PR = $\sqrt{(1^2 + 1^2)} = \sqrt{2}$.

Fig. 19.15   **Fig. 19.16**

Hence

$$\sin 45° = \frac{1}{\sqrt{2}} \quad \cos 45° = \frac{1}{\sqrt{2}} \quad \tan 45° = 1$$

A quantity which is not exactly expressible as a rational number is called a **surd**. For example, $\sqrt{2}$ and $\sqrt{3}$ are called surds because they cannot be expressed as a fraction and the decimal part may be continued indefinitely. For example, $\sqrt{2} = 1.414\ 213\ 5....$
From above, sin 30° = cos 60°, sin 45° = cos 45° and sin 60° = cos 30°.

In general, $\quad \mathbf{\sin \theta = \cos (90° - \theta)}$
and $\qquad \mathbf{\cos \theta = \sin (90° - \theta)}$

For example, it may be checked by calculator that sin 25° = cos 65°, sin 42° = cos 48°, cos 84°10' = sin 5°50', and so on.

> *Problem 20* Using surd forms, evaluate
> $$\frac{3 \tan 60° - 2 \cos 30°}{\tan 30°}$$

From above, tan 60° = $\sqrt{3}$, cos 30° = $\sqrt{3}/2$ and tan 30° = $1/\sqrt{3}$. Hence

$$\frac{3 \tan 60° - 2 \cos 30°}{\tan 30°} = \frac{3(\sqrt{3}) - 2(\frac{\sqrt{3}}{2})}{\frac{1}{\sqrt{3}}} = \frac{3\sqrt{3} - \sqrt{3}}{\frac{1}{\sqrt{3}}}$$

$$= \frac{2\sqrt{3}}{\frac{1}{\sqrt{3}}} = 2\sqrt{3}\left(\frac{\sqrt{3}}{1}\right)$$

$$= 2(3) = \mathbf{6}$$

A further problem on the fractional and surd forms of trigonometric ratios may be found in section 19.10, Problem 26, page 186.

19.8 Solution of right-angled triangles

To 'solve a right-angled triangle' means 'to find the unknown sides and angles'. This is achieved by using (i) the theorem of Pythagoras, and/or (ii) trigonometric ratios.

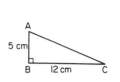

Fig. 19.17

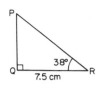

Fig. 19.18

Problem 21 Sketch a right-angled triangle ABC such that $\angle B = 90°$, AB = 5 cm and BC = 12 cm. Determine the length of AC and hence evaluate sin C, cos C and tan C.

Triangle ABC is shown in *Fig. 19.17*. By Pythagoras' theorem, AC = $\sqrt{(5^2 + 12^2)}$ = 13. By definition:

$$\sin C = \frac{\text{opposite side}}{\text{hypotenuse}} = \frac{5}{13} \text{ or } \mathbf{0.384\ 6}$$

$$\cos C = \frac{\text{adjacent side}}{\text{hypotenuse}} = \frac{12}{13} \text{ or } \mathbf{0.923\ 1}$$

$$\tan C = \frac{\text{opposite side}}{\text{adjacent side}} = \frac{5}{12} \text{ or } \mathbf{0.416\ 7}$$

Problem 22 In triangle PQR shown in *Fig. 19.18*, find the lengths of PQ and PR.

---

tan 38° = PQ/QR = PQ/7.5. Hence

$$PQ = 7.5 \tan 38° = 7.5(0.781\ 3) = \mathbf{5.860\ cm}$$

cos 38° = QR/PR = 7.5/PR. Hence

$$PR = \frac{7.5}{\cos 38°} = \frac{7.5}{0.788\ 0} = \mathbf{9.518\ cm}$$

(Check: Using Pythagoras' theorem $(7.5)^2 + (5.860)^2 = 90.59 = (9.518)^2$.)

Problem 23 Solve the triangle ABC shown in *Fig. 19.19*.

To 'solve triangle ABC' means 'to find the length AC and angles B and C'.

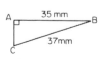

Fig. 19.19

Fig. 19.20

$$\sin C = \frac{35}{37} = 0.945\ 9$$

Hence

$$C = \arcsin 0.945\ 9 = \mathbf{71°4'}$$

$B = 180° - 90° - 71°4' = \mathbf{18°56'}$ (since angles in a triangle add up to 180°)

$$\sin B = \frac{AC}{37}$$

hence

$$AC = 37 \sin 18°56' = 37(0.324\ 5) = \mathbf{12.0\ mm}$$

(Check: Using Pythagoras' theorem $37^2 = 35^2 + 12^2$.)

Problem 24 Solve triangle XYZ given $\angle X = 90°$, $\angle Y = 23°17'$ and YZ = 20.0 mm. Determine also its area.

It is always advisable to make a reasonably accurate sketch so as to visualize the expected magnitudes of unknown sides and angles. Such a sketch is shown in *Fig. 19.20*.

$$\angle Z = 180° - 90° - 23°17' = \mathbf{66°43'}$$

$$\sin 23°17' = \frac{XZ}{20.0}$$

Hence

$$XZ = 20.0 \sin 23°17'$$
$$= 20.0(0.395\ 3) = \mathbf{7.906\ mm}$$

$$\cos 23°17' = \frac{XY}{20.0}$$

Hence

$$XY = 20.0 \cos 23°17'$$
$$= 20.0(0.918\ 5) = \mathbf{18.37\ mm}$$

(Check: Using Pythagoras' theorem $(18.37)^2 + (7.906)^2 = 400.0 = (20.0)^2$.)

Area of triangle XYZ = $\frac{1}{2}$(base)(perpendicular height)

$$= \frac{1}{2}(XY)(XZ)$$

$$= \frac{1}{2}(18.37)(7.906)$$

$$= \mathbf{72.62\ mm^2}$$

Further problems on the solution of right-angled triangles may be found in section 19.10, Problems 27 to 29, page 186.

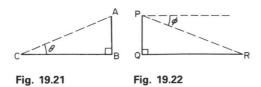

Fig. 19.21   **Fig. 19.22**

19.9 Angles of elevation and depression

(a) If, in *Fig. 19.21*, BC represents horizontal ground and AB a vertical flagpole, then the **angle of elevation** of the top of the flagpole, A, from the point C is the angle that the imaginary straight line AC must be raised (or elevated) from the horizontal CB, i.e. angle θ.

(b) If, in *Fig. 19.22*, PQ represents a vertical cliff and R a ship at sea, then the **angle of depression** of the ship from point P is the angle through which the imaginary straight line PR must be lowered (or depressed) from the horizontal to the ship, i.e. angle ϕ. (Note, $\angle$PRQ is also ϕ – alternate angles between parallel lines.)

Problem 25 An electricity pylon stands on horizontal ground. At a point 80 m from the base of the pylon, the angle of elevation of the top of the pylon is 23°. Calculate the height of the pylon to the nearest metre.

Figure 19.23 shows the pylon AB and the angle of elevation of A from point C is 23°.

$$\tan 23° = \frac{AB}{BC} = \frac{AB}{80}$$

Hence height of pylon AB = 80 tan 23°

$\qquad\qquad\qquad\qquad = 80(0.424\ 5) = 33.96$ m

$\qquad\qquad\qquad\qquad = $ **34 m to the nearest metre**

Problem 26 A surveyor measures the angle of elevation of the top of a perpendicular building as 19°. He moves 120 m nearer the building and finds the angle of elevation is now 47°. Determine the height of the building.

Fig. 19.23

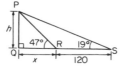

Fig. 19.24

The building PQ and the angles of elevation are shown in *Fig. 19.24*. In triangle PQS

$$\tan 19° = \frac{h}{x + 120}$$

hence $h = \tan 19°(x + 120)$, i.e.

$$h = 0.344\ 3(x + 120) \qquad (1)$$

In triangle PQR

$$\tan 47° = \frac{h}{x}$$

hence $h = \tan 47°(x)$, i.e.

$$h = 1.072\ 4x \qquad (2)$$

Equating equations (1) and (2) gives:

$\qquad 0.344\ 3(x + 120) = 1.072\ 4x$

$\qquad 0.344\ 3x + (0.344\ 3)(120) = 1.072\ 4x$

$\qquad (0.344\ 3)(120) = (1.072\ 4 - 0.344\ 3)x$

$\qquad 41.316 = 0.728\ 1x$

$$x = \frac{41.316}{0.728\ 1} = 56.74 \text{ m}$$

From equation (2), height of building $h = 1.072\ 4x = 1.072\ 4(56.74) = $ **60.85 m**.

Problem 27 A surveyor measures the angle of elevation of the top of a building as 17°15′. He moves 80.0 m nearer to the building and measures the angle of elevation as 29°35′. Determine the perpendicular height of the building.

In *Fig. 19.25*, AB represents the perpendicular height of the building, C is the initial position and D the final position of the surveyor.

This is a similar problem to the previous one but is determined using cotangents instead of tangents. Either method is acceptable.

Using trigonometric ratios, cot 17°15′ = AC/h, from which AC = h cot 17°15′.

Similarly, AD = h cot 29°35′, and since DC = AC − AD

$$DC = h \cot 17°15′ - h \cot 29°35′$$
$$= h(\cot 17°15′ - \cot 29°35′) = h(1.459\ 4)$$

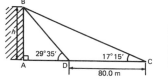

Fig. 19.25

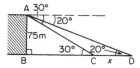

Fig. 19.26

Hence the perpendicular height of the building,

$$h = \frac{80.0}{1.459\,4} = \textbf{54.82 m}$$

Problem 28 The angle of depression of a ship viewed at a particular instant from the top of a 75 m vertical cliff is 30°. Find the distance of the ship from the base of the cliff at this instant. The ship is sailing away from the cliff at constant speed and 1 minute later its angle of depression from the top of the cliff is 20°. Determine the speed of the ship in km/h.

Fig 19.26 shows the cliff AB, the initial position of the ship at C and the final position at D. Since the angle of depression is initially 30° then ∠ACB = 30° (alternate angles between parallel lines).

$$\tan 30° = \frac{AB}{BC} = \frac{75}{BC}$$

Hence

$$BC = \frac{75}{\tan 30°} = \frac{75}{0.577\,4}$$

$$= \textbf{129.9 m} = \textbf{initial position of ship}$$

In triangle ABD,

$$\tan 20° = \frac{AB}{BD} = \frac{75}{BC + CD} = \frac{75}{129.9 + x}$$

Hence

$$129.9 + x = \frac{75}{\tan 20°} = \frac{75}{0.364\,0} = 206.0 \text{ m}$$

from which $x = 206.0 - 129.9 = 76.1$ m

Thus the ship sails 76.1 m in 1 minute, i.e. 60 s, hence speed of ship

$$= \frac{\text{distance}}{\text{time}} = \frac{76.1}{60} \text{ m/s} = \frac{76.1 \times 60 \times 60}{60 \times 1000} \text{ km/h}$$

$$= \textbf{4.566 km/h}$$

Further problems on angles of elevation and depression may be found in section 19.10, Problem 30 to 38, page 187.

19.10 Further problems on introduction to trigonometry

Trigonometric ratios of acute angles

1 Sketch a triangle XYZ such that ∠Y = 90°, XY = 9 cm and YZ = 40 cm. Determine sin Z, cos Z, tan X and cos X.

$$\left[\sin Z = \frac{9}{41}, \cos Z = \frac{40}{41}, \tan X = \frac{40}{9}, \cos X = \frac{9}{41}\right]$$

2 In triangle ABC shown in *Fig. 19.27*, find sin A, cos A, tan A, sin B, cos B and tan B.

$$\left[\sin A = \frac{3}{5}, \cos A = \frac{4}{5}, \tan A = \frac{3}{4},\right.$$
$$\left.\sin B = \frac{4}{5}, \cos B = \frac{3}{5}, \tan B = \frac{4}{3}\right]$$

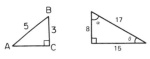

Fig. 19.27   **Fig. 19.28**

3 If $\cos A = \frac{15}{17}$ find sin A and tan A, in fraction form.

$$\left[\sin A = \frac{8}{17}, \tan A = \frac{8}{15}\right]$$

4 If $\tan X = \frac{15}{112}$, find sin X and cos X, in fraction form.

$$\left[\sin X = \frac{15}{113}, \cos X = \frac{112}{113}\right]$$

5 For the right-angled triangle shown in *Fig. 19.28*, find (a) sin α (b) cos θ (c) sec θ (d) cosec α (e) tan θ (f) cot θ (g) cot α.

$$\left[(a) \frac{15}{17} (b) \frac{15}{17} (c) 1\frac{2}{15} (d) 1\frac{2}{15} (e) \frac{8}{15} (f) 1\frac{7}{8} (g) \frac{8}{15}\right]$$

6 If $\tan θ = \frac{7}{24}$, find the other five trigonometric ratios in fraction form.

$$\left[\sin θ = \frac{7}{25}, \cos θ = \frac{24}{25}, \sec θ = 1\frac{1}{24},\right.$$
$$\left.\text{cosec } θ = 3\frac{4}{7}, \cot θ = 3\frac{3}{7}\right]$$

7 In a right-angled triangle PQR, PQ = 11.6 cm, QR = 5.3 cm and R = 90°. Evaluate, correct to 4 significant figures, (a) sec Q (b) cot P (c) tan Q (d) cosec Q (e) cot Q.
[(a) 2.189 (b) 1.947 (c) 1.947 (d) 1.124 (e) 0.513 7]

Evaluating trigonometric ratios

In *Problems 8* to *15*, evaluate correct to 4 decimal places:

8   (a) sine 27°   (b) sine 172.41°   (c) sine 302°52′
    [(a) 0.454 0 (b) 0.132 1 (c) −0.839 9]
9   (a) cosine 124°   (b) cosine 21.46°   (c) cosine 284°10′
    [(a) −0.559 2 (b) 0.930 7 (c) 0.244 7]
10 (a) tangent 145°   (b) tangent 310.59°
    (c) tangent 49°16′
    [(a) −0.700 2 (b) −1.167 1 (c) 1.161 2]
11 (a) secant 73°   (b) secant 286.45°   (c) secant 155°41′
    [(a) 3.420 3 (b) 3.531 3 (c) −1.097 4]
12 (a) cosecant 213°   (b) cosecant 15.62°
    (c) cosecant 311°50′
    [(a) −1.836 1 (b) 3.713 9 (c) −1.342 1]
13 (a) cotangent 71°   (b) cotangent 151.62°
    (c) cotangent 321°23′
    [(a) 0.344 3 (b) −1.851 0 (c) −1.251 9]
14 (a) sine $\dfrac{2\pi}{3}$ (b) cos 1.681   (c) tan 3.672

    [(a) 0.866 0 (b) −0.101 0 (c) 0.586 5]

15 (a) sec $\dfrac{\pi}{8}$ (b) cosec 2.961   (c) cot 2.612

    [(a) 1.082 4 (b) 5.567 5 (c) −1.708 3]

In *Problems 16* and *17*, determine the acute angle in degrees (correct to 2 decimal places), degrees and minutes, and in radians (correct to 3 decimal places).

16 (a) arcsin 0.234 1   (b) arccos 0.827 1
    (c) arctan 0.810 6
    $\begin{bmatrix} \text{(a) } 13.54°,\ 13°32′,\ 0.236\ \text{rad} \\ \text{(b) } 34.20°,\ 34°12′,\ 0.597\ \text{rad} \\ \text{(c) } 39.03°,\ 39°2′,\ 0.681\ \text{rad} \end{bmatrix}$
17 (a) arcsec 1.621 4   (b) arccosec 2.489 1
    (c) arccot 1.961 4
    $\begin{bmatrix} \text{(a) } 51.92°,\ 51°55′,\ 0.906\ \text{rad} \\ \text{(b) } 23.69°,\ 23°41′,\ 0.413\ \text{rad} \\ \text{(c) } 27.01°,\ 27°1′,\ 0.417\ \text{rad} \end{bmatrix}$
18 Evaluate the following, each correct to 4 significant figures:
    (a) 4 cos 56°19′ − 3 sin 21°57′
    (b) $\dfrac{11.5 \tan 49°11′ - \sin 90°}{3 \cos 45°}$

    (c) $\dfrac{5 \sin 86°3′}{3 \tan 14°29′ - 2 \cos 31°9′}$
    [(a) 1.097 (b) 5.805 (c) −5.325]
19 Determine the acute angle, in degrees and minutes, correct to the nearest minute, given by
    $\arcsin\left(\dfrac{4.32 \sin 42°16′}{7.86}\right)$

    [21°42′]

20 Evaluate correct to 4 decimal places
    $\dfrac{(\sin 34°27′)(\cos 69°2′)}{(2 \tan 53°39′)}$
    [0.074 5]
21 If tan x = 1.527 6 determine sec x, cosec x and cot x.
    (Assume x is an acute angle.)
    [1.825 8, 1.195 2, 0.654 6]
22 Evaluate, using 4 figure tables, correct to 4 significant figures:
    (a) 3 cot 14°15′ sec 23°9′
    (b) $\dfrac{\operatorname{cosec} 27°19′ + \sec 45°29′}{1 - \operatorname{cosec} 27°19′ \sec 45°29′}$

    (c) $\dfrac{30 \tan 61° \sec 54° - 15 \cot 14°}{2 \operatorname{cosec} 24°}$
    [(a) 12.85 (b) −1.710 (c) 6.490]
23 Evaluate (a) cosec (−125°)   (b) tan (−241°)
    (c) sec (−49°15′).
    [(a) −1.220 8 (b) −1.804 0 (c) 1.532 0]

Evaluating trigonometric ratios of any magnitude

24 Find all the angles between 0° and 360°:
    (a) whose sine is −0.732 1
    (b) whose cosecant is 2.531 7
    (c) whose cotangent is −0.631 2.
    [(a) 227°4′ or 312°56′ (b) 23°16′ or 156°44′
    (c) 122°16′ or 302°16′]
25 Solve for all values of θ between 0° and 360°:
    (a) arccos (− 0.531 6) = θ (b) arcsec 2.316 2 = θ
    (c) arctan 0.831 4 = θ.
    [(a) 122°7′ or 237°53′ (b) 64°25′ or 295°35′
    (c) 39°44′ or 219°44′]

Fractional and surd forms of trigonometric ratios

26 Evaluate the following without using tables or calculators, leaving, where necessary, in surd form:
    (a) 3 sin 30° − 2 cos 60°   (b) 5 tan 60° − 3 sin 60°
    (c) $\dfrac{\tan 60°}{3 \tan 30°}$
    (d) (tan 45°)(4 cos 60° − 2 sin 60°)
    (e) $\dfrac{\tan 60° - \tan 30°}{1 + \tan 30° \tan 60°}$

    $\left[\text{(a) } \dfrac{1}{2} \ \text{(b) } \dfrac{7}{2}\sqrt{3} \ \text{(c) } 1 \ \text{(d) } 2 - \sqrt{3} \ \text{(e) } \dfrac{1}{\sqrt{3}} \right]$

Solution of right-angled triangles

27 Solve the triangles shown in *Fig. 19.29*.
    $\begin{bmatrix} \text{(i) } BC = 3.50 \text{ cm},\ AB = 6.10 \text{ cm},\ \angle B = 55° \\ \text{(ii) } FE = 5 \text{ cm},\ \angle E = 53°8′,\ \angle F = 36°52′ \\ \text{(iii) } GH = 9.841 \text{ mm},\ GI = 11.32 \text{ mm},\ \angle H = 49° \end{bmatrix}$

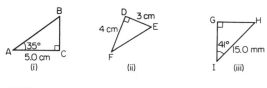

Fig. 19.29

28 Solve the triangles shown in *Fig. 19.30* and find their areas.

> (i) KL = 5.43 cm, JL = 8.62 cm, ∠J = 39°,
>   area = 18.19 cm²
> (ii) MN = 28.86 mm, NO = 13.82 mm, ∠O = 64°25′,
>   area = 199.4 mm²
> (iii) PR = 7.934 m, ∠Q = 65°3′, ∠R = 24°57′,
>   area = 14.64 m²

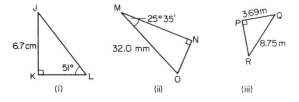

Fig. 19.30

29 A ladder rests against the top of the perpendicular wall of a building and makes an angle of 67° with the ground. If the foot of the ladder is 12 m from the wall, calculate the height of the building.
[28.27 m]

Angles of elevation and depression

30 A vertical tower stands on level ground. At a point 105 m from the foot of the tower the angle of elevation of the top is 19°. Find the height of the tower.
[36.15 m]

31 If the angle of elevation of the top of a vertical 30 m high aerial is 32°, how far is it to the aerial?
[48 m]

32 From the top of a vertical cliff 90 m high the angle of depression of a boat is 19°50′. Determine the distance of the boat from the cliff.
[249.5 m]

33 From the top of a vertical cliff 80.0 m high the angles of depression of two buoys lying due west of the cliff are 23° and 15°, respectively. How far are the buoys apart?
[110.1 m]

34 From a point on horizontal ground a surveyor measures the angle of elevation of the top of a flagpole as 18°40′. He moves 50 m nearer to the flagpole and measures the angle of elevation as 26°22′. Determine the height of the flagpole.
[53.0 m]

35 A flagpole stands on the edge of the top of a building. At a point 200 m from the building the angles of elevation of the top and bottom of the pole are 32° and 30°, respectively. Calculate the height of the flagpole.
[9.50 m]

36 From a ship at sea, the angles of elevation of the top and bottom of a vertical lighthouse standing on the edge of a vertical cliff are 31° and 26°, respectively. If the lighthouse is 25.0 m high, calculate the height of the cliff.
[107.8 m]

37 From a window 4.2 m above horizontal ground the angle of depression of the foot of a building across the road is 24° and the angle of elevation of the top of the same building is 34°. Determine, correct to the nearest centimetre, the width of the road and the height of the building.
[9.43 m, 10.56 m]

38 The elevation of a tower from two points, one due east of the tower and the other due west of it are 20° and 24°, respectively, and the two points of observation are 300 m apart. Find the height of the tower to the nearest metre.
[60 m]

20

The solution of triangles and their areas

20.1 Sine and cosine rules

To 'solve a triangle' means 'to find the values of unknown sides and angles'.

If a triangle is **right angled**, trigonometric ratios and the theorem of Pythagoras may be used for its solution. However, for a **non-right-angled triangle**, trigonometric ratios and Pythagoras' theorem **cannot** be used. Instead, two rules, called the sine rule and the cosine rule, are used.

Sine rule

With reference to triangle ABC of *Fig. 20.1*, the **sine rule** states:

$$\frac{a}{\sin A} = \frac{b}{\sin B} = \frac{c}{\sin C}$$

The rule may be used only when:

(i) 1 side and any 2 angles are initially given, or
(ii) 2 sides and an angle (not the included angle) are initially given.

Cosine rule

With reference to triangle ABC of *Fig. 20.1*, the **cosine rule** states:

$$a^2 = b^2 + c^2 - 2bc \cos A$$
$$\text{or } b^2 = a^2 + c^2 - 2ac \cos B$$
$$\text{or } c^2 = a^2 + b^2 - 2ab \cos C$$

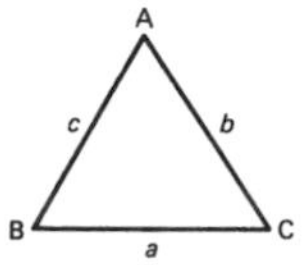

Fig. 20.1

The rule may be used only when:

(i) 2 sides and the included angle are initially given, or
(ii) 3 sides are initially given.

20.2 Area of any triangle

The **area of any triangle** such as ABC of *Fig. 20.1* is given by:

(i) $\frac{1}{2} \times$ base $\times$ perpendicular height, or

(ii) $\frac{1}{2} ab \sin C$, or

   $\frac{1}{2} ac \sin B$ or $\frac{1}{2} bc \sin A$, or

(iii) $\sqrt{[s(s - a)(s - b)(s - c)]}$,

   where $s = \frac{a + b + c}{2}$

20.3 Worked problems on the solution of triangles and their areas

> *Problem 1* In the triangle XYZ, $X = 51°$, $Y = 67°$ and YZ = 15.2 cm. Solve the triangle and find its area.

The triangle XYZ is shown in *Fig. 20.2*. Since the angles in a triangle add up to 180°, then $Z = 180° - 51° - 67° = 62°$. Applying the sine rule:

$$\frac{15.2}{\sin 51°} = \frac{y}{\sin 67°} = \frac{z}{\sin 62°}$$

Using $\dfrac{15.2}{\sin 51°} = \dfrac{y}{\sin 67°}$ and transposing gives:

$$y = \frac{15.2 \sin 67°}{\sin 51°} = \textbf{18.00 cm} = \textbf{XZ}$$

Using $\dfrac{15.2}{\sin 51°} = \dfrac{z}{\sin 62°}$ and transposing gives:

$$z = \frac{15.2 \sin 62°}{\sin 51°} = \textbf{17.27 cm} = \textbf{XY}$$

Area of triangle XYZ = $\dfrac{1}{2}xy \sin Z$

$$= \frac{1}{2}(15.2)(18.00) \sin 62°$$

$$= \textbf{120.8 cm}^2$$

or area = $\dfrac{1}{2}xz \sin Y = \dfrac{1}{2}(15.2)(17.27) \sin 67° = \textbf{120.8 cm}^2$).

It is always worth checking with triangle problems that the longest side is opposite the largest angle, and vice-versa. In this problem, Y is the largest angle and thus XZ should be the longest of the three sides.

> *Problem 2* Solve the triangle ABC given $B = 78°51'$, AC = 22.31 mm and AB = 17.92 mm. Find also its area.

Triangle ABC is shown in *Fig. 20.3*. Applying the sine rule:

$$\frac{22.31}{\sin 78°51'} = \frac{17.92}{\sin C}$$

from which, $\sin C = \dfrac{17.92 \sin 78°51'}{22.31} = 0.788\,1$

Hence $C = \arcsin 0.788\,1 = 52°0'$ or $128°0'$ (see Chapter 19). Since $B = 78°51'$, C cannot be $128°0'$, since $128°0' + 78°51'$ is greater than 180°. Thus only $C = 52°0'$ is valid. Angle $A = 180° - 78°51' - 52°0' = 49°9'$. Applying the sine rule:

$$\frac{a}{\sin 49°9'} = \frac{22.31}{\sin 78°51'}$$

from which,

$$a = \frac{22.31 \sin 49°9'}{\sin 78°51'} = 17.20 \text{ mm}$$

Hence $A = \textbf{49°9'}$, $C = \textbf{52°0'}$ and $\textbf{BC} = \textbf{17.20 mm}$.

Area of triangle ABC = $\dfrac{1}{2}ac \sin B =$

$\dfrac{1}{2}(17.20)(17.92) \sin 78°51' = \textbf{151.2 mm}^2$.

> *Problem 3* Solve the triangle PQR and find its area given that QR = 36.5 mm, PR = 29.6 mm and $Q = 36°$.

Triangle PQR is shown in *Fig. 20.4*. Applying the sine rule:

$$\frac{29.6}{\sin 36°} = \frac{36.5}{\sin P}$$

from which,

$$\sin P = \frac{36.5 \sin 36°}{29.6} = 0.724\,8$$

Hence $P = \arcsin 0.724\,8 = 46°27'$ or $133°33'$.
When $P = 46°27'$ and $Q = 36°$ then $R = 180° - 46°27' - 36° = 97°33'$.

Fig. 20.2

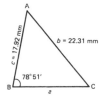

Fig. 20.3

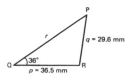

Fig. 20.4

When $P = 133°33'$ and $Q = 36°$ then $R = 180° - 133°33' - 36° = 10°27'$.

Thus, in this problem, there are **two** separate sets of results and both are feasible solutions. Such a situation is called the **ambiguous case**.

Case 1 $P = 46°27'$, $Q = 36°$, $R = 97°33'$, $p = 36.5$ mm and $q = 29.6$ mm.
From the sine rule:

$$\frac{r}{\sin 97°33'} = \frac{29.6}{\sin 36°}$$

from which,

$$r = \frac{29.6 \sin 97°33'}{\sin 36°} = \textbf{49.92 mm}$$

$$\text{Area} = \frac{1}{2}pq \sin R = \frac{1}{2}(36.5)(29.6) \sin 97°33' = \textbf{535.5 mm}^2.$$

Case 2 $P = 133°33'$, $Q = 36°$, $R = 10°27'$, $p = 36.5$ mm and $q = 29.6$ mm.
From the sine rule:

$$\frac{r}{\sin 10°27'} = \frac{29.6}{\sin 36°}$$

from which,

$$r = \frac{29.6 \sin 10°27'}{\sin 36°} = \textbf{9.134 mm}$$

$$\text{Area} = \frac{1}{2}pq \sin R = \frac{1}{2}(36.5)(29.6) \sin 10°27' = \textbf{97.98 mm}^2.$$

Triangle PQR for case 2 is shown in *Fig. 20.5*.

Problem 4 Solve triangle DEF and find its area given that EF = 35.0 mm, DE = 25.0 mm and $E = 64°$.

Triangle DEF is shown in *Fig. 20.6*. Applying the cosine rule:

$$e^2 = d^2 + f^2 - 2df \cos E$$
i.e. $e^2 = (35.0)^2 + (25.0)^2 - \{2(35.0)(25.0) \cos 64°\}$
$$= 1225 + 625 - 767.1 = 1083$$
$$e = \sqrt{1083} = \textbf{32.91 mm}$$

Applying the sine rule:

$$\frac{32.91}{\sin 64°} = \frac{25.0}{\sin F}$$

from which,

$$\sin F = \frac{25.0 \sin 64°}{32.91°} = 0.682\ 8$$

Thus

$$F = \arcsin 0.682\ 8 = 43°4' \text{ or } 136°56'$$

$F = 136°56'$ is not possible in this case since $136°56' + 64°$ is greater than $180°$. Thus only $F = \textbf{43°4'}$ is valid.

$$D = 180° - 64° - 43°4' = \textbf{72°56'}$$

$$\text{Area of triangle DEF} = \frac{1}{2}df \sin E = \frac{1}{2}(35.0)(25.0) \sin 64°$$

$$= \textbf{393.2 mm}^2.$$

Problem 5 A triangle ABC has sides $a = 9.0$ cm, $b = 7.5$ cm and $c = 6.5$ cm. Determine its three angles and its area.

Triangle ABC is shown in *Fig. 20.7*. It is usual first to calculate the largest angle to determine whether the triangle is acute or obtuse. In this case the largest angle is A (i.e. opposite the longest side).
Applying the cosine rule:

$$a^2 = b^2 + c^2 - 2bc \cos A$$

from which,

$$2bc \cos A = b^2 + c^2 - a^2$$

and

$$\cos A = \frac{b^2 + c^2 - a^2}{2bc} = \frac{7.5^2 + 6.5^2 - 9.0^2}{2(7.5)(6.5)}$$

$$= 0.179\ 5$$

Hence $A = \arccos 0.179\ 5 = \textbf{79°40'}$ (or $280°20'$, which is obviously impossible). The triangle is thus acute angled.

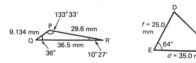

Fig. 20.5   **Fig. 20.6**

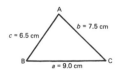

Fig. 20.7

since cos *A* is positive. (If cos *A* had been negative, angle *A* would be obtuse, i.e. lie between 90° and 180°.) Applying the sine rule:

$$\frac{9.0}{\sin 79°40'} = \frac{7.5}{\sin B}$$

from which,

$$\sin B = \frac{7.5 \sin 79°40'}{9.0} = 0.819\,8$$

Hence

$$B = \arcsin 0.819\,8 = \mathbf{55°4'}$$

$C = 180° - 79°40' - 55°4' = \mathbf{45°16'}$.

Area $= \sqrt{[s(s-a)(s-b)(s-c)]}$, where

$$s = \frac{a+b+c}{2} = \frac{9.0 + 7.5 + 6.5}{2} = 11.5 \text{ cm}$$

Hence area $= \sqrt{[11.5(11.5 - 9.0)(11.5 - 7.5)(11.5 - 6.5)]}$

$$= \sqrt{[11.5(2.5)(4.0)(5.0)]}$$

$$= \mathbf{23.98 \text{ cm}^2}$$

Alternatively, area $= \dfrac{1}{2}ab \sin C = \dfrac{1}{2}(9.0)(7.5) \sin 45°16'$

$$= \mathbf{23.98 \text{ cm}^2}.$$

Problem 6 Solve triangle XYZ (*Fig. 20.8*) and find its area given that $Y = 128°$, XY = 7.2 cm and YZ = 4.5 cm.

Applying the cosine rule:

$$\begin{aligned} y^2 &= x^2 + z^2 - 2xz \cos Y \\ &= (4.5)^2 + (7.2)^2 - \{2(4.5)(7.20) \cos 128°\} \\ &= 20.25 + 51.84 - \{-39.89\} \\ &= 20.25 + 51.84 + 39.89 = 112.0 \end{aligned}$$

$$y = \sqrt{(112.0)} = \mathbf{10.58 \text{ cm}}$$

Applying the sine rule:

$$\frac{10.58}{\sin 128°} = \frac{7.2}{\sin Z}$$

from which,

$$\sin Z = \frac{7.2 \sin 128°}{10.58} = 0.536\,3$$

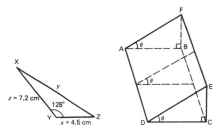

Fig. 20.8   **Fig. 20.9**

Hence $Z = \arcsin 0.536\,3 = \mathbf{32°26'}$ (or 147°34' which, here, is impossible). $X = 180° - 128° - 32°26' = \mathbf{19°34'}$.

Area $= \dfrac{1}{2}xz \sin Y = \dfrac{1}{2}(4.5)(7.2) \sin 128° = \mathbf{12.77 \text{ cm}^2}$.

Further problems on the solution of triangles and their areas may be found in section 20.6, Problems 1 to 5, page 194.

20.4 Lengths and areas on an inclined plane

In *Fig. 20.9*, rectangle ADEF is a plane inclined at an angle of θ to the horizontal plane ABCD.

$$\cos \theta = \frac{DC}{DE}, \text{ from which } DE = \frac{DC}{\cos \theta}$$

Hence the line of greatest slope on an inclined plane is given by:

$$\left(\frac{1}{\cos \theta}\right) \textbf{(its projection on to the horizontal plane)}$$

Area of ADEF $= (AD)(DE)$

$$= (AD)\left(\frac{DC}{\cos \theta}\right)$$

$$= \left(\frac{1}{\cos \theta}\right)\textbf{(area of horizontal plane)}$$

Problem 7 A vertical, cylindrical ventilation shaft of diameter 36.0 cm has its end at an angle of 20° to the horizontal as shown in *Fig. 20.10*. Determine the area of the end cover plate.

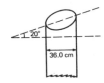

Fig. 20.10

Cover plate area

$$= \left(\frac{1}{\cos \theta}\right) \text{(horizontal plane area), from above}$$

$$= \left(\frac{1}{\cos 20°}\right)\left[\pi\left(\frac{36.0}{2}\right)^2\right] = \mathbf{1083 \ cm^2}$$

Problem 8   A rectangular chimney stack having dimensions of 1.2 m by 0.70 m passes through a roof that has a pitch of 35°. Determine the area of the void in the roof through which the stack passes.

Area of void in roof

$$= \left(\frac{1}{\cos \theta}\right) \text{(area of horizontal plane)}$$

$$= \left(\frac{1}{\cos 35°}\right)(1.2 \times 0.70) = \mathbf{1.025 \ m^2}$$

Further problems on the lengths and areas on an inclined plane may be found in section 20.6, Problems 6 and 7, page 195.

20.5   Practical situations involving trigonometry

There are a number of **practical situations** where the use of trigonometry is needed to find unknown sides and angles of triangles. This is demonstrated in *Problems 9 to 15.*

Problem 9   A room 8.0 m wide has a span roof which slopes at 33° on one side and 40° on the other. Find the length of the roof slopes, correct to the nearest centimetre.

A section of the roof is shown in *Fig. 20.11.*
Angle at ridge, $B = 180° - 33° - 40° = 107°$. From the sine rule:

$$\frac{8.0}{\sin 107°} = \frac{a}{\sin 33°}$$

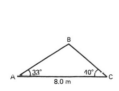

Fig. 20.11

from which,

$$a = \frac{8.0 \sin 33°}{\sin 107°} = 4.556 \ m$$

Also from the sine rule:

$$\frac{8.0}{\sin 107°} = \frac{c}{\sin 40°}$$

from which,

$$c = \frac{8.0 \sin 40°}{\sin 107°} = 5.377 \ m$$

Hence the roof slopes are 4.56 m and 5.38 m, correct to the nearest centimetre.

Problem 10   A man leaves a point walking at 6.5 km/h in a direction E 20° N (i.e. a bearing of 70°). A cyclist leaves the same point at the same time in a direction E 40° S (i.e. a bearing of 130°) travelling at a constant speed. Find the average speed of the cyclist if the walker and cyclist are 80 km apart after 5 hours.

After 5 hours the walker has travelled $5 \times 6.5 = 32.5$ km (shown as AB in *Fig. 20.12*). If AC is the distance the cyclist travels in 5 hours then BC = 80 km. Applying the sine rule:

$$\frac{80}{\sin 60°} = \frac{32.5}{\sin C}$$

from which,

$$\sin C = \frac{32.5 \sin 60°}{80} = 0.351 \ 8$$

Hence $C = \arcsin 0.351 \ 8 = 20°36'$ (or 159°24' which is impossible in this case). $B = 180° - 60° - 20°36' = 99°24'$. Applying the sine rule:

$$\frac{80}{\sin 60°} = \frac{b}{\sin 99°24'}$$

from which,

$$b = \frac{80 \sin 90°24'}{\sin 60°} = 91.14 \ km$$

Since the cyclist travels 91.14 km in 5 hours then:

$$\text{average speed} = \frac{\text{distance}}{\text{time}} = \frac{91.14}{5} = \mathbf{18.23 \ km/h}$$

Fig. 20.12

Problem 11 Two voltage phasors are shown in *Fig. 20.13*. If $V_1 = 40$ V and $V_2 = 100$ V determine the value of their resultant (i.e. length OA) and the angle the resultant makes with V_1.

Angle OBA $= 180° - 45° = 135°$. Applying the cosine rule:

$$OA^2 = V_1^2 + V_2^2 - 2V_1V_2 \cos OBA$$
$$= 40^2 + 100^2 - \{2(40)(100) \cos 135°\}$$
$$= 1600 + 10\,000 - \{-5657\}$$
$$= 1600 + 10\,000 + 5657 = 17\,257$$

The resultant OA $= \sqrt{(17\,257)} = 131.4$ V

Applying the sine rule:

$$\frac{131.4}{\sin 135°} = \frac{100}{\sin AOB}$$

from which,

$$\sin AOB = \frac{100 \sin 135°}{131.4} = 0.538\,1$$

Hence angle AOB $= \arcsin 0.538\,1 = 32°33'$ (or $147°27'$ which is impossible in this case).
Hence the resultant voltage is 131.4 volts at $32°33'$ to V_1.

Problem 12 In *Fig. 20.14*, PR represents the inclined jib of a crane and is 10.0 m long. PQ is 4.0 m long. Determine the length of tie QR and the inclination of the jib to the vertical.

Applying the sine rule:

$$\frac{PR}{\sin 120°} = \frac{PQ}{\sin R}$$

from which,

$$\sin R = \frac{PQ \sin 120°}{PR} = \frac{(4.0) \sin 120°}{10.0} = 0.346\,4$$

Hence $R = \arcsin 0.346\,4 = 20°16'$ (or $159°44'$, which is impossible in this case).
$P = 180° - 120° - 20°16' = \mathbf{39°44'}$, **which is the inclination of the jib to the vertical.**
Applying the sine rule:

$$\frac{10.0}{\sin 120°} = \frac{QR}{\sin 39°44'}$$

from which,

$$QR = \frac{10.0 \sin 39°44'}{\sin 120°} = \mathbf{7.38\ m = length\ of\ tie}$$

Problem 13 A vertical aerial stands on horizontal ground. A surveyor positioned due east of the aerial measures the elevation of the top as 48°. He moves due south 30.0 m and measures the elevation as 44°. Determine the height of the aerial.

In *Fig. 20.15*, DC represents the aerial, A is the initial position of the surveyor and B his final position. From triangle ACD, $\tan 48° = DC/AC$, from which, $AC = DC/\tan 48°$ $= DC \cot 48°$.
Similarly, from triangle BCD, $BC = DC \cot 44°$.
For triangle ABC, using Pythagoras' theorem:

$$BC^2 = AB^2 + AC^2$$

$$(DC \cot 44°)^2 = (30.0)^2 + (DC \cot 48°)^2$$

$$DC^2(\cot^2 44° - \cot^2 48°) = 30.0^2$$

$$DC^2 = \frac{30.0^2}{\cot^2 44° - \cot^2 48°} = 3440$$

Hence, height of aerial, DC $= \sqrt{3440} = 58.65$ m.

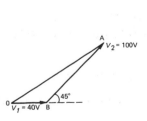

Fig. 20.13   **Fig. 20.14**

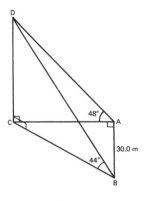

Fig. 20.15

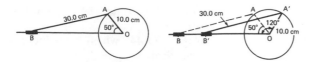

Fig. 20.16   **Fig. 20.17**

Problem 14   A crank mechanism of a petrol engine is shown in *Fig. 20.16*. Arm OA is 10.0 cm long and rotates clockwise about 0. The connecting rod AB is 30.0 cm long and end B is constrained to move horizontally.

(a) For the position shown in *Fig. 20.16* determine the angle between the connecting rod AB and the horizontal and the length of OB.
(b) How far does B move when angle AOB changes from 50° to 120°?

(a) Applying the sine rule:

$$\frac{AB}{\sin 50°} = \frac{AO}{\sin B}$$

from which,

$$\sin B = \frac{AO \sin 50°}{AB} = \frac{10.0 \sin 50°}{30.0} = 0.255\,3$$

Hence B = arcsin 0.255 3 = 14°47′ (or 165°13′, which is impossible in this case)
Hence the connecting rod AB makes an angle of 14°47′ with the horizontal.
Angle OAB = 180° − 50° − 14°47′ = 115°13′. Applying the sine rule:

$$\frac{30.0}{\sin 50°} = \frac{OB}{\sin 115°13′}$$

from which,

$$\mathbf{OB} = \frac{30.0 \sin 115°13′}{\sin 50°} = \mathbf{35.43\ cm}$$

(b) *Figure 20.17* shows the initial and final positions of the crank mechanism. In triangle OA′B′, applying the sine rule:

$$\frac{30.0}{\sin 120°} = \frac{10.0}{\sin A′B′O}$$

from which,

$$\sin A′B′O = \frac{10.0 \sin 120°}{30.0} = 0.288\,7$$

Hence A′B′O = arcsin 0.288 7 = 16°47′ (or 163°13′ which is impossible in this case)
Angle OA′B′ = 180° − 120° − 16°47′ = 43°13′.
Applying the sine rule:

$$\frac{30.0}{\sin 120°} = \frac{OB′}{\sin 43°13′}$$

from which,

$$OB′ = \frac{30.0 \sin 43°13′}{\sin 120°} = 23.72\ cm$$

Since OB = 35.43 cm and OB′ = 23.72 cm then BB′ = 35.43 − 23.72 = 11.68 cm.
Hence B moves 11.68 cm when angle AOB changes from 50° to 120°.

Problem 15   The area of a field is in the form of a quadrilateral ABCD as shown in *Fig. 20.18*. Determine its area.

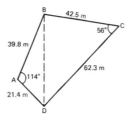

Fig. 20.18

A diagonal drawn from B to D divides the quadrilateral into two triangles. Area of quadrilateral ABCD

$$= \text{area of triangle ABD} + \text{area of triangle BCD}$$
$$= \frac{1}{2}(39.8)(21.4) \sin 114′ + \frac{1}{2}(42.5)(62.3) \sin 56°$$
$$= 389.04 + 1097.5$$
$$= \mathbf{1487\ m^2}$$

Further problems on practical situations involving trigonometry may be found in section 20.6, Problems 8 to 20, page 195.

20.6   Further problems on the solution of triangles and their areas

Solution of triangles and their areas

1   Use the sine rule to solve the following triangles ABC and find their areas.
   (a) $A = 29°$, $B = 68°$, $b = 27$ mm

(b) $B = 71°26'$, $C = 56°32'$, $b = 8.60$ cm
(c) $A = 117°$, $C = 24°30'$, $a = 15.2$ mm
$\left[\begin{array}{l}\text{(a) } C = 83°,\ a = 14.1\text{ mm},\ c = 28.9\text{ mm,}\\ \quad\text{area} = 189\text{ mm}^2\\ \text{(b) } A = 52°2',\ c = 7.568\text{ cm},\ a = 7.152\text{ cm,}\\ \quad\text{area} = 25.65\text{ cm}^2\\ \text{(c) } B = 38°30',\ b = 10.62\text{ mm},\ c = 7.074\text{ mm,}\\ \quad\text{area} = 33.47\text{ mm}^2\end{array}\right]$

2 Use the sine rule to solve the following triangles DEF and find their areas.
(a) $d = 17$ cm, $f = 22$ cm, $F = 26°$
(b) $e = 4.20$ m, $f = 7.10$ m, $F = 81°$
(c) $d = 32.6$ mm, $e = 25.4$ mm, $D = 104°22'$
$\left[\begin{array}{l}\text{(a) } D = 19°48',\ E = 134°12',\ e = 36.0\text{ cm,}\\ \quad\text{area} = 134\text{ cm}^2\\ \text{(b) } E = 35°45',\ D = 63°15',\ d = 6.419\text{ m,}\\ \quad\text{area} = 13.31\text{ m}^2\\ \text{(c) } E = 49°0',\ F = 26°38',\ f = 15.09\text{ mm,}\\ \quad\text{area} = 185.6\text{ mm}^2\end{array}\right]$

3 Use the sine rule to solve the following triangles JKL and find their areas.
(a) $j = 3.85$ cm, $k = 3.23$ cm, $K = 36°$
(b) $k = 46$ mm, $l = 36$ mm, $L = 35°$
(c) $j = 2.92$ m, $l = 3.24$ m, $J = 27°30'$
$\left[\begin{array}{l}\text{(a) } J = 44°29',\ L = 99°31',\ l = 5.420\text{ cm,}\\ \quad\text{area} = 6.133\text{ cm}^2\\ \text{OR } J = 135°31',\ L = 8°29',\ l = 0.811\text{ cm}\\ \quad\text{area} = 0.917\text{ cm}^2\\ \text{(b) } K = 47°8',\ J = 97°52',\ j = 62.2\text{ mm,}\\ \quad\text{area} = 820.6\text{ mm}^2\\ \text{OR } K = 132°52',\ J = 12°8',\ j = 13.19\text{ mm,}\\ \quad\text{area} = 174.0\text{ mm}^2\\ \text{(c) } L = 30°49',\ K = 121°41',\ k = 5.381\text{ m,}\\ \quad\text{area} = 4.025\text{ m}^2\\ \text{OR } L = 149°11',\ K = 3°19',\ k = 0.366\text{ m,}\\ \quad\text{area} = 0.274\text{ m}^2\end{array}\right]$

4 Use the cosine and sine rules to solve the following triangles PQR and find their areas.
(a) $q = 12$ cm, $r = 16$ cm, $P = 54°$
(b) $p = 56$ mm, $q = 38$ mm, $R = 64°$

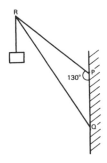

Fig. 20.19

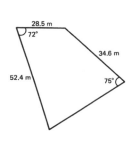

Fig. 20.20

(c) $q = 3.25$ m, $r = 4.42$ m, $P = 105°$
$\left[\begin{array}{l}\text{(a) } p = 13.2\text{ cm},\ Q = 47°21',\ R = 78°39',\\ \quad\text{area} = 77.7\text{ cm}^2\\ \text{(b) } r = 52.1\text{ mm},\ Q = 40°58',\ P = 75°2',\\ \quad\text{area} = 956\text{ mm}^2\\ \text{(c) } p = 6.127\text{ m},\ Q = 30°49',\ R = 44°11',\\ \quad\text{area} = 6.938\text{ m}^2\end{array}\right]$

5 Use the cosine and sine rules to solve the following triangles XYZ and find their areas.
(a) $x = 10.0$ cm, $y = 8.0$ cm, $z = 7.0$ cm
(b) $x = 2.4$ m, $y = 3.6$ m, $z = 1.5$ m
(c) $x = 21$ mm, $y = 34$ mm, $z = 42$ mm
$\left[\begin{array}{l}\text{(a) } X = 83°20',\ Y = 52°37',\ Z = 44°3',\\ \quad\text{area} = 27.8\text{ cm}^2\\ \text{(b) } X = 28°57',\ Y = 133°26',\ Z = 17°37',\\ \quad\text{area} = 1.31\text{ m}^2\\ \text{(c) } X = 29°46',\ Y = 53°31',\ Z = 96°43',\\ \quad\text{area} = 355\text{ mm}^2\end{array}\right]$

Lengths and areas on an inclined plane

6 A vertical 35.0 cm by 35.0 cm ventilation shaft has an end covered by a plate that makes an angle of 21°15' with the horizontal. Determine the area of the plate. [1314 cm²]

7 A chimney stack has a diameter of 1.5 m and passes through a roof that has a pitch of 36°30'. Determine the area of the resulting void in the roof covering. [2.20 m²]

Practical situations involving trigonometry

8 A ship P sails at a steady speed of 45 km/h in a direction of W 32° N (i.e. a bearing of 302°) from a port. At the same time another ship Q leaves the port at a steady speed of 35 km/h in a direction N 15° E (i.e. a bearing of 015°). Determine their distance apart after 4 hours. [193 km]

9 Two sides of a triangular plot of land are 52.0 m and 34.0 m, respectively. If the area of the plot is 620 m² find (a) the length of fencing required to enclose the plot and (b) the angles of the triangular plot. [(a) 122.6 m (b) 94°49', 40°39', 44°32']

10 A jib crane is shown in *Fig. 20.19*. If the tie rod PR is 8.0 long and PQ is 4.5 m long determine (a) the length of jib RQ and (b) the angle between the jib and the tie rod. [(a) 11.4 m (b) 17°33']

11 A building site is in the form of a quadrilateral as shown in *Fig. 20.20*, and its area is 1510 m². Determine the length of the perimeter of the site. [163.4 m]

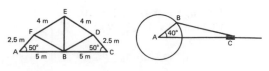

Fig. 20.21   **Fig. 20.22**

12 Determine the length of members BF and EB in the roof truss shown in *Fig. 20.21*.
[BF = 3.9 m, EB = 4.0 m]

13 A laboratory 9.0 m wide has a span roof which slopes at 36° on one side and 44° on the other. Determine the lengths of the roof slopes.
[6.35 m, 5.37 m]

14 PQ and QR are the phasors representing the alternating currents in two branches of a circuit. Phasor PQ is 20.0 A and is horizontal. Phasor QR (which is joined to the end of PQ to form triangle PQR) is 14.0 A and is at an angle of 35° to the horizontal. Determine the resultant phasor PR and the angle it makes with phasor PQ.
[32.48 A, 14°19′]

15 A vertical aerial AB, 9.60 m high, stands on ground which is inclined 12° to the horizontal. A stay connects the top of the aerial A to a point C on the ground 10.0 m downhill from B, the foot of the aerial. Determine (a) the length of the stay and (b) the angle the stay makes with the ground.
[(a) 15.23 m (b) 38°4′]

16 Three forces acting on a fixed point are represented by the sides of a triangle of dimensions 7.2 cm, 9.6 cm and 11.0 cm. Determine the angles between the lines of action and the three forces.
[80°25′, 59°23′, 40°12′]

17 A reciprocating engine mechanism is shown in *Fig. 20.22*. The crank AB is 12.0 cm long and the connecting rod BC is 32.0 cm long. For the position shown determine the length of AC and the angle between the crank and the connecting rod.
[40.25 cm, 126°3′]

18 From *Fig. 20.22*, determine how far C moves, correct to the nearest millimetre when angle CAB changes from 40° to 160°, B moving in an anticlockwise direction.
[19.8 cm]

19 A surveyor, standing W 25° S of a tower measures the angle of elevation of the top of the tower as 46°30′. From a position E 23° S from the tower the elevation of the top is 37°15′. Determine the height of the tower if the distance between the two observations is 75 m.
[36.2 m]

20 An aeroplane is sighted due east from a radar station at an elevation of 40° and a height of 8000 m and later at an elevation of 35° and height 5500 m in a direction E 70° S. If it is descending uniformly, find the angle of descent. Determine also the speed of the aeroplane in km/h if the time between the two observations is 45 s.
[13°57′, 829.9 km/h]

21

Trigonometric graphs and the combination of waveforms

21.1 Sine and cosine curves

Graphs of sine and cosine waveforms

(i) A graph of $y = \sin A$ is shown by the broken line in *Fig. 21.1* and is obtained by drawing up a table of values as in section 19.5. A similar table may be produced for $y = \sin 2A$.

$A°$	0	30	45	60	90	120
$2A$	0	60	90	120	180	240
$\sin 2A$	0	0.866	1.0	0.866	0	−0.866

$A°$	135	150	180	210	225	240
$2A$	270	300	360	420	450	480
$\sin 2A$	−1.0	−0.866	0	0.866	1.0	0.866

$A°$	270	300	315	330	360
$2A$	540	600	630	660	720
$\sin 2A$	0	−0.866	−1.0	−0.866	0

A graph of $y = \sin 2A$ is shown in *Fig. 21.1*.

(ii) A graph of $y = \sin \frac{1}{2}A$ is shown in *Fig. 21.2* using the following table of values.

$A°$	0	30	60	90	120	150	180
$\frac{1}{2}A$	0	15	30	45	60	75	90
$\sin \frac{1}{2}A$	0	0.259	0.50	0.707	0.866	0.966	1.0

$A°$	210	240	270	300	330	360
$\frac{1}{2}A$	105	120	135	150	165	180
$\sin \frac{1}{2}A$	0.966	0.866	0.707	0.50	0.259	0

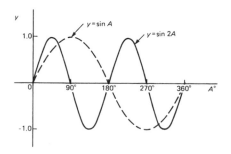

Fig. 21.1

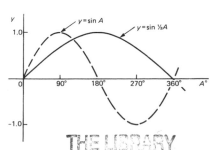

Fig. 21.2

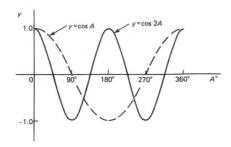

Fig. 21.3

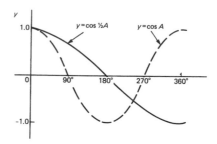

Fig. 21.4

(iii) A graph of $y = \cos A$ is shown by the broken line in *Fig. 21.3* and is obtained by drawing up a table of values as in section 5 of Chapter 19. A similar table may be produced for $y = \cos 2A$.

$A°$	0	30	45	60	90	120
$2A$	0	60	90	120	180	240
$\cos 2A$	1.0	0.50	0	−0.50	−1.0	−0.50

$A°$	135	150	180	210	225	240
$2A$	270	300	360	420	450	480
$\cos 2A$	0	0.50	1.0	0.50	0	−0.50

$A°$	270	300	315	330	360
$2A$	540	600	630	660	720
$\cos 2A$	−1.0	−0.50	0	0.50	1.0

A graph of $y \cos 2A$ is shown in *Fig. 21.3*.

(iv) A graph of $y = \cos \dfrac{1}{2}A$ is shown in *Fig. 21.4* using the following table of values.

$A°$	0	30	60	90	120	150	180
$\frac{1}{2}A$	0	15	30	45	60	75	90
$\cos \frac{1}{2}A$	1.0	0.966	0.866	0.707	0.50	0.259	0

$A°$	210	240	270	300	330	360
$\frac{1}{2}A$	105	120	135	150	165	180
$\cos \frac{1}{2}A$	−0.259	−0.50	−0.707	−0.866	−0.966	−1.0

Periodic time and period

(i) Each of the graphs shown in *Figs 21.1* to *21.4* will repeat themselves as angle A increases and are thus called **periodic functions**.

(ii) $y = \sin A$ and $y = \cos A$ repeat themselves every 360° (or 2π radians); thus 360° is called the **period** of these waveforms. $y = \sin 2A$ and $y = \cos 2A$ repeat themselves every 180° (or π radians); thus 180° is the period of these waveforms.

(iii) In general, if $y = \sin pA$ or $y = \cos pA$ (where p is a constant) then the period of the waveform is $360°/p$ (or $2\pi/p$ rad). Hence if $y = \sin 3A$ then the period is $360/3$, i.e. 120°, and if $y = \cos 4A$ then the period is $360/4$, i.e. 90°.

Amplitude

Amplitude is the name given to the maximum or peak value of a sine wave. Each of the graphs shown in *Figs 21.1* to *21.4* has an amplitude of +1 (i.e. they oscillate between +1 and −1). However, if $y = 4 \sin A$, each of the values in the table is multiplied by 4 and the maximum value, and thus amplitude, is 4. Similarly, if $y = 5 \cos 2A$, the amplitude is 5 and the period is $360°/2$, i.e. 180°.

Lagging and leading angles

(i) A sine or cosine curve may not always start at 0°. To show this a periodic function is represented by $y = \sin (A \pm \alpha)$ or $y = \cos (A \pm \alpha)$, where α is a phase displacement compared with $y = \sin A$ or $y = \cos A$.

(ii) By drawing up a table of values, a graph of $y = \sin (A − 60°)$ may be plotted as shown in *Fig. 21.5*. If $y = \sin A$ is assumed to start at 0° then $y = \sin (A − 60°)$ starts 60° later (i.e. has a zero value 60° later). Thus $y = \sin (A − 60°)$ is said to **lag** $y = \sin A$ by 60°.

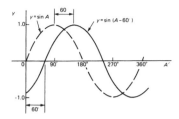

Fig. 21.5

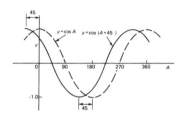

Fig. 21.6

(iii) By drawing up a table of values, a graph of $y = \cos(A + 45°)$ may be plotted as shown in *Fig. 21.6*. If $y = \cos A$ is assumed to start at $0°$ then $y = \cos(A + 45°)$ starts $45°$ earlier (i.e. has a zero value $45°$ earlier). Thus $y = \cos(A + 45°)$ is said to **lead** $y = \cos A$ by $45°$.

(iv) Generally, a graph of $y = \sin(A - \alpha)$ lags $y = \sin A$ by angle α, and a graph of $y = \sin(A + \alpha)$ leads $y = \sin A$ by angle α.

(v) A cosine curve is the same shape as a sine curve but starts $90°$ earlier, i.e. leads by $90°$. Hence $\cos A = \sin(A + 90°)$.

Graphs of sin² A and cos² A

(i) A graph of $y = \sin^2 A$ is shown in *Fig. 21.7* using the following table of values.

$A°$	0	30	60	90	120
$\sin A$	0	0.50	0.866	1.0	0.866
$(\sin A)^2 = \sin^2 A$	0	0.25	0.75	1.0	0.75

$A°$	150	180	210	240	270
$\sin A$	0.50	0	-0.50	-0.866	-1.0
$(\sin A)^2 = \sin^2 A$	0.25	0	0.25	0.75	1.0

$A°$	300	330	360
$\sin A$	-0.866	-0.50	0
$(\sin A)^2 = \sin^2 A$	0.75	0.25	0

(ii) A graph of $y = \cos^2 A$ is shown in *Fig. 21.8* using the following table of values.

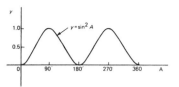

Fig. 21.7

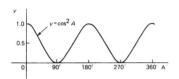

Fig. 21.8

$A°$	0	30	60	90	120
$\cos A$	1.0	0.866	0.50	0	-0.50
$(\cos A)^2 = \cos^2 A$	1.0	0.75	0.25	0	0.25

$A°$	150	180	210	240	270
$\cos A$	-0.866	-1.0	-0.866	-0.50	0
$(\cos A)^2 = \cos^2 A$	0.75	1.0	0.75	0.25	0

$A°$	300	330	360
$\cos A$	0.50	0.866	1.0
$(\cos A)^2 = \cos^2 A$	0.25	0.75	1.0

(iii) $y = \sin^2 A$ and $y = \cos^2 A$ are both periodic functions of period $180°$ (or π rad) and both contain only positive values. Thus a graph of $y = \sin^2 2A$ has a period $180°/2$, i.e. $90°$. Similarly, a graph of $y = 4\cos^2 3A$ has a maximum value of 4 and a period of $180°/3$, i.e. $60°$.

Problem 1 Sketch $y = \sin 3A$ between $A = 0°$ and $A = 360°$.

Amplitude = 1; period $= 360°/3 = 120°$.
A sketch of $y = \sin 3A$ is shown in *Fig. 21.9*.

Fig. 21.9

Problem 2   Sketch $y = 3 \sin 2A$ from $A = 0$ to $A = 2\pi$ radians.

Amplitude = 3; period = $2\pi/2 = \pi$ rads (or 180°). A sketch of $y = 3 \sin 2A$ is shown in *Fig. 21.10*.

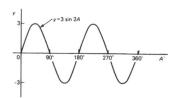

Fig. 21.10

Problem 3   Sketch $y = 4 \cos 2x$ from $x = 0°$ to $x = 360°$.

Amplitude = 4; period = $360°/2 = 180°$. A sketch of $y = 4\cos 2x$ is shown in *Fig. 21.11*.

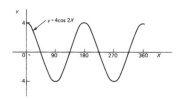

Fig. 21.11

Problem 4   Sketch $y = 2 \sin \dfrac{3}{5}A$ over one cycle.

Amplitude = 2; period = $\dfrac{360°}{\dfrac{3}{5}} = \dfrac{360° \times 5}{3} = 600°$

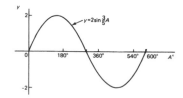

Fig. 21.12

A sketch of $y = 2 \sin \dfrac{3}{5}A$ is shown in *Fig. 21.12*.

Problem 5   Sketch $y = 5 \sin (A + 30°)$ from $A = 0°$ to $A = 360°$.

Amplitude = 5; period = $360°/1 = 360°$.
$5 \sin (A + 30°)$ leads $5 \sin A$ by 30° (i.e. starts 30° earlier). A sketch of $y = 5 \sin (A + 30°)$ is shown in *Fig. 21.13*.

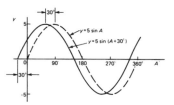

Fig. 21.13

Problem 6   Sketch $y = 7 \sin (2A - \pi/3)$ over one cycle.

Amplitude = 7; period = $2\pi/2 = \pi$ radians.
In general, $y = \sin (pt - \alpha)$ **lags** $y = \sin pt$ by α/p, hence $7 \sin (2A - \pi/3)$ lags $7 \sin 2A$ by $\pi/(3/2)$, i.e. $\pi/6$ rad or 30°.

A sketch of $y = 7 \sin (2A - \pi/3)$ is shown in *Fig. 21.14*.

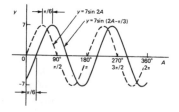

Fig. 21.14

Problem 7   Sketch $y = 2 \cos (\omega t - 3\pi/10)$ over one cycle.

Amplitude = 2; period = $2\pi/\omega$ rad,
$2 \cos (\omega t - 3\pi/10)$ lags $2 \cos \omega t$ by $3\pi/10\omega$ radians.
A sketch of $y = 2 \cos (\omega t - 3\pi/10)$ is shown in *Fig. 21.15*.

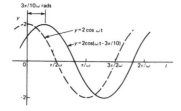

Fig. 21.15

Problem 8 Sketch $y = 3 \sin^2 \dfrac{1}{2}A$ in the range $0 \leqslant A \leqslant 360°$.

Maximum value = 3; period = $180°/(1/2) = 360°$. A sketch of $3 \sin^2 \dfrac{1}{2}A$ is shown in *Fig. 21.16*.

Fig. 21.16

Problem 9 Sketch $y = 7 \cos^2 2A$ between $A = 0°$ and $A = 360°$.

Maximum value = 7; period = $180°/2 = 90°$.
A sketch of $y = 7 \cos^2 2A$ is shown in *Fig. 21.17*.

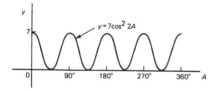

Fig. 21.17

Further problems on sine and cosine curves may be found in section 21.4, Problems 1 to 6, page 206.

21.2 Sinusoidal form *A* sin (*ωt* ± *α*)

In *Fig. 21.18*, let OR represent a vector that is free to rotate anticlockwise about 0 at a velocity of ω rad/s. A rotating vector is called a **phasor**. After a time t seconds OR will

have turned through an angle ωt radians (shown as angle TOR in *Fig. 21.18*). If ST is constructed perpendicular to OR, then sin ωt = ST/OT, i.e. ST = OT sin ωt.

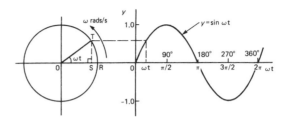

Fig. 21.18

If all such vertical components are projected on to a graph of y against ωt, a sine wave results of amplitude OR (as shown in Chapter 19).

If phasor OR makes one revolution (i.e. 2π radians) in T seconds, then the angular velocity, $\omega = 2\pi/T$ rad/s, from which, $T = 2\pi/\omega$ **seconds**. T is known as the **periodic time**.

The number of complete cycles occurring per second is called the **frequency, f**.

$$\text{Frequency} = \frac{\text{number of cycles}}{\text{second}}$$

$$= \frac{1}{T} = \frac{\omega}{2\pi} \text{ Hz, i.e. } \boldsymbol{f = \frac{\omega}{2\pi} \text{ Hz}}$$

Hence angular velocity, $\boldsymbol{\omega = 2\pi f \text{ rad/s}}$.

Given a general sinusoidal periodic function $y = A \sin(\omega t \pm \alpha)$, then A = amplitude, ω = angular velocity, $2\pi/\omega$ = periodic time, T, $\omega/2\pi$ = frequency, f, and α = angle of lead or lag (compared with $y = A \sin \omega t$).

Problem 10 An alternating current is given by $i = 30 \sin(100\pi t + 0.27)$ amperes. Find the amplitude, periodic time, frequency and phase angle (in degrees and minutes).

$i = 30 \sin(100\pi t + 0.27)$ A
Amplitude = **30 A**.
Angular velocity $\omega = 100\pi$. Hence periodic time,

$$T = \frac{2\pi}{\omega} = \frac{2\pi}{100\pi} = \frac{1}{50}$$

$$= \textbf{0.02 s or 20 ms}$$

Frequency, $f = \dfrac{1}{T} = \dfrac{1}{0.02} = \textbf{50 Hz}$

Phase angle, $\alpha = 0.27 \text{ rad} = \left(0.27 \times \dfrac{180}{\pi}\right)^{\circ}$

$$= \textbf{15°28′ leading } i = \textbf{30 sin (100}\pi t\textbf{)}$$

Problem 11 An oscillating mechanism has a maximum displacement of 2.5 m and a frequency of 60 Hz. At time $t = 0$ the displacement is 90 cm. Express the displacement in the general form $A \sin(\omega t \pm \alpha)$.

Amplitude = maximum displacement = 2.5 m. Angular velocity, $\omega = 2\pi f = 2\pi(60) = 120\pi \text{ rad/s}$. Hence displacement = $2.5 \sin(120\pi t + \alpha)$ m.
When $t = 0$, displacement = 90 cm = 0.90 m. Hence

$$0.90 = 2.5 \sin(0 + \alpha)$$

i.e. $\sin \alpha = \dfrac{0.90}{2.5} = 0.36$

Hence $\alpha = \arcsin 0.36 = 21°6′ = 0.368 \text{ rad}$

Thus **displacement = 2.5 sin (120πt + 0.368) m**.

Problem 12 The instantaneous value of voltage in an a.c. circuit at any time t seconds is given by $v = 340 \sin(50\pi t - 0.541)$ volts. Determine:

(a) the amplitude, periodic time, frequency and phase angle (in degrees),
(b) the value of the voltage when $t = 0$,
(c) the value of the voltage when $t = 10$ ms,
(d) the time when the voltage first reaches 200 V, and
(e) the time when the voltage is a maximum.

Sketch one cycle of the waveform.

(a) Amplitude = **340 V**
   Angular velocity, $\omega = 50\pi$. Hence periodic time:

$$T = \dfrac{2\pi}{\omega} = \dfrac{2\pi}{50\pi} = \dfrac{1}{25} = \textbf{0.04 s or 40 ms}$$

Frequency $f = \dfrac{1}{T} = \dfrac{1}{0.04} = \textbf{25 Hz}$

Phase angle = 0.541 rad

$$= \left(0.541 \times \dfrac{180}{\pi}\right)^{\circ}$$

$$= \textbf{31° lagging } v = 340 \sin(50\pi t)$$

(b) When $t = 0$, $v = 340 \sin(0 - 0.541) = 340 \sin(-31°)$
   = **−175.1 V**.
(c) When $t = 10$ ms then

$$v = 340 \sin\left(50\pi\dfrac{10}{10^3} - 0.541\right)$$

$$= 340 \sin(1.029\ 8)$$
$$= 340 \sin 59° = \textbf{291.4 volts}$$

(d) When $v = 200$ volts then

$$200 = 340 \sin(50\pi t - 0.541)$$

$$\dfrac{200}{340} = \sin(50\pi t - 0.541)$$

Hence $(50\pi t - 0.541) = \arcsin \dfrac{200}{340}$

$$= 36.03° \text{ or } 0.628\ 8 \text{ rad}$$
$$50\pi t = 0.628\ 8 + 0.541 = 1.169\ 8$$

Hence time, $t = \dfrac{1.169\ 8}{50\pi} = \textbf{7.447 ms}$

(e) When the voltage is a maximum, $v = 340$ V. Hence

$$340 = 340 \sin(50\pi t - 0.541)$$
$$1 = \sin(50\pi t - 0.541)$$
$$50\pi t - 0.541 = \arcsin 1 = 90° \text{ or } 1.570\ 8 \text{ rad}$$
$$50\pi t = 1.570\ 8 + 0.541 = 2.111\ 8$$

Hence time, $t = \dfrac{2.111\ 8}{50\pi} = \textbf{13.44 ms}$

A sketch of $v = 340 \sin(50\pi t - 0.541)$ volts is shown in *Fig. 21.19*.

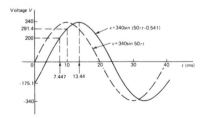

Fig. 21.19

Further problems on the sinusoidal form A sin ($\omega t \pm \alpha$) may be found in section 21.4, Problems 7 to 9, page 206.

21.3 Combination of two periodic functions of the same frequency

There are a number of instances in engineering and science where waveforms combine and where it is required to determine the single phasor (called the resultant) which could replace two or more separate phasors. Uses are found in electrical alternating current theory, in mechanical vibrations, in the addition of forces and with sound waves. There are several methods of determining the resultant and two such methods are shown below.

(i) **Plotting the periodic functions graphically**
This may be achieved by sketching the separate functions on the same axes and then adding (or subtracting) ordinates at regular intervals (see *Problems 13 to 15*). Alternatively, a table of values may be drawn up before plotting the resultant waveforms (see *Problem 16*).

(ii) **Resolution of phasors by drawing or calculation**
The resultant of two periodic functions may be found from their relative positions when the time is zero. For example, if $y_1 = 4 \sin \omega t$ and $y_2 = 3 \sin (\omega t - \pi/3)$ then each may be represented as phasors as shown in *Fig. 21.20*, y_1 being 4 units long and drawn horizontally and y_2 being 3 units long, lagging y_1 by $\pi/3$ radians or 60°. To determine the resultant of $y_1 + y_2$, y_1 is drawn horizontally as shown in *Fig. 21.21* and y_2 is joined to the end of y_1 at 60° to the horizontal. The resultant is given by y_R. This is the same as the diagonal of a parallelogram which is shown completed in *Fig. 21.22*. Resultant y_R, in *Figs 21.21* and *21.22*, is determined either by:

(a) scaled drawing and measurement, or
(b) by use of the cosine rule (and then sine rule to calculate angle ϕ), or
(c) by determining horizontal and vertical components of lengths oa and ab in *Fig. 21.21*, and then using Pythagoras' theorem to calculate ob.

In this case, by calculation, $y_R = 6.083$ and angle $\phi = 25.28°$ or 0.441 rad. Thus the resultant may be expressed in sinusoidal form as $y_R = 6.083 \sin (\omega t - 0.441)$. If the resultant phasor, $y_R = y_1 - y_2$ is required, then y_2 is still 3 units long but is drawn in the opposite direction, as shown in *Fig. 21.23*, and y_R is determined by measurement or calculation. (See *Problems 17 to 19*.)

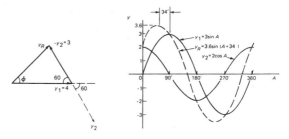

Fig. 21.23   **Fig. 21.24**

> *Problem 13* Plot the graph of $y_1 = 3 \sin A$ from $A = 0°$ to $A = 360°$. On the same axes plot $y_2 = 2 \cos A$. By adding ordinates plot $y_R = 3 \sin A + 2 \cos A$ and obtain a sinusoidal expression for this resultant waveform.

$y_1 = 3 \sin A$ and $y_2 = 2 \cos A$ are shown plotted in *Fig. 21.24*. Ordinates may be added at, say, 15° intervals. For example,

at   0°, $y_1 + y_2 = 0 + 2 = 2$
at 15°, $y_1 + y_2 = 0.78 + 1.93 = 2.71$
at 120°, $y_1 + y_2 = 2.6 + -1 = 1.6$
at 210°, $y_1 + y_2 = -1.5 - 1.73 = -3.23$, and so on.

The resultant waveform, shown by the broken line, has the same period, i.e. 360°, and thus the same frequency as the single phasors. The maximum value, or amplitude, of the resultant is 3.6. The resultant waveform leads $y_1 = 3 \sin A$ by 34° or 0.593 rad. The sinusoidal expression for the resultant waveform is:

$$y_R = 3.6 \sin (A + 34°) \text{ or } y_R = 3.6 \sin (A + 0.593)$$

> *Problem 14* Plot the graphs of $y_1 = 4 \sin \omega t$ and $y_2 = 3 \sin (\omega t - \pi/3)$ on the same axis, over one cycle. By adding ordinates at intervals plot $y_R = y_1 + y_2$ and obtain a sinusoidal expression for the resultant waveform.

$y_1 = 4 \sin \omega t$ and $y_2 = 3 \sin (\omega t - \pi/3)$ are shown plotted in *Fig. 21.25*. Ordinates are added at 15° intervals and the

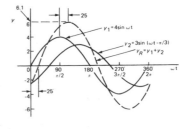

Fig. 21.25

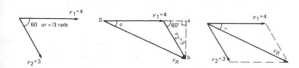

Fig. 21.20   **Fig. 21.21**   **Fig. 21.22**

resultant is shown by the broken line. The amplitude of the resultant is 6.1 and it lags y_1 by 25° or 0.436 rad. Hence the sinusoidal expression for the resultant waveform is:

$$y_R = 6.1 \sin (\omega t - 0.436)$$

Problem 15   Determine a sinusoidal expression for $y_1 - y_2$ when $y_1 = 4 \sin \omega t$ and $y_2 = 3 \sin (\omega t - \pi/3)$.

---

y_1 and y_2 are shown plotted in *Fig. 21.26*. At 15° intervals y_2 is subtracted from y_1. For example:

at   0°, $y_1 - y_2 = 0 - (-2.6) = +2.6$
at   30°, $y_1 - y_2 = 2 - (-1.5) = +3.5$
at 150°, $y_1 - y_2 = 2 - 3 = -1$, and so on.

The amplitude, or peak value of the resultant (shown by the broken line), is 3.6 and it leads y_1 by 45° or 0.79 rad. Hence

$$y_1 - y_2 = 3.6 \sin (\omega t + 0.79)$$

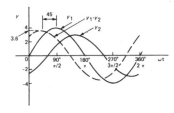

Fig. 21.26

Problem 16   Draw a graph to represent current $i = 2.4 \sin t + 3.2 \sin (t + 40°)$ amperes and express i in the general form $i = A \sin (t \pm \alpha)$.

A table of values may be drawn up as shown below.

$t°$	0	30	60	90	120
$\sin t$	0	0.5	0.866	1.0	0.866
$2.4 \sin t$	0	1.2	2.08	2.4	2.08
$(t + 40°)$	40	70	100	130	160
$\sin (t + 40°)$	0.64	0.94	0.98	0.77	0.34
$3.2 \sin (t + 40°)$	2.05	3.01	3.14	2.46	1.09
$i = 2.4 \sin t + 3.2 \sin (t + 40°)$	2.05	4.21	5.22	4.86	3.17

$t°$	150	180	210	240	270
$\sin t$	0.5	0	-0.5	-0.866	-1.0
$2.4 \sin t$	1.2	0	-1.2	-2.08	-2.4
$(t + 40°)$	190	220	250	280	310
$\sin (t + 40°)$	-0.17	-0.64	-0.94	-0.98	-0.77
$3.2 \sin (t + 40°)$	-0.54	-2.05	-3.01	-3.14	-2.46
$i = 2.4 \sin t + 3.2 \sin (t + 40°)$	0.66	-2.05	-4.21	-5.22	-4.48

$t°$	300	330	360
$\sin t$	-0.866	-0.5	0
$2.4 \sin t$	- 2.08	-1.2	0
$(t + 40°)$	340	370	400
$\sin (t + 40°)$	-0.34	0.17	0.64
$3.2 \sin (t + 40°)$	-1.09	0.54	2.05
$i = 2.4 \sin t + 3.2 \sin (t + 40°)$	-3.17	-0.66	2.05

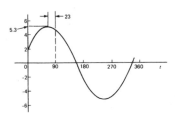

Fig. 21.27

A graph of $i = 2.4 \sin t + 3.2 \sin (t + 40°)$ is shown in *Fig. 21.27*. The amplitude is 5.3 amperes and i is 23°, i.e. 0.40 rad, ahead of a sine wave starting at 0°. Hence

$$i = 5.3 \sin (t + 0.40) \text{ amperes}$$

Problem 17   Given $y_1 = 2 \sin \omega t$ and $y_2 = 3\sin (\omega t + \pi/4)$, obtain an expression of the resultant $y_R = y_1 + y_2$, (a) by drawing and (b) by calculation.

(a) When time $t = 0$ the position of phasors y_1 and y_2 are as shown in *Fig. 21.28(a)*. To obtain the resultant, y_1 is drawn horizontally, 2 units long, y_2 is drawn 3 units long at an angle of $\pi/4$ rads or 45° and joined to the end of

y_1 as shown in *Fig. 21.28(b)*. y_R is measured as 4.6 units long and angle ϕ is measured as 27° or 0.47 rad. Alternatively, y_R is the diagonal of the parallelogram formed as shown in *Fig. 21.28(c)*.

Hence, by drawing, $y_R = \mathbf{4.6 \sin(\omega t + 0.47)}$

(b) From *Fig. 21.28(b)*, and using the cosine rule:

$$y_R{}^2 = 2^2 + 3^2 - [2(2)(3) \cos 135°]$$
$$= 4 + 9 - [-8.485] = 21.49$$

Hence $y_R = \sqrt{(21.49)} = 4.64$

Using the sine rule: $\dfrac{3}{\sin \phi} = \dfrac{4.64}{\sin 135°}$ from which

$$\sin \phi = \frac{3 \sin 135°}{4.64} = 0.457\,2$$

Hence $\phi = \arcsin 0.457\,2 = 27°12'$ or 0.475 rad
By calculation $y_R = \mathbf{4.64 \sin(\omega t + 0.475)}$

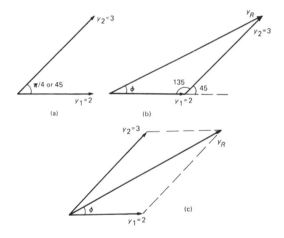

Fig. 21.28

Problem 18 Two alternating voltages are given by $v_1 = 15 \sin \omega t$ volts and $v_2 = 25 \sin(\omega t - \pi/6)$ volts. Determine a sinusoidal expression for the resultant $v_R = v_1 + v_2$ by finding horizontal and vertical components.

The relative positions of v_1 and v_2 at time $t = 0$ are shown in *Fig. 21.29(a)* and the phasor diagram is shown in *Fig. 21.29(b)*.

The horizontal component of $v_R = oa + ab$
$$= 15 + 25 \cos 30°$$
$$= 36.65 \text{ V}$$

The vertical component of $v_R = bc = 25 \sin 30°$
$$= 12.50 \text{ V}$$

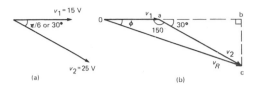

Fig. 21.29

Hence

$$v_R \;(= oc) = \sqrt{[(36.65)^2 + (12.50)^2]} \text{ by Pythagoras'}$$
theorem
$$= 38.72 \text{ volts}$$

$$\tan \phi = \frac{bc}{ob} = \frac{12.50}{36.65} = 0.341\,1$$

from which, $\phi = \arctan 0.341\,1 = 18°50'$ or 0.329 radians.
Hence

$$v_R = v_1 + v_2 = \mathbf{38.72 \sin(\omega t - 0.329) \text{ V}}$$

Problem 19 For the voltages in *Problem 18*, determine the resultant $v_R = v_1 - v_2$.

To find the resultant $v_R = v_1 - v_2$, the phasor v_2 of *Fig. 21.29(b)* is reversed in direction as shown in *Fig. 21.30*. Using the cosine rule:

$$v_R{}^2 = 15^2 + 25^2 - 2(15)(25) \cos 30°$$
$$= 225 + 625 - 649.5 = 200.5$$
$$v_R = \sqrt{(200.5)} = 14.16 \text{ volts}$$

Using the sine rule: $\dfrac{25}{\sin \phi} = \dfrac{14.16}{\sin 30°}$ from which

$$\sin \phi = \frac{25 \sin 30°}{14.16} = 0.882\,8$$

Hence $\phi = \arcsin 0.882\,8 = 61.98°$ or 118.02°. From *Fig. 21.30*, ϕ is obtuse, hence $\phi = 118.02°$ or 2.06 radians. Hence

$$v_R = v_1 - v_2 = \mathbf{14.16 \sin(\omega t + 2.06) \text{ V}}$$

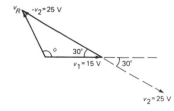

Fig. 21.30

Further problems on the combination of two periodic functions of the same frequency may be found in section 21.4, Problems 10 to 14, page 206.

21.4 Further problems on trigonometric graphs and the combination of waveforms

Sine and cosine curves

In *Problems 1* to *3* state the amplitude and period of the waveform and sketch the curve between 0° and 360°.

1 (a) $y = \cos 3A$ (b) $y = 2 \sin \dfrac{5x}{2}$ (c) $y = 3 \sin 4t$

[(a) 1, 120° (b) 2, 144° (c) 3, 90°]

2 (a) $y = 3 \cos \dfrac{\theta}{2}$ (b) $y = \dfrac{7}{2} \sin \dfrac{3x}{8}$

(c) $y = 6 \sin (t - 45°)$

[(a) 3, 720° (b) $\dfrac{7}{2}$, 960° (c) 6, 360°]

3 (a) $y = 4 \cos (2\theta + 30°)$ (b) $y = 2 \sin^2 2t$

(c) $y = 5 \cos^2 \dfrac{3}{2}\theta$

[(a) 4, 180° (b) 2, 90° (c) 5, 120°]

In *Problems 4* to *6* state the amplitude, periodic time, frequency and phase angle (in degrees).

4 $v = 200 \sin (200\pi t + 0.29)$ V
[200 V, 10 ms, 100 Hz, 16°37′ leading]

5 $i = 32 \sin (400\pi t - 0.42)$ A
[32 A, 5 ms, 200 Hz, 24°4′ lagging]

6 $x = 5 \sin (314.2t + 0.33)$ cm
[5 cm, 20 ms, 50 Hz, 18°54′ leading]

Sinusoidal form A sin (ωt ± α)

7 A sinusoidal voltage has a maximum value of 120 V and a frequency of 50 Hz. At time $t = 0$, the voltage is (a) zero, and (b) 50 V. Express the instantaneous voltage v in the form $v = A \sin (\omega t \pm \alpha)$.
[(a) $v = 120 \sin 100\pi t$ volts
 (b) $v = 120 \sin (100\pi t + 0.43)$ volts]

8 An alternating current has a periodic time of 25 ms and a maximum value of 20 A. When time $t = 0$, current $i = -10$ amperes. Express the current i in the form $i = A \sin (\omega t \pm \alpha)$.

[$i = 20 \sin \left(80\pi t - \dfrac{\pi}{6} \right)$ amperes]

9 The current in an a.c. circuit at any time t seconds is given by:

$i = 5 \sin (100\pi t - 0.432)$ amperes

Determine (a) the amplitude, periodic time, frequency and phase angle (in degrees) (b) the value of current at $t = 0$ (c) the value of current at $t = 8$ ms (d) the time when the current is first a maximum (e) the time when the current first reaches 3A. Sketch one cycle of the waveform showing relevant points.
[(a) 5 A, 20 ms, 50 Hz, 24°45′ lagging (b) −2.093 A
 (c) 4.363 A (d) 6.375 ms (e) 3.423 ms]

Combination of two periodic functions of the same frequency

10 Plot the graph of $y = 2 \sin A$ from $A = 0°$ to $A = 360°$. On the same axes plot $y = 4 \cos A$. By adding ordinates at intervals plot $y = 2 \sin A + 4 \cos A$ and obtain a sinusoidal expression for the waveform.
[4.5 sin $(A + 63°26′)$]

11 Two alternating voltages are given by $v_1 = 10 \sin \omega t$ volts and $v_2 = 14 \sin (\omega t + \pi/3)$ volts. By plotting v_1 and v_2 on the same axes over one cycle obtain a sinusoidal expression for (a) $v_1 + v_2$ (b) $v_1 - v_2$.
[(a) 20.9 sin $(\omega t + 0.62)$ volts
 (b) 12.5 sin $(\omega t - 1.33)$ volts]

12 Draw up a table of values for the waveform $y = 5 \sin (\omega t + \pi/4) + 3 \sin (\omega t - \pi/3)$ and plot a graph of y against ωt. Express y in the form $y = A \sin (\omega t \pm \alpha)$.
[5.1 sin $(\omega t + 0.184)$]

In *Problems 13* and *14*, express the combination of periodic functions in the form $A \sin (\omega t \pm \alpha)$ either by drawing or by calculation.

13 (a) $12 \sin \omega t + 5 \cos \omega t$

(b) $7 \sin \omega t + 5 \sin \left(\omega t + \dfrac{\pi}{4} \right)$

(c) $6 \sin \omega t + 3 \sin \left(\omega t - \dfrac{\pi}{6} \right)$

[(a) 13 sin $(\omega t + 0.395)$
 (b) 11.11 sin $(\omega t + 0.324)$
 (c) 8.73 sin $(\omega t - 0.173)$]

14 (a) $i = 25 \sin \omega t - 15 \sin \left(\omega t + \dfrac{\pi}{3} \right)$

(b) $v = 8 \sin \omega t - 5 \sin \left(\omega t - \dfrac{\pi}{4} \right)$

(c) $x = 9 \sin \left(\omega t + \dfrac{\pi}{3} \right) - 7 \sin \left(\omega t - \dfrac{3\pi}{8} \right)$

[(a) $i = 21.79 \sin (\omega t - 0.639)$
 (b) $v = 5.695 \sin (\omega t + 0.670)$
 (c) $x = 14.38 \sin (\omega t + 1.444)$]

22

Trigonometric identities and the solution of equations

22.1 Trigonometric identities

A **trigonometric identity** is an expression that is true for all values of the unknown variable.

$$\tan\theta = \frac{\sin\theta}{\cos\theta}, \cot\theta = \frac{\cos\theta}{\sin\theta}, \sec\theta = \frac{1}{\cos\theta},$$

$$\operatorname{cosec}\theta = \frac{1}{\sin\theta} \text{ and } \cot\theta = \frac{1}{\tan\theta}$$

are examples of trigonometric identities (see Chapter 19).

Applying Pythagoras' theorem to the right-angled triangle shown in *Fig. 22.1* gives:

$$a^2 + b^2 = c^2 \qquad (1)$$

Dividing each term of equation (1) by c^2 gives:

$$\frac{a^2}{c^2} + \frac{b^2}{c^2} = \frac{c^2}{c^2}, \text{ i.e. } \left(\frac{a}{c}\right)^2 + \left(\frac{b}{c}\right)^2 = 1$$

$$(\cos\theta)^2 + (\sin\theta)^2 = 1$$

Hence $\cos^2\theta + \sin^2\theta = 1$ \qquad (2)

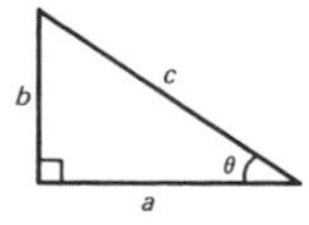

Fig. 22.1

Dividing each term of equation (1) by a^2 gives:

$$\frac{a^2}{a^2} + \frac{b^2}{a^2} = \frac{c^2}{a^2}, \text{ i.e. } 1 + \left(\frac{b}{a}\right)^2 = \left(\frac{c}{a}\right)^2$$

Hence $1 + \tan^2\theta = \sec^2\theta$ \qquad (3)

Dividing each term of equation (1) by b^2 gives:

$$\frac{a^2}{b^2} + \frac{b^2}{b^2} = \frac{c^2}{b^2}, \text{ i.e. } \left(\frac{a}{b}\right)^2 + 1 = \left(\frac{c}{b}\right)^2$$

Hence $\cot^2\theta + 1 = \operatorname{cosec}^2\theta$ \qquad (4)

Equations (2), (3) and (4) are three further examples of trigonometric identities. For the proof of further trigonometric identities, see the following worked problem.

Problem 1 Prove the identity $\sin^2\theta \cot\theta \sec\theta = \sin\theta$.

With trigonometric identities it is necessary to start with the left-hand side (LHS) and attempt to make it equal to the right-hand side (RHS) or vice-versa. It is often useful to change all of the trigonometric ratios into sines and cosines where possible. Thus

$$\text{LHS} = \sin^2\theta \cot\theta \sec\theta = \sin^2\theta\left(\frac{\cos\theta}{\sin\theta}\right)\left(\frac{1}{\cos\theta}\right)$$

$$= \sin\theta \text{ (by cancelling)} = \text{RHS}$$

Problem 2 Prove that $\dfrac{\tan x + \sec x}{\sec x\left(1 + \dfrac{\tan x}{\sec x}\right)} = 1$

$$\text{LHS} = \frac{\tan x + \sec x}{\sec x\left(1 + \dfrac{\tan x}{\sec x}\right)} = \frac{\dfrac{\sin x}{\cos x} + \dfrac{1}{\cos x}}{\left(\dfrac{1}{\cos x}\right)\left(1 + \dfrac{\dfrac{\sin x}{\cos x}}{\dfrac{1}{\cos x}}\right)}$$

$$= \frac{\dfrac{\sin x + 1}{\cos x}}{\left(\dfrac{1}{\cos x}\right)\left[1 + \left(\dfrac{\sin x}{\cos x}\right)\left(\dfrac{\cos x}{1}\right)\right]}$$

$$= \frac{\dfrac{\sin x + 1}{\cos x}}{\dfrac{1}{\cos x}[1 + \sin x]} = \left(\frac{\sin x + 1}{\cos x}\right)\left(\frac{\cos x}{1 + \sin x}\right)$$

$$= 1 \text{ (by cancelling)} = \text{RHS}$$

Problem 3 Prove that $\dfrac{1 + \cot\theta}{1 + \tan\theta} = \cot\theta$

$$\text{LHS} = \frac{1 + \cot\theta}{1 + \tan\theta} = \frac{1 + \dfrac{\cos\theta}{\sin\theta}}{1 + \dfrac{\sin\theta}{\cos\theta}} = \frac{\dfrac{\sin\theta + \cos\theta}{\sin\theta}}{\dfrac{\cos\theta + \sin\theta}{\cos\theta}}$$

$$= \left(\frac{\sin\theta + \cos\theta}{\sin\theta}\right)\left(\frac{\cos\theta}{\cos\theta + \sin\theta}\right) = \frac{\cos\theta}{\sin\theta}$$

$$= \cot\theta = \text{RHS}$$

Problem 4 Show that $\cos^2\theta - \sin^2\theta = 1 - 2\sin^2\theta$

From equation (2), $\cos^2\theta + \sin^2\theta = 1$, from which, $\cos^2\theta = 1 - \sin^2\theta$. Hence,

$$\text{LHS} = \cos^2\theta - \sin^2\theta = (1 - \sin^2\theta) - \sin^2\theta$$
$$= 1 - \sin^2\theta - \sin^2\theta = 1 - 2\sin^2\theta = \text{RHS}$$

Problem 5 Prove that $\sqrt{\left(\dfrac{1 - \sin x}{1 + \sin x}\right)} = \sec x - \tan x$

$$\text{LHS} = \sqrt{\left(\frac{1 - \sin x}{1 + \sin x}\right)} = \sqrt{\left\{\frac{(1 - \sin x)(1 - \sin x)}{(1 + \sin x)(1 - \sin x)}\right\}}$$

$$= \sqrt{\left\{\frac{(1 - \sin x)^2}{(1 - \sin^2 x)}\right\}}$$

Since $\cos^2 x + \sin^2 x = 1$ then $1 - \sin^2 x = \cos^2 x$,

$$\text{LHS} = \sqrt{\left\{\frac{(1 - \sin x)^2}{(1 - \sin^2 x)}\right\}} = \sqrt{\left\{\frac{(1 - \sin x)^2}{\cos^2 x}\right\}}$$

$$= \frac{1 - \sin x}{\cos x} = \frac{1}{\cos x} - \frac{\sin x}{\cos x}$$

$$= \sec x - \tan x = \text{RHS}$$

Problem 6 Use a calculator to show that the identity $1 + \cot^2\theta = \text{cosec}^2\theta$ is true when $\theta = 223°$.

$$\cot 223° = \frac{1}{\tan 223°} = 1.072\,368\,7\ldots$$

$$\cot^2 223° = (\cot 223°)^2 = (1.072\,368\,7\ldots)^2$$

$$= 1.149\,974\,6\ldots$$

Hence $\quad \text{LHS} = 1 + \cot^2 223° = 2.149\,974\,6\ldots$

$$\text{cosec } 223° = \frac{1}{\sin 223°} = -1.466\,279\,18\ldots$$

$$\text{RHS} = \text{cosec}^2 223° = (\text{cosec } 223°)^2$$

$$= (-1.466\,279\,18\ldots)^2 = 2.149\,974\,6\ldots$$

Since LHS = RHS the identity is shown to be true when $\theta = 223°$.

Further problems on trigonometric identities may be found in section 22.3, Problems 1 to 7, page 212.

22.2 Trigonometric equations

Equations which contain trigonometric ratios are called **trigonometric equations**. There are usually an infinite number of solutions to such equations: however, solutions are often restricted to those between 0° and 360°. A knowledge of angles of any magnitude is essential in the solution of trigonometric equations (see Chapter 19) and calculators cannot be relied upon to give all the solutions (as shown on page 181). *Fig. 22.2* shows a summary for angles of any magnitude.

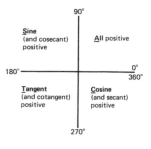

Fig. 22.2

Equations of the type $a \sin^2 A + b \sin A + c = 0$

(i) **When $a = 0$,** $b \sin A + c = 0$, hence

$$\sin A = -\frac{c}{b} \text{ and } A = \textbf{arcsin}\left(-\frac{c}{b}\right)$$

There are two values of A between $0°$ and $360°$ which satisfy such an equation, provided $-1 \leqslant c/b \leqslant 1$ (see *Problems 7 to 10*).

(ii) **When $b = 0$,** $a \sin^2 A + c = 0$, hence

$$\sin^2 A = -\frac{c}{a}$$

$$\sin A = \sqrt{\left(-\frac{c}{a}\right)}$$

and $\quad A = \textbf{arcsin}\,\sqrt{\left(-\frac{c}{a}\right)}$

If either a or c is a negative number, then the value within the square root sign is positive. Since when a square root is taken there is a positive and negative answer there are four values of A between $0°$ and $360°$ which satisfy such an equation, provided $-1 \leqslant c/a \leqslant 1$ (see *Problems 11 to 13*).

(iii) **When a, b and c are all non-zero:**
$a \sin^2 A + b \sin A + c = 0$ is a quadratic equation in which the unknown is $\sin A$. The solution of a quadratic equation is obtained either by factorizing (if possible) or by using the quadratic formula:

$$\sin A = \frac{-b \pm \sqrt{(b^2 - 4ac)}}{2a} \text{ (see *Problems 14 to 16*)}$$

(iv) Often the trigonometric identities $\cos^2 A + \sin^2 A = 1$, $1 + \tan^2 A = \sec^2 A$ and $\cot^2 A + 1 = \text{cosec}^2 A$ need to be used to reduce equations to one of the above forms (see *Problems 17 to 20*).

> **Problem 7** Solve the trigonometric equation $5 \sin \theta + 3 = 0$ for values of θ from $0°$ to $360°$.

$5 \sin \theta + 3 = 0$, from which $\sin \theta = -3/5 = -0.600\,0$. Hence $\theta = \arcsin\,(-0.600\,0)$. Sine is negative in the third and fourth quadrants (see *Fig. 22.3*). The acute angle $\arcsin\,(0.600\,0) = 36°52'$ (shown as α in *Fig. 22.3(b)*). Hence

$\theta = 180° + 36°52'$, i.e. **216°52'** or
$\theta = 360° - 36°52'$, i.e. **323°8'**

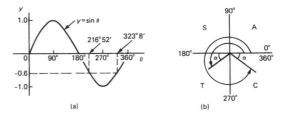

Fig. 22.3

> **Problem 8** Solve $1.5 \tan x - 1.8 = 0$ for $0° \leqslant x \leqslant 360°$.

$1.5 \tan x - 1.8 = 0$, from which $\tan x = 1.8/1.5 = 1.200\,0$. Hence $x = \arctan 1.200\,0$. Tangent is positive in the first and third quadrants (see *Fig. 22.4*). The acute angle $\arctan 1.200\,0 = 50°12'$. Hence

$$x = \textbf{50°12'} \text{ or } 180° + 50°12' = \textbf{230°12'}$$

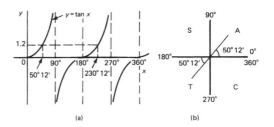

Fig. 22.4

> **Problem 9** Solve $4 \sec t = 5$ for values of t between $0°$ and $360°$.

$4 \sec t = 5$, from which, $\sec t = \dfrac{5}{4} = 1.250\,0$. Hence $t = \text{arcsec } 1.250\,0$. Secant $= 1/\text{cosine}$ is positive in the first and fourth quadrants (see *Fig. 22.5*). The acute angle arcsec $1.250\,0 = 36°52'$. Hence

$$t = \textbf{36°52'} \text{ or } 360° - 36°52' = \textbf{323°8'}$$

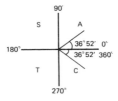

Fig. 22.5

Problem 10   Solve 3.2 (cot θ – 1) = –12 for values of θ between 0° and 360°.

3.2 (cot θ –1) = –12, hence

$$\cot \theta - 1 = \frac{-12}{3.2} = -3.750\ 0$$

$$\cot \theta = -3.750\ 0 + 1 = -2.750\ 0$$

$$\theta = \text{arccot}\ (-2.750\ 0)$$

Cotangent = 1/tangent is negative in the second and fourth quadrants (see *Fig. 22.6*). The acute angle arccot (2.750 0) = 19°59′. Hence

$$\theta = 180° - 19°59' = \mathbf{160°1'}$$
$$\text{or} \quad \theta = 360° - 19°59' = \mathbf{340°1'}$$

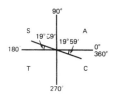

Fig. 22.6

Problem 11   Solve 2 – 4 cos² A = 0 for values of A in the range 0° ≤ A ≤ 360°.

2 – 4 cos² A = 0, from which cos² $A = \dfrac{2}{4} = 0.500\ 0$. Hence

cos $A = \sqrt{(0.500\ 0)} = \pm 0.707\ 1$ and A = arccos ($\pm 0.707\ 1$). Cosine is positive in quadrants one and four and negative in quadrants two and three. Thus in this case there are four solutions, one in each quadrant (see *Fig. 22.7*). The acute angle arccos 0.707 1 = 45°. Hence

A = 45°, 135°, 225° or 315°

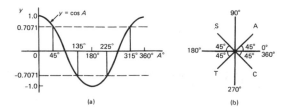

Fig. 22.7

Problem 12   Solve 0.75 sec² x – 1.2 = 0 for values of x between 0° and 360°.

0.75 sec² x – 1.2 = 0, from which, sec² x = 1.2/0.75 = 1.600 0. Hence sec $x = \sqrt{(1.600\ 0)} = \pm 1.264\ 9$ and x =

arcsec ($\pm 1.264\ 9$). There are four solutions, one in each quadrant: the acute angle arcsec 1.264 9 = 37°46′. Hence

x = 37°46′, 142°14′, 217°46′ or 322°14′

Problem 13   Solve $\dfrac{1}{2}$ cot² y = 1.3 for 0° ≤ y ≤ 360°.

$\dfrac{1}{2}$ cot² y = 1.3, from which, cot² y = 2(1.3) = 2.6. Hence cot $y = \sqrt{2.6} = \pm 1.612\ 5$, and y = arccot ($\pm 1.612\ 5$). There are four solutions, one in each quadrant: the acute angle arccot 1.612 5 = 31°48′. Hence

y = 31°48′, 148°12′, 211°48′ or 328°12′

Problem 14   Solve the equation 8 sin² θ + 2 sin θ – 1 = 0, for all values of θ between 0° and 360°.

Factorizing 8 sin² θ + 2 sin θ – 1 = 0 gives (4 sin θ – 1)(2 sin θ + 1) = 0. Hence 4 sin θ – 1 = 0, from which, sin $\theta = \dfrac{1}{4} = 0.250\ 0$, or 2 sin θ + 1 = 0, from which, sin $\theta = -\dfrac{1}{2} = -0.500\ 0$.

θ = arcsin 0.250 0 = 14°29′ or 165°31′, since sine is positive in the first and second quadrants, or θ = arcsin (–0.500 0) = 210° or 330°, since sine is negative in the third and fourth quadrants. Hence

θ = 14°29′, 165°31′, 210° or 330°

Problem 15   Solve 6 cos² θ + 5 cos θ – 6 = 0 for values of θ from 0° to 360°.

Factorizing 6 cos² θ + 5 cos θ – 6 = 0 gives (3 cos θ – 2)(2cos θ + 3) = 0. Hence 3 cos θ – 2 = 0, from which, cos $\theta = \dfrac{2}{3} = 0.666\ 7$, or 2 cos θ + 3 = 0, from which, cos $\theta = -\dfrac{3}{2} = -1.500\ 0$.

The minimum value of a cosine is –1, hence the latter expression has no solution and is thus neglected. Hence

θ = arccos 0.666 7 = 48°11′ or 311°49′

since cosine is positive in the first and fourth quadrants.

Problem 16   Solve 9 tan² x + 16 = 24 tan x in the range 0° ≤ x ≤ 360°.

Rearranging gives 9 tan² x – 24 tan x + 16 = 0 and factorizing gives (3 tan x – 4)(3 tan x – 4) = 0, i.e. (3 tan x – 4)²

= 0. Hence tan $x = \dfrac{4}{3} = 1.333\ 3$ and

$x = \arctan 1.333\ 3 = \textbf{53°8′ or 233°8′}$

since tangent is positive in the first and third quadrants.

Problem 17 Solve $5 \cos^2 t + 3 \sin t - 3 = 0$ for values of t from $0°$ to $360°$.

Since $\cos^2 t + \sin^2 t = 1$, $\cos^2 t = 1 - \sin^2 t$. Substituting for $\cos^2 t$ in $5 \cos^2 t + 3 \sin t - 3 = 0$ gives

$$5(1 - \sin^2 t) + 3 \sin t - 3 = 0$$
$$5 - 5 \sin^2 t + 3 \sin t - 3 = 0$$
$$-5 \sin^2 t + 3 \sin t + 2 = 0$$
$$5 \sin^2 t - 3 \sin t - 2 = 0$$

Factorizing gives $(5 \sin t + 2)(\sin t - 1) = 0$. Hence $5 \sin t + 2 = 0$, from which, $\sin t = -\dfrac{2}{5} = -0.400\ 0$, or $\sin t - 1 = 0$, from which, $\sin t = 1$.
$t = \arcsin (-0.400\ 0) = 203°35′$ or $336°25′$, since sine is negative in the third and fourth quadrants, or $t = \arcsin 1 = 90°$. Hence

$t = \textbf{90°, 203°35′ or 336°25′}$ as shown in *Fig. 22.8*.

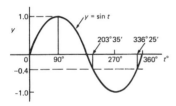

Fig. 22.8

Problem 18 Solve $18 \sec^2 A - 3 \tan A = 21$ for values of A between $0°$ and $360°$.

$1 + \tan^2 A = \sec^2 A$. Substituting for $\sec^2 A$ in $18 \sec^2 A - 3 \tan A = 21$ gives $18(1 + \tan^2 A) - 3 \tan A = 21$, i.e.

$$18 + 18 \tan^2 A - 3 \tan A - 21 = 0$$
$$18 \tan^2 A - 3 \tan A - 3 = 0$$

Factorizing gives $(6 \tan A - 3)(3 \tan A + 1) = 0$. Hence $6 \tan A - 3 = 0$, from which, $\tan A = \dfrac{3}{6} = 0.500\ 0$ or $3 \tan A + 1 = 0$, from which, $\tan A = -\dfrac{1}{3} = -0.333\ 3$. Thus $A = \arctan (0.500\ 0) = 26°34′$ or $206°34′$, since tangent is positive in the first and third quadrants, or $A = \arctan (-0.333\ 3) = 161°34′$ or $341°34′$, since tangent is negative in the second and fourth quadrants. Hence

$A = \textbf{26°34′, 161°34′, 206°34′ or 341°34′}$

Problem 19 Solve $3 \operatorname{cosec}^2 \theta - 5 = 4 \cot \theta$ in the range $0 \leqslant \theta \leqslant 360°$.

$\cot^2 \theta + 1 = \operatorname{cosec}^2 \theta$. Substituting for $\operatorname{cosec}^2 \theta$ in $3 \operatorname{cosec}^2 \theta - 5 = 4 \cot \theta$ gives

$$3 (\cot^2 \theta + 1) - 5 = 4 \cot \theta$$
$$3 \cot^2 \theta + 3 - 5 = 4 \cot \theta$$
$$3 \cot^2 \theta - 4 \cot \theta - 2 = 0$$

Since the left-hand side does not factorize the quadratic formula is used. Thus

$$\cot \theta = \frac{-(-4) \pm \sqrt{[(-4)^2 - 4(3)(-2)]}}{2(3)}$$

$$= \frac{4 \pm \sqrt{(16 + 24)}}{6} = \frac{4 \pm \sqrt{40}}{6}$$

$$= \frac{10.324\ 6}{6} \text{ or } -\frac{2.324\ 6}{6}$$

Hence $\cot \theta = 1.720\ 8$ or $-0.387\ 4$. $\theta = \operatorname{arccot} 1.720\ 8 = 30°10′$ or $210°10′$, since cotangent is positive in the first and third quadrants, or $\theta = \operatorname{arccot} (-0.387\ 4) = 111°11′$ or $291°11′$, since cotangent is negative in the second and fourth quadrants. Hence

$\theta = \textbf{30°10′, 111°11′, 210°10′ or 291°11′}$

Problem 20 Solve $7 \sin^2 \theta - 4 \cos \theta = 5$ for values of θ between $0°$ and $360°$.

Since $\cos^2 \theta + \sin^2 \theta = 1$, $\sin^2 \theta = 1 - \cos^2 \theta$. Substituting for $\sin^2 \theta$ in $7 \sin^2 \theta - 4 \cos \theta = 5$ gives $7(1 - \cos^2 \theta) - 4 \cos \theta = 5$, i.e.

$$7 - 7 \cos^2 \theta - 4 \cos \theta = 5$$
$$-7 \cos^2 \theta - 4 \cos \theta + 2 = 0$$
$$7 \cos^2 \theta + 4 \cos \theta - 2 = 0$$

Since the left-hand side does not factorize the quadratic formula is used. Thus

$$\cos \theta = \frac{-4 \pm \sqrt{\{(4)^2 - 4(7)(-2)\}}}{2(7)} = \frac{-4 \pm \sqrt{(16 + 56)}}{14}$$

$$= \frac{-4 \pm \sqrt{72}}{14} = \frac{4.485\ 3}{14} \text{ or } \frac{-12.485\ 3}{14}$$

$$= 0.320\ 4 \text{ or } -0.891\ 8$$
$$\theta = \arccos 0.320\ 4 = 71°19′ \text{ or } 228°41′$$

or $\theta = \arccos(-0.891\ 8) = 153°6'$ or $206°54'$

Hence $\theta = \mathbf{71°19',\ 153°6',\ 206°54'}$ or $\mathbf{288°41'}$

Further problems on trigonometric equations may be found in section 22.3, Problems 8 to 17.

22.3   Further problems on trigonometric identities and the solution of equations

Trigonometric identities

In *Problems 1* to *6* prove the trigonometric identities.

1   $\sin x \cot x = \cos x$

2   $\dfrac{1}{\sqrt{(1 - \cos^2 \theta)}} = \operatorname{cosec} \theta$

3   $2 \cos^2 A - 1 = \cos^2 A - \sin^2 A$

4   $\dfrac{\cos x - \cos^3 x}{\sin x} = \sin x \cos x$

5   $(1 + \cot \theta)^2 + (1 - \cot \theta)^2 = 2 \operatorname{cosec}^2 \theta$

6   $\dfrac{\sin^2 x (\sec x + \operatorname{cosec} x)}{\cos x \tan x} = 1 + \tan x$

7   Show that the trigonometric identities $\cos^2 \theta + \sin^2 \theta = 1$, $1 + \tan^2 \theta = \sec^2 \theta$ and $\cot^2 \theta + 1 = \operatorname{cosec}^2 \theta$ are valid when θ is (a) 124°, (b) 231° and (c) 312°46′.

Trigonometric equations

In *Problems 8* to *17* solve the equations for angles between 0° and 360°.

8   (a) $4 - 7 \sin \theta = 0$ (b) $2.5 \cos x + 1.75 = 0$
   [(a) $\theta = 34°51'$ or $145°9'$
   (b) $x = 134°26'$ or $225°34'$]

9   (a) $3 \operatorname{cosec} A + 5.5 = 0$ (b) $4(2.32 - 5.4 \cot t) = 0$
   [(a) $A = 213°3'$ or $326°57'$
   (b) $t = 66°45'$ or $246°45'$]

10   (a) $5 \sin^2 y = 3$ (b) $3 \tan^2 \phi - 2 = 0$
   [(a) $y = 50°46', 129°14', 230°46'$ or $309°14'$
   (b) $\phi = 39°14', 140°46', 219°14'$ or $320°46'$]

11   (a) $5 + 3 \operatorname{cosec}^2 D = 8$ (b) $2 \cot^2 \theta = 5$
   [(a) $D = 90°$ or $270°$
   (b) $\theta = 32°19', 147°41', 212°19'$ or $327°41'$]

12   (a) $15 \sin^2 A + \sin A - 2 = 0$
   (b) $8 \tan^2 \theta + 2 \tan \theta = 15$
   [(a) $A = 19°28', 160°32', 203°35'$ or $336°25'$
   (b) $\theta = 51°20', 123°41', 231°20'$ or $303°41'$]

13   (a) $\sec x + 6 = \sec^2 x$ (b) $2 \operatorname{cosec}^2 t - 5 \operatorname{cosec} t = 12$
   [(a) $x = 70°32', 120°, 240°$ or $289°28'$
   (b) $t = 14°29', 165°31', 221°49'$ or $318°11'$]

14   (a) $12 \sin^2 \theta - 6 = \cos \theta$ (b) $16 \sec x - 2 = 14 \tan^2 x$
   [(a) $\theta = 48°11', 138°35', 221°25'$ or $311°49'$
   (b) $x = 52°56'$ or $307°4'$]

15   (a) $4 \cot^2 A - 6 \operatorname{cosec} A + 6 = 0$
   (b) $2 \cos^2 y + 3 \sin y = 3$
   [(a) 90°
   (b) 30°, 90° or 150°]

16   (a) $3.2 \sin^2 x + 2.5 \cos x - 1.8 = 0$
   (b) $5 \sec t + 2 \tan^2 t = 3$
   [(a) $112°11'$ or $247°49'$
   (b) $107°50'$ or $252°10'$]

17   (a) $2.9 \cos^2 \alpha - 7 \sin \alpha + 1 = 0$
   (b) $3 \operatorname{cosec}^3 \beta = 8 - 7 \cot \beta$
   [(a) $\alpha = 27°50'$ or $152°10'$
   (b) $\beta = 60°10', 161°1', 240°10'$ or $341°1'$]

23

Cartesian and polar coordinates

23.1 Introduction

There are two ways in which the position of a point in a plane can be represented. These are:

(a) by **Cartesian coordinates**, i.e. (x,y), and
(b) by **polar coordinates**, i.e. (r,θ), where r is a 'radius' from a fixed point and θ is an angle from a fixed point.

From plotting graphs in Chapters 10 to 13, we are familiar with Cartesian coordinates. Polar coordinates provide us with another method of plotting points.

23.2 Changing from Cartesian into polar coordinates

In *Fig. 23.1*, if lengths x and y are known, then the length of r can be obtained from Pythagoras' theorem (see Chapter 14) since OPQ is a right-angled triangle. Hence

$$r^2 = (x^2 + y^2)$$

from which $\boxed{r = \sqrt{(x^2 + y^2)}}$

From trigonometric ratios (see Chapter 19),

$$\tan \theta = \frac{y}{x}$$

from which $\boxed{\theta = \arctan \frac{y}{x}}$

$r = \sqrt{(x^2 + y^2)}$ and $\theta = \arctan (y/x)$ are the two formulae we need to change from Cartesian to polar coordinates. The angle θ, which may be expressed in degrees or radians, must **always** be measured from the positive x axis, i.e. measured from the line OQ in *Fig. 23.1*. It is suggested that when changing from Cartesian to polar coordinates a diagram should always be sketched.

Problem 1 Change the Cartesian coordinates (3,4) into polar coordinates.

A diagram representing the point (3,4) is shown in *Fig. 23.2*.

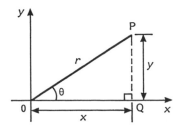

Fig. 23.1

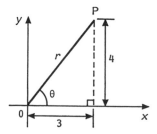

Fig. 23.2

From Pythagoras' theorem, $r = \sqrt{(3^2 + 4^2)} = 5$ (note that -5 has no meaning in this context). By trigonometric ratios, $\theta = \arctan \dfrac{4}{3} = 53.13°$ or 0.927 rad (note that $53.13°$ $= 53.13 \times (\pi/180)$ rad $= 0.927$ rad).

Hence (3,4) in Cartesian coordinates corresponds to (5,53.13°) or (5,0.927 rad) in polar coordinates.

Problem 2 Express in polar coordinates the position $(-4,3)$.

A diagram representing the point using the Cartesian coordinates $(-4,3)$ is shown in *Fig. 23.3*.

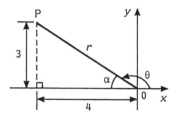

Fig. 23.3

From Pythagoras' theorem, $r = \sqrt{(4^2 + 3^2)} = 5$. By trigonometric ratios, $\alpha = \arctan \dfrac{3}{4} = 36.87°$ or 0.644 rad.

Hence

$$\theta = 180° - 36.87° = 143.13°$$
or $\quad \theta = \pi - 0.644 = 2.498$ rad

Hence the position of point P in polar coordinates form is (5,143.13°) or (5,2.498 rad).

Problem 3 Express $(-5,-12)$ in polar coordinates.

A sketch showing the position $(-5,-12)$ is shown in *Fig. 23.4*.

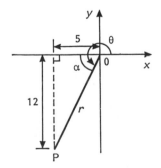

Fig. 23.4

$r = \sqrt{(5^2 + 12^2)} = 13$

$$\alpha = \arctan \dfrac{12}{5} = 67.38° \text{ or } 1.176 \text{ rad}$$

Hence $\quad \theta = 180° + 67.38° = 247.38°$
or $\quad \theta = \pi + 1.176 = 4.318$ rad

Thus $(-5,-12)$ in Cartesian coordinates corresponds to (13,247.38°) or (13,4.318 rad) in polar coordinates.

Problem 4 Express $(2,-5)$ in polar coordinates.

A sketch showing the position $(2,-5)$ is shown in *Fig. 23.5*.

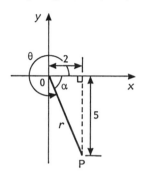

Fig. 23.5

$$r = \sqrt{(2^2 + 5^2)} = \sqrt{29} = 5.385 \text{ correct to 3 decimal places}$$

$$\alpha = \arctan \dfrac{5}{2} = 68.20° \text{ or } 1.190 \text{ rad}$$

Hence $\quad \theta = 360° - 68.20° = 291.80°$
or $\quad \theta = 2\pi - 1.190 = 5.093$ rad

Thus $(2,-5)$ in Cartesian coordinates corresponds to (5.385,291.80°) or (5.385,5.093 rad) in polar coordinates.

23.3 Changing from polar into Cartesian coordinates

From the right-angled triangle OPQ in *Fig. 23.6*

$$\cos \theta = \frac{x}{r} \text{ and } \sin \theta = \frac{y}{r}, \text{ from trigonometric ratios}$$

Hence $\quad \boxed{x = r \cos \theta} \quad$ and $\quad \boxed{y = r \sin \theta}$

If length r and angle θ are known then $x = r \cos \theta$ and $y = r \sin \theta$ are the two formulae we need to change from polar to Cartesian coordinates.

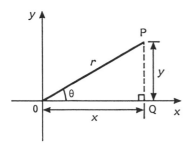

Fig. 23.6

Problem 5   Change (4,32°) into Cartesian coordinates.

A sketch showing the position (4,32°) is shown in *Fig. 23.7*.

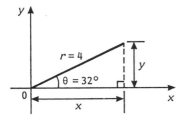

Fig. 23.7

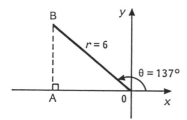

Fig. 23.8

Now $x = r \cos \theta = 4 \cos 32° = 3.39$
and $y = r \sin \theta = 4 \sin 32° = 2.12$

Hence (4,32°) in polar coordinates corresponds to (3.39,2.12) in Cartesian coordinates.

Problem 6   Express (6,137°) in Cartesian coordinates.

A sketch showing the position (6,137°) is shown in *Fig. 23.8*.

$x = r \cos \theta = 6 \cos 137° = -4.388$

which corresponds to length OA in *Fig. 23.8*.

$y = r \sin \theta = 6 \sin 137° = 4.092$

which corresponds to length AB in Fig. 23.8.
Thus (6,137°) in polar coordinates corresponds to (−4.388,4.092) in Cartesian coordinates.

(Note that when changing from polar to Cartesian coordinates it is not quite so essential to draw a sketch. Use of $x = r \cos \theta$ and $y = r \sin \theta$ automatically produces the correct signs.)

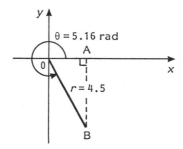

Fig. 23.9

Problem 7   Express (4.5,5.16 rad) in Cartesian coordinates.

A sketch showing the position (4.5,5.16 rad) is shown in *Fig. 23.9*.

$x = r \cos \theta = 4.5 \cos 5.16 = 1.948$

which corresponds to length OA in *Fig. 23.9*.

$y = r \sin \theta = 4.5 \sin 5.16 = -4.057$

which corresponds to length AB in *Fig. 23.9*.
Thus (1.948,−4.057) in Cartesian coordinates corresponds to (4.5,5.16 rad) in polar coordinates.

23.4   Use of R → P and P → R functions on calculators

Another name for Cartesian coordinates is **rectangular** coordinates. Many scientific notation calculators possess R → P and P → R functions. The R is the first letter of the word rectangular and the P is the first letter of the word polar. Check the operation manual for your particular calculator to determine how to use these two functions. They make changing from Cartesian to polar coordinates, and vice-versa, so much quicker and easier.

23.5   Further problems on Cartesian and polar coordinates

In *Problems 1 to 8*, express the given Cartesian coordinates as polar coordinates, correct to 2 decimal places, in both degrees and in radians.

1   (3,5)
    [(5.83,59.04°) or (5.83,1.03 rad)]

2 (6.18,2.35)
  [(6.61,20.82°) or (6.61,0.36 rad)]
3 (−2,4)
  [(4.47,116.57°) or (4.47,2.03 rad)]
4 (−5.4,3.7)
  [(6.55,145.58°) or (6.55,2.54 rad)]
5 (−7,−3)
  [(7.62,203.20°) or (7.62,3.55 rad)]
6 (−2.4,−3.6)
  [(4.33,236.31°) or (4.33,4.12 rad)]
7 (5,−3)
  [(5.83,329.04°) or (5.83,5.74 rad)]
8 (9.6,−12.4)
  [(15.8,307.75°) or (15.68,5.37 rad)]

In *Problems 9* to *16*, express the given polar coordinates as Cartesian coordinates, correct to 3 decimal places.

9 (5,75°)
  [1.294,4.830)]
10 (4.4,1.12 rad)
  [(1.197,3.960)]
11 (7,140°)
  [(−5.362,4.500)]
12 (3.6,2.5 rad)
  [(−2.884,2.154)]
13 (10.8,210°)
  [(−9.353,−5.400)]
14 (4,4 rad)
  [(−2.615,−3.027)]
15 (1.5,300°)
  [(0.750,−1.299)]
16 (6,5.5 rad)
  [(4.252,−4.233)]

24

Vectors

24.1 Introduction

Some physical quantities are entirely defined by a numerical value and are called **scalar quantities** or **scalars**. Examples of scalars include time, mass, temperature, energy and volume. Other physical quantities are defined by both a numerical value and a direction in space and these are called **vector quantities** or **vectors**. Examples of vectors include force, velocity, moment and displacement.

24.2 Vector addition

A vector may be represented by a straight line, the length of the line being directly proportional to the magnitude of the quantity and the direction of the line being in the same direction as the line of action of the quantity. An arrow is used to denote the sense of the vector, that is, for a horizontal vector, say, whether it acts from left to right or vice-versa. The arrow is positioned at the end of the vector and this position is called the 'nose' of the vector. *Fig. 24.1* shows a velocity of 20 m/s at an angle of 45° to the horizontal and may be depicted by *oa* = 20 m/s at 45° to the horizontal.

To distinguish between vector and scalar quantities, various ways are used, and the one adopted in this text is to denote vector quantities in bold print. Thus, *oa* represents a vector

quantity, but *oa* is the magnitude of vector *oa*. Also, positive angles are measured in an anticlockwise direction from a horizontal, right facing line and negative angles in a clockwise direction from this line. Thus 90° is a line vertically upwards and −90° is a line vertically downwards.

The resultant of adding two vectors together, say V_1 at angle θ_1 and V_2 at angle $(-\theta_2)$, as shown in *Fig. 24.2(a)*, can be obtained by drawing *oa* to represent V_1 and then drawing *ar* to represent V_2. The resultant of $V_1 + V_2$ is given by *or*. This is shown in *Fig. 24.2(b)*, the vector equation being *oa* + *ar* = *or*. This is called the **'nose-to-tail' method** of vector addition.

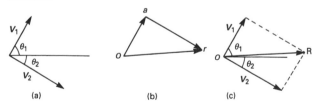

Fig. 24.2

Alternatively, by drawing lines parallel to V_1 and V_2 from the noses of V_2 and V_1, respectively, and letting the point of intersection of these parallel lines be R, gives **OR** as the magnitude and direction of the resultant of adding V_1 and V_2, as shown in *Fig. 24.2(c)*. This is called the **'parallelogram' method** of vector addition.

Problem 1 A force of 4 N is inclined at an angle of 45° to a second force of 7 N, both forces acting at a point. Find the magnitude of the resultant of these two forces and the direction of the resultant with respect to the 7 N force by both the 'triangle' and the 'parallelogram' methods.

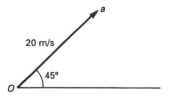

Fig. 24.1

The forces are shown in *Fig. 24.3(a)*. Although the 7 N force is shown as a horizontal line, it could have been drawn in any direction. Using the 'nose-to-tail' method, a line 7 units long is drawn horizontally to give vector *oa* in *Fig. 24.3(b)*. To the nose of this vector *ar* is drawn 4 units long at an angle of 45° to *oa*. The resultant of vector addition is *or* and by measurement is **10.2 units long** and at an angle of **16°** to the 7 N force.

Fig. 24.3(c) uses the 'parallelogram' method in which lines are drawn parallel to the 7 N and 4 N forces from the noses of the 4 N and 7 N forces, respectively. These intersect at R. Vector **OR** gives the magnitude and direction of the resultant of vector addition and as obtained by the 'nose-to-tail' method is 10.2 units long at an angle of 16° to the 7 N force. Thus by both methods, the resultant of vector addition is

a force of 10.2 N at an angle of 16° to the 7 N force

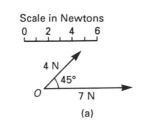

Scale in Newtons

(a)

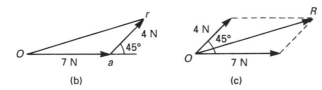

(b)   (c)

Fig. 24.3

Problem 2   Use a graphical method to determine the magnitude and direction of the resultant of the three velocities shown in *Fig. 24.4*.

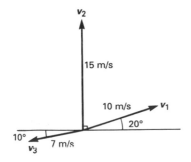

Fig. 24.4

Often it is easier to use the 'nose-to-tail' method when more than two vectors are being added. The order in which the vectors are added is immaterial. In this case the order taken is v_1, then v_2, then v_3 but just the same result would have been obtained if the order had been, say, v_1, v_3 and finally v_2. v_1 is drawn 10 units long at an angle of 20° to the horizontal, shown by *oa* in *Fig. 24.5*. v_2 is added to v_1 by drawing a line 15 units long vertically upwards from *a*, shown as *ab*. Finally, v_3 is added to $v_1 + v_2$ by drawing a line 7 units long at an angle at 190° from *b*, shown as *br*. The resultant of vector addition is *or* and by measurement is 17.5 units long at an angle of 82° to the horizontal. Thus

$$v_1 + v_2 + v_3 = \textbf{17.5 m/s at 82° to the horizontal}$$

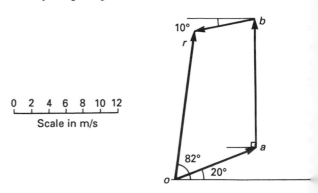

Scale in m/s

Fig. 24.5

24.3   Resolution of vectors

A vector can be resolved into two component parts such that the vector addition of the component parts is equal to the original vector. The two components usually taken are a horizontal component and a vertical component. For the vector shown as F in *Fig. 24.6*, the horizontal component is $F\cos\theta$ and the vertical component is $F\sin\theta$.

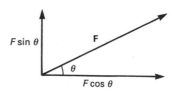

Fig. 24.6

For the vectors F_1 and F_2 shown in *Fig. 24.7*, the horizontal component of vector addition is

$$H = F_1 \cos\theta_1 + F_2 \cos\theta_2$$

and the vertical component of vector addition is

$$V = F_1 \sin\theta_1 + F_2 \sin\theta_2$$

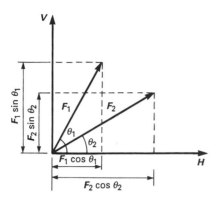

Fig. 24.7

Having obtained H and V, the magnitude of the resultant vector R is given by $\sqrt{(H^2 + V^2)}$ and its angle to the horizontal is given by arctan (V/H).

Problem 3   Resolve the acceleration vector of 17 m/s² at an angle of 120° to the horizontal into a horizontal and a vertical component.

For a vector A at angle θ to the horizontal, the horizontal component is given by $A \cos \theta$ and the vertical component by $A \sin \theta$. Any convention of signs may be adopted, in this case horizontally from left to right is taken as positive and vertically upwards is taken as positive.
Horizontal component $H = 17 \cos 120° = \mathbf{-8.5\ m/s^2}$, acting from right to left.
Vertical component $V = 17 \sin 120° = \mathbf{14.72\ m/s^2}$, acting vertically upwards.
These component vectors are shown in *Fig. 24.8*.

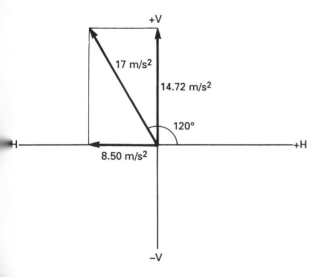

Fig. 24.8

Problem 4   Calculate the resultant force of the two forces given in *Problem 1*.

Reference to *Fig. 24.3* shows that there are horizontal components due to both the 7 N and the 4 N forces. Horizontal component of force,

$$H = 7 \cos 0° + 4 \cos 45°$$
$$= 7 + 2.828 = 9.828\ N$$

Vertical component of force,

$$V = 7 \sin 0° + 4 \sin 45°$$
$$= 0 + 2.828 = 2.828\ N$$

The magnitude of the resultant of vector addition

$$= \sqrt{(H^2 + V^2)}$$
$$= \sqrt{(9.828^2 + 2.828^2)}$$
$$= \sqrt{104.59} = 10.23\ N$$

The direction of the resultant of vector addition

$$= \arctan\left(\frac{V}{H}\right)$$
$$= \arctan\left(\frac{2.828}{9.828}\right) = 16.05°$$

Thus, the resultant of the two forces is a single vector of 10.23 N at 16.05° to the 7 N vector.

Problem 5   Calculate the resultant velocity of the three velocities given in *Problem 2*.

With reference to *Fig. 24.4*:
Horizontal component of the velocity,

$$H = 10 \cos 20° + 15 \cos 90° + 7 \cos 190°$$
$$= 9.397 + 0 + (-6.894)$$
$$= 2.503\ m/s$$

Vertical component of the velocity,

$$V = 10 \sin 20° + 15 \sin 90° + 7 \sin 190°$$
$$= 3.420 + 15 + (-1.216)$$
$$= 17.204\ m/s$$

Magnitude of the resultant of vector addition

$$= \sqrt{(H^2 + V^2)}$$
$$= \sqrt{(2.503^2 + 17.204^2)} = \sqrt{302.24} = 17.39\ m/s$$

Direction of the resultant of vector addition

$$= \arctan\left(\frac{V}{H}\right)$$

$$= \arctan\left(\frac{17.204}{2.503}\right)$$

$$= \arctan 6.873\,4 = 81.72°$$

Thus, the resultant of the three velocities is a single vector of 17.39 m/s at 81.72° to the horizontal.

24.4   Vector subtraction

In *Fig. 24.9*, a force vector *F* is represented by *oa*. The vector (−*oa*) can be obtained by drawing a vector from *o* in the opposite sense to *oa* but having the same magnitude, shown as *ob* in *Fig. 24.9*, i.e. *ob* = (−*oa*).

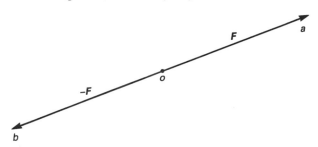

Fig. 24.9

For two vectors acting at a point, as shown in *Fig. 24.10(a)*, the resultant of vector addition is *os* = *oa* + *ob*. *Fig. 24.10(b)* shows vectors *ob* + (−*oa*), that is, *ob* − *oa* and the vector equation is *ob* − *oa* = *od*. Comparing *od* in *Fig. 24.10(b)* with the broken line ab in *Fig. 24.10(a)* shows that the second diagonal of the 'parallelogram' method of vector addition gives the magnitude and direction of vector subtraction of *oa* from *ob*.

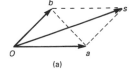
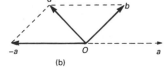

(a)   (b)

Fig. 24.10

Problem 6   Accelerations of a_1 = 1.5 m/s² at 90° and a_2 = 2.6 m/s² at 145° act at a point. Find $a_1 + a_2$ and $a_1 - a_2$ by (i) drawing a scale vector diagram and (ii) by calculation.

(i)   The scale vector diagram is shown in *Fig. 24.11*. By measurement,

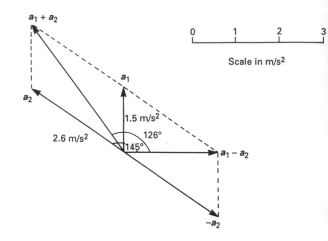

Fig. 24.11

$$a_1 + a_2 = \textbf{3.7 m/s}^2 \textbf{ at 126}°$$
$$a_1 - a_2 = \textbf{2.1 m/s}^2 \textbf{ at 0}°$$

(ii)   Resolving horizontally and vertically gives:

Horizontal component of $a_1 + a_2$,
$H = 0 + 2.6 \cos 145° = -2.13$
Vertical component of $a_1 + a_2$,
$V = 1.5 + 2.6 \sin 145° = 2.99$

Magnitude of $a_1 + a_2 = \sqrt{(-2.13^2 + 2.99^2)}$
$$= 3.67 \text{ m/s}^2$$

Direction of $a_1 + a_2 = \arctan\left(\frac{2.99}{-2.13}\right)'$

and must lie in the second quadrant since H is negative and V is positive.

Arctan (2.99/−2.13) = −54.5°, and for this to be in the second quadrant, the true angle is 180° displaced, i.e. 180° − 54.5° or 125.5°. Thus

$$a_1 + a_2 = \textbf{3.67 m/s}^2 \textbf{ at 125.5}°$$

Horizontal component of $a_1 - a_2$, that is $a_1 + (-a_2)$

$$= 0 + 2.6 \cos (145° - 180°)$$
$$= 2.6 \cos (-35°) = 2.13$$

Vertical component of $a_1 - a_2$, that is $a_1 + (-a_2)$

$$= 1.5 + 2.6 \sin (-35°) = 0$$

Magnitude of $a_1 - a_2 = \sqrt{(2.13^2 + 0^2)} = 2.13$ m/s²

Direction of $a_1 - a_2 = \arctan\left(\dfrac{0}{2.13}\right) = 0°$

Thus $a_1 - a_2 = \mathbf{2.13\ m/s^2\ at\ 0°}$

Problem 7 Calculate the resultant of (i) $v_1 - v_2 + v_3$ and (ii) $v_1 - v_2 - v_3$ when $v_1 = 22$ units at 140°, $v_2 = 40$ units at 190° and $v_3 = 15$ units at 290°.

(i) The vectors are shown in *Fig. 24.12* with their angles referred to the horizontal axis, When this is done, the sign is determined by reference to the horizontal and vertical axes. Thus for horizontal resolution:

$v_1 = -22 \cos 40°$, $v_2 = -40 \cos 10°$ and $v_3 = 15 \cos 70°$

The horizontal component of $v_1 - v_2 + v_3$

$= (-22 \cos 40°) - (-40 \cos 10°) + (15 \cos 70°)$
$= -16.85 + 39.39 + 5.13$
$= 27.67$ units

For vertical resolution:

$v_1 = 22 \sin 40°$, $v_2 = -40 \sin 10°$
and $v_3 = -15 \sin 70°$

The vertical component of $v_1 - v_2 + v_3$

$= (22 \sin 40°) - (-40 \sin 10°) + (-15 \sin 70°)$
$= 14.14 + 6.95 - 14.10$
$= 6.99$ units

The magnitude of the resultant, R, which can be represented by the mathematical symbol for 'the **modulus** of' as $|v_1 - v_2 + v_3|$ is given by

$|R| = \sqrt{(27.67^2 + 6.99^2)} = 28.5$ units

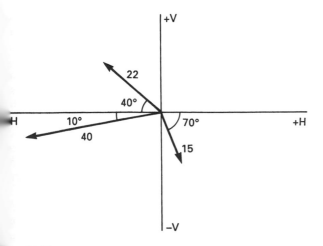

g. 24.12

The direction of the resultant, R, which can be represented by the mathematical symbol for 'the **argument** of' as arg $(v_1 - v_2 + v_3)$ is given by

$\arg R = \arctan\left(\dfrac{6.99}{27.67}\right) = 14.2°$

Thus, $v_1 - v_2 + v_3 = \mathbf{28.5\ units\ at\ 14.2°}$.

(ii) The horizontal component of $v_2 - v_1 - v_3$

$= (-40 \cos 10°) - (22 \cos 40°) - (15 \cos 70°)$
$= -39.39 + 16.85 - 5.13$
$= -27.67$ units

The vertical component of $v_2 - v_1 - v_3$

$= (-40 \sin 10°) - (-22 \sin 40°) - (-15 \sin 70°)$
$= -6.95 - 14.14 + 14.10$
$= -6.99$ units

Let $R = v_2 - v_1 - v_3$. Then $|R| = \sqrt{[(-27.67)^2 + (-6.99)^2]}$
$= 28.5$ units

and, $\arg R = \arctan\left(\dfrac{-6.99}{-27.67}\right)$

and must lie in the third quadrant since both H and V are negative quantities.
Arctan (6.99/27.67) = 14.2°, hence the required angle is 180° + 14.2° = 194.2°. Thus

$v_2 - v_1 - v_3 = \mathbf{28.5\ units\ at\ 194.2°}$

This result is as expected, since $v_2 - v_1 - v_3 = -(v_1 - v_2 + v_3)$ and the vector 28.5 units at 194.2° is minus times the vector 28.5 units at 14.2°.

Problem 8 Two cars, P and Q, are travelling towards the junction of two roads which are at right angles to one another. Car P has a velocity of 45 km/h due east and car Q a velocity of 55 km/h due south. Calculate (i) the velocity of car P relative to car Q, and (ii) the velocity of car Q relative to car P.

For relative velocity problems, some fixed datum point should be selected. This is often a fixed point on the earth's surface. In any vector equation, only the start and finish points affect the resultant vector of a system. Two different systems are shown in *Fig. 24.13*, but in each of the systems, the resultant vector is ad.
The vector equation of the system shown in *Fig. 24.13(a)* is:

$$ad = ab + bd$$

and that for the system shown in *Fig. 24.13(b)* is:

$$ad = ab + bc + cd$$

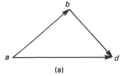

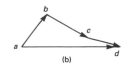

Fig. 24.13

Thus in vector equations of this form, only the first and last letters, *a* and *d*, respectively, fix the magnitude and direction of the resultant vector. This principle is used in relative velocity problems.

(i)   The directions of the cars are shown in *Fig. 24.14(a)*, called a **space diagram**. The velocity diagram is shown in *Fig. 24.14(b)*, in which *pe* is taken as the velocity of car *P* relative to point *e* on the earth's surface. The velocity of *P* relative to *Q* is vector *pq* and the vector equation is *pq = pe + eq*. Hence the vector directions are as shown, *eq* being in the opposite direction to *qe*. From the geometry of the vector triangle,

$$|pq| = \sqrt{(45^2 + 55^2)} = 71.1 \text{ km/h}$$

$$\arg \boldsymbol{pq} = \arctan\left(\frac{55}{45}\right) = 50.7°$$

That is, the velocity of car *P* relative to car *Q* is 71.1 km/h at 50.7°.

(ii)   The velocity of car *Q* relative to car *P* is given by the vector equation *qp = qe + ep* and the vector diagram is as shown in *Fig. 24.14(c)*, having *ep* opposite in direction to *pe*. From the geometry of this vector triangle:

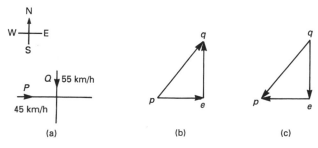

Fig. 24.14

$$|qp| = \sqrt{(45^2 + 55^2)} = 71.1 \text{ m/s}$$

$$\arg \boldsymbol{qp} = \arctan\left(\frac{55}{45}\right) = 50.7°$$

but must lie in the third quadrant, that is, the required angle is 180° + 50.7° = 230.7°.
Thus the velocity of car *Q* relative to car *P* is 71.1 m/s at 230.7°.

24.5   Further problems on vectors

1   Forces of 23 N and 41 N act at a point and are inclined at 90° to each other. Find, by drawing, the resultant force and its direction relative to the 41 N force.
[47 N at 29°]

2   Forces *A*, *B* and *C* are coplanar and act at a point. Force *A* is 12 kN at 90°, *B* is 5 kN at 180° and *C* is 13 kN at 293°. Determine graphically the resultant force.
[Zero]

3   Calculate the magnitude and direction of velocities of 3 m/s at 18° and 7 m/s at 115° when acting simultaneously on a point.
[7.27 m/s at 90.8°]

4   Three forces of 2 N, 3 N and 4 N act as shown in *Fig. 24.15*. Calculate the magnitude of the resultant force and its direction relative to the 2 N force.
[6.24 N at 76.18°]

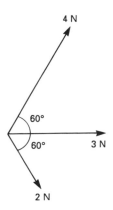

Fig. 24.15

5   A load of 5.89 N is lifted by two strings, making angles of 20° and 35° with the vertical. Calculate the tensions in the strings.
[For a system such as this, the vectors representing the forces form a closed triangle when the system is in equilibrium.]
[2.46 N, 4.12 N]

6   The acceleration of a body is due to four component coplanar accelerations. These are 2 m/s² due north, 3 m/s² due east, 4 m/s² to the south-west and 5 m/s² to the south-east. Calculate the resultant acceleration and its direction
[5.7 m/s² at 310°]

7   A car is moving along a straight horizontal road at 79.2 km/h and rain is falling vertically downwards at 26.4 km/h. Find the velocity of the rain relative to the driver of the car.
[83.5 km/h at 71.6° to the vertical]

8   Calculate the time needed to swim across a river 142 m wide when the swimmer can swim at 2 km/h in still water and the river is flowing at 1 km/h.
[4 minutes 55 seconds]

25

Presentation of statistical data

25.1 Some statistical terminology

Data are obtained largely by two methods:

(a) by counting – for example, the number of stamps sold by a post office in equal periods of time, and
(b) by measurement – for example, the heights of a group of people.

When data are obtained by counting and only whole numbers are possible, the data are called **discrete**. Measured data can have any value within certain limits and are called **continuous** (see *Problem 1*).

A **set** is a group of data and an individual value within the set is called a **member** of the set. Thus, if the masses of five people are measured correct to the nearest 0.1 kilogram and are found to be 53.1 kg, 59.4 kg, 62.1 kg, 77.8 kg and 64.4 kg, then the set of masses in kilograms for these five people is:

$$\{53.1,\ 59.4,\ 62.1,\ 77.8,\ 64.4\}$$

and one of the members of the set is 59.4.

A set containing all the members is called a **population**. Some members selected at random from a population are called a **sample**. Thus all car registration numbers form a population, but the registration numbers of, say, 20 cars taken at random throughout the country are a sample drawn from that population.

The number of times that the value of a member occurs in a set is called the **frequency** of that member. Thus in the set: $\{2, 3, 4, 5, 4, 2, 4, 7, 9\}$, member 4 has a frequency of three, member 2 has a frequency of 2 and the other members have a frequency of one.

The **relative frequency** with which any member of a set occurs is given by the ratio:

$$\frac{\text{frequency of member}}{\text{total frequency of all members}}$$

For the set: $\{2, 3, 5, 4, 7, 5, 6, 2, 8\}$, the relative frequency of member 5 is $\dfrac{2}{9}$.

Often, relative frequency is expressed as a percentage and the **percentage relative frequency** is: (relative frequency $\times\ 100$)%.

Problem 1 Data are obtained on the topics given below. State whether they are discrete or continuous data.

(a) The number of days on which rain falls in a month for each month of the year.
(b) The mileage travelled by each of a number of salesmen.
(c) The time that each of a batch of similar batteries lasts.
(d) The amount of money spent by each of several families on food.

(a) The number of days on which rain falls in a given month must be an integer value and is obtained by **counting** the number of days. Hence, these data are **discrete**.

(b) A salesman can travel any number of miles (and parts of a mile) between certain limits and these data are **measured**. Hence the data are **continuous**.

(c) The time that a battery lasts is **measured** and can have any value between certain limits. Hence these data are **continuous**.

(d) The amount of money spent on food can only be expressed correct to the nearest half pence, the amount being **counted**. Hence, these data are **discrete**.

Further problems on discrete and continuous data may be found in section 25.4, Problems 1 and 2, page 231.

25.2 Presentation of ungrouped data

Ungrouped data can be presented diagrammatically in several ways and these include:

(a) **pictograms**, in which pictorial symbols are used to represent quantities (see *Problem 2*),

(b) **horizontal bar charts**, having data represented by equally spaced horizontal rectangles (see *Problem 3*), and

(c) **vertical bar charts**, in which data are represented by equally spaced vertical rectangles (see *Problem 4*).

Trends in ungrouped data over equal periods of time can be presented diagrammatically by a **percentage component bar chart**. In such a chart, equally spaced rectangles of any width, but whose height corresponds to 100%, are constructed. The rectangles are then subdivided into values corresponding to the percentage relative frequencies of the members (see *Problem 5*).

A **pie diagram** is used to show diagrammatically the parts making up the whole. In a pie diagram, the area of a circle represents the whole, and the areas of the sectors of the circle are made proportional to the parts which make up the whole (see *Problem 6*).

Problem 2 The number of television sets repaired in a workshop by a technician in six, one-month periods is as shown below. Present these data as a pictogram.

Month	January	February	March	April	May	June
Number repaired	11	6	15	9	13	8

Each symbol shown in *Fig. 25.1* represents two television sets repaired. Thus, in January, $5\frac{1}{2}$ symbols are used to represent the 11 sets repaired, in February, 3 symbols are used to represent the 6 sets repaired, and so on.

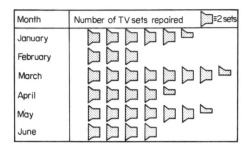

Fig. 25.1

Problem 3 The distance in miles travelled by four salesmen in a week are as shown below.

Salesmen	P	Q	R	S
Distance travelled (miles)	413	264	597	143

Use a horizontal bar chart to represent these data diagrammatically.

Equally spaced horizontal rectangles of any width, but whose length is proportional to the distance travelled, are used. Thus, the length of the rectangle for salesman P is proportional to 413 miles, and so on. The horizontal bar chart depicting these data is shown in *Fig. 25.2*.

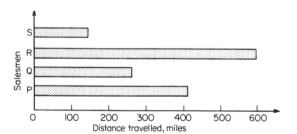

Fig. 25.2

Problem 4 The number of issues of tools or materials from a store in a factory is observed for seven, one-hour periods in a day, and the results of the survey are as follows:

Period	1	2	3	4	5	6	7
Number of issues	34	17	9	5	27	13	6

Present these data on a vertical bar chart.

In a vertical bar chart, equally spaced vertical rectangles of any width, but whose height is proportional to the quantity being represented, are used. Thus the height of the rectangle for period 1 is proportional to 34 units, and so on. The vertical bar chart depicting these data is shown in *Fig. 25.3*.

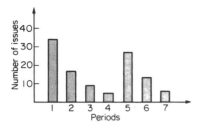

Fig. 25.3

Problem 5 The numbers of various types of dwell-ings sold by a company annually over a three-year period are as shown below. Draw percentage compo-nent bar charts to present these data.

	Year 1	Year 2	Year 3
4-roomed bungalows	24	17	7
5-roomed bungalows	38	71	118
4-roomed houses	44	50	53
5-roomed houses	64	82	147
6-roomed houses	30	30	25

A table of percentage relative frequency values, correct to the nearest 1%, is the first requirement. Since,

percentage relative frequency

$$= \frac{\text{frequency of member} \times 100}{\text{total frequency}}$$

then for 4-roomed bungalows in year 1:

percentage relative frequency

$$= \frac{24 \times 100}{24 + 38 + 44 + 64 + 30} = 12\%$$

The percentage relative frequencies of the other types of dwellings for each of the three years are similarly calcu-lated and the results are as shown in the table below.

	Year 1	Year 2	Year 3
4-roomed bungalows	12%	7%	2%
5-roomed bungalows	19%	28%	34%
4-roomed houses	22%	20%	15%
5-roomed houses	32%	33%	42%
6-roomed houses	15%	12%	7%

The percentage component bar chart is produced by con-structing three equally spaced rectangles of any width, cor-responding to the three years. The heights of the rectangles correspond to 100% relative frequency, and are subdivided into the values in the table of percentages shown above. A

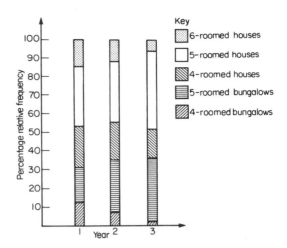

Fig. 25.4

key is used (different types of shading or different colour schemes) to indicate corresponding percentage values in the rows of the table of percentages. The percentage com-ponent bar chart is shown in *Fig. 25.4*.

Problem 6 The retail price of a product costing £2 is made up as follows: materials 10p, labour 20p, research and development 40p, overheads 70p, profit 60p. Present these data on a pie diagram.

A circle of any radius is drawn, and the area of the circle represents the whole, which in this case is £2. The circle is subdivided into sectors so that the areas of the sectors are proportional to the parts, i.e. the parts which make up the total retail price. For the area of a sector to be proportional to a part, the angle at the centre of the circle must be pro-portional to that part. The whole, £2 or 200p, corresponds to 360°. Therefore,

10p corresponds to $360 \times \dfrac{10}{200}$ degrees, i.e. 18°

20p corresponds to $360 \times \dfrac{20}{200}$ degrees, i.e. 36°

Fig. 25.5

and so on, giving the angles at the centre of the circle for the parts of the retail price as: 18°, 36°, 72°, 126° and 108°, respectively.

The pie diagram is shown in *Fig. 25.5*.

Problem 7

(a) Using the data given in *Fig. 25.2* only, calculate the amount of money paid to each salesman for travelling expenses, if they are paid an allowance of 27p per mile.
(b) Using the data presented in *Fig. 25.4*, comment on the housing trends over the three-year period.
(c) Determine the profit made by selling 700 units of the product shown in *Fig. 25.5*.

(a) By measuring the length of rectangle P the mileage covered by salesman P is equivalent to 413 miles. Hence salesman P receives a travelling allowance of

$$\frac{£413 \times 27}{100}, \text{ i.e. } £111.51$$

Similarly, for salesman Q, miles travelled are 264 and his allowance is

$$\frac{£264 \times 27}{100}, \text{ i.e. } £71.28$$

Salesman R travels 597 miles and he receives

$$\frac{£597 \times 27}{100}, \text{ i.e. } £161.19$$

Finally, salesman S receives

$$\frac{£143 \times 27}{100}, \text{ that is, } £38.61$$

(b) An analysis of *Fig. 25.4* shows that 5-roomed bungalows and 5-roomed houses are becoming more popular, the greatest change in the three years being a 15% increase in the sales of 5-roomed bungalows.
(c) Since 1.8° corresponds to 1p and the profit occupies 108° of the pie diagram, then the profit per unit is

$$\frac{108 \times 1}{1.8}, \text{ that is, 60p}$$

The profit when selling 700 units of the product is

$$£\frac{700 \times 60}{100}, \text{ that is, } £420$$

Further problems on presentation of ungrouped data may be found in section 25.4, Problems 3 to 19, page 23.

25.3 Presentation of grouped data

When the number of members in a set is small, say ten or less, the data can be represented diagrammatically without further analysis, by means of pictograms, bar charts, percentage components bar charts or pie diagrams (as shown in section 25.2).

For sets having more than ten members, those members having similar values are grouped together in **classes** to form a **frequency distribution**. To assist in accurately counting members in the various classes, a **tally diagram** is used (see *Problems 8* and *12*).

A frequency distribution is merely a table showing classes and their corresponding frequencies (see *Problems 8* and *12*).

The new set of values obtained by forming a frequency distribution is called **grouped data**.

The terms used in connection with grouped data are shown in *Fig. 25.6(a)*. The size or range of a class is given by the **upper class boundary value** minus the **lower class boundary value**, and in *Fig. 25.6* is 7.65 – 7.35, i.e., 0.3. The **class interval** for the class shown in *Fig. 25.6(b)* is 7.4 to 7.6 and the class mid-point value is given by

$$\frac{\text{(upper class boundary value)} - \text{(lower class boundary value)}}{2}$$

and in *Fig. 25.6* is $\dfrac{7.65 - 7.35}{2}$, i.e. 7.5

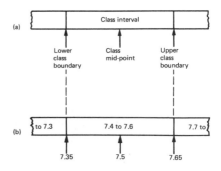

Fig. 25.6

One of the principal ways of presenting grouped data diagrammatically is by using a **histogram**, in which the **areas** of vertical, adjacent rectangles are made proportional to frequencies of the classes (see *Problem 9*). When class intervals are equal, the heights of the rectangles of a histogram are equal to the frequencies of the classes. For histograms having unequal class intervals, the area must be proportional to the frequency. Hence, if the class interval of class A is twice the class interval of class B, then for equal frequencies, the height of the rectangle representing A is half that of B (see *Problem 11*).

Another method of presenting grouped data diagrammatically is by using a **frequency polygon**, which is the graph produced by plotting frequency against class mid-point values and joining the co-ordinates with straight lines (see *Problem 12*).

A **cumulative frequency distribution** is a table showing the cumulative frequency for each value of upper class boundary. The cumulative frequency for a particular value of upper class boundary is obtained by adding the frequency of the class to the sum of the previous frequencies. A cumulative frequency distribution is formed in *Problems 13 and 14*.

The curve obtained by joining the coordinates of cumulative frequency (vertically) against upper class boundary (horizontally) is called an **ogive** or a **cumulative frequency distribution curve** (see *Problems 13 and 15*).

Problem 8 The data given below refer to the gain of each of a batch of 40 transistors, expressed correct to the nearest whole number. Form a frequency distribution for these data having seven classes.

81	83	87	74	76	89	82	84
86	76	77	71	86	85	87	88
84	81	80	81	73	89	82	79
81	79	78	80	85	77	84	78
83	79	80	83	82	79	80	77

The **range** of the data is the value obtained by taking the value of the largest member from that of the smallest member. Inspection of the set of data shows that, range = 89 − 71 = 18. The size of each class is given approximately by range divided by the number of classes. Since 7 classes are required, the size of each class is 18/7, that is, approximately 3. To achieve seven equal classes spanning a range of values from 71 to 89, the class intervals are selected as: 70–72, 73–75, and so on.

To assist with accurately determining the number in each class, a **tally diagram** is produced, as shown in *Table 25.1(a)*. This is obtained by listing the classes in the left-hand column, and then inspecting each of the 40 members of the set in turn and allocating them to the appropriate classes by putting '1s' in the appropriate rows. Every fifth '1' allocated to a particular row is shown as an oblique line crossing the four previous '1s', to help with final counting.

Table 25.1(a)

Class	Tally
70–72	l
73–75	ll
76–78	⊬⊬ ll
79–81	⊬⊬ ⊬⊬ ll
82–84	⊬⊬ llll
85–87	⊬⊬ l
88–90	lll

Table 25.1(b)

Class	Class mid-point	Frequency
70–72	71	1
73–75	74	2
76–78	77	7
79–81	80	12
82–84	83	9
85–87	86	6
88–90	89	3

A frequency distribution for the data is shown in *Table 25.1(b)* and lists classes and their corresponding frequencies, obtained from the tally diagram. (Class mid-point values are also shown in the table, since they are used for constructing the histogram for these data (see *Problem 9*).)

Problem 9 Construct a histogram for the data given in *Table 25.1(b)*.

The histogram is shown in *Fig. 25.7*. The width of the rectangles correspond to the upper class boundary values minus the lower class boundary values and the heights of the rectangles correspond to the class frequencies. The easiest way to draw a histogram is to mark the class mid-point values on the horizontal scale and draw the rectangles symmetrically about the appropriate class mid-point values and touching one another.

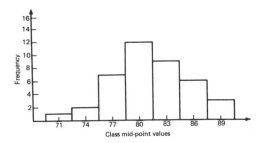

Fig. 25.7

Problem 10 The amount of money earned weekly by 40 people working part-time in a factory, correct to the nearest £10, is shown below. Form a frequency distribution having 6 classes for these data.

80	90	70	110	90	160	110	80
140	30	90	50	100	110	60	100
80	90	110	80	100	90	120	70
130	170	80	120	100	110	40	110
50	100	110	90	100	70	110	80

Inspection of the set given shows that the majority of the members of the set lie between £80 and £110 and that there are a much smaller number of extreme values ranging from £30 to £180. If equal class intervals are selected, the frequency distribution obtained does not give as much information as one with unequal class intervals. Since the majority of members are between £80 and £100, the class intervals in this range are selected to be smaller than those outside of this range. There is no unique solution and one possible solution is shown in *Table 25.2*.

Table 25.2

Class	Frequency
20–40	2
50–70	6
80–90	12
100–110	14
120–140	4
150–170	2

Problem 11 Draw a histogram for the data given in *Table 25.2*.

When dealing with unequal class intervals, the histogram must be drawn so that the areas (and not the heights of the rectangles), are proportional to the frequencies of the classes. The data given are shown in columns 1 and 2 of *Table 25.3*. Columns 3 and 4 give the upper and lower class boundaries, respectively. In column 5, the class ranges (i.e. upper class boundary minus lower class boundary values) are listed.

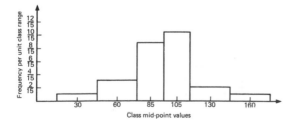

Fig. 25.8

The heights of the rectangles are proportional to the ratio (frequency/class range), as shown in column 6. The histogram is shown in *Fig. 25.8*.

Problem 12 The masses of 50 ingots in kilograms are measured correct to the nearest 0.1 kg and the results are as shown below. Produce a frequency distribution having about 7 classes for these data and then present the grouped data as (a) a frequency polygon and (b) a histogram.

8.0	8.6	8.2	7.5	8.0	9.1	8.5	7.6	8.2	7.8
8.3	7.1	8.1	8.3	8.7	7.8	8.7	8.5	8.4	8.5
7.7	8.4	7.9	8.8	7.2	8.1	7.8	8.2	7.7	7.5
8.1	7.4	8.8	8.0	8.4	8.5	8.1	7.3	9.0	8.6
7.4	8.2	8.4	7.7	8.3	8.2	7.9	8.5	7.9	8.0

The **range** of the data is the member having the largest value minus the member having the smallest value. Inspection of the set of data shows that:

$$\text{range} = 9.1 - 7.1 = 2.0$$

The size of each class is given approximately by

$$\frac{\text{range}}{\text{number of classes}}$$

Since about seven classes are required, the size of each class is 2.0/7, that is approximately 0.3.

The terms used in connection with grouped data are shown in *Fig. 25.9(a)*. The size of a class is given by the **upper class boundary** value minus the **lower class boundary** value, and in this problem the size is 0.3. To achieve about seven classes spanning the range of values from 7.1 to 9.1, the **class limits** are selected as 7.1 to 7.3, 7.4 to 7.6, 7.7 to 7.9, and so on. The **class mid-point** value is given by

$$\frac{(\text{upper class boundary value}) - (\text{lower class boundary value})}{2}$$

Thus the class mid-point for the 7.4 to 7.6 class is (7.65 − 7.35)/2, i.e. 7.5. The various values associated with the 7.4 to 7.6 class are shown in *Fig. 25.9(b)*.

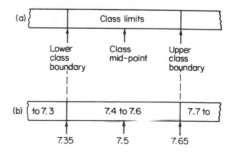

Fig. 25.9

Table 25.3

1 Class	2 Frequency	3 Upper class boundary	4 Lower class boundary	5 Class range	6 Height of rectangle
20–40	2	45	15	30	$\dfrac{2}{30}=\dfrac{1}{15}$
50–70	6	75	45	30	$\dfrac{6}{30}=\dfrac{3}{15}$
80–90	12	95	75	20	$\dfrac{12}{20}=\dfrac{9}{15}$
100–110	14	115	95	20	$\dfrac{14}{20}=\dfrac{10\frac{1}{2}}{15}$
120–140	4	145	115	30	$\dfrac{4}{30}=\dfrac{2}{15}$
150–170	2	175	145	30	$\dfrac{2}{30}=\dfrac{1}{15}$

To assist with accurately determining the number in each class, a **tally diagram** is produced as shown in *Table 25.4(a)*. This is obtained by listing the classes in the left-hand column and then inspecting each of the 50 members of the set of data in turn and allocating it to the appropriate class by putting a '1' in the appropriate row. Each fifth '1' allocated to a particular row is marked as an oblique line to help with final counting.

A **frequency distribution** for the data is shown in *Table 25.4(b)* and lists classes and their corresponding frequencies. Class mid-points are also shown in this table, since they are used when constructing the frequency polygon and histogram.

A **frequency polygon** is shown in *Fig. 25.10(a)*, the co-ordinates corresponding to the class mid-point/frequency values, given in *Table 25.4(b)*. The coordinates are joined by straight lines and the polygon is 'anchored-down' at each end by joining to the next class mid-point value and zero frequency.

A **histogram** is shown in *Fig. 25.10(b)*, the width of a rectangle corresponding to (upper class boundary value – lower class boundary value) and height corresponding to

Table 25.4(a) *Tally diagram*

Class	Tally
7.1 to 7.3	|||
7.4 to 7.6	₩
7.7 to 7.9	₩ ||||
8.0 to 8.2	₩ ₩ ||||
8.3 to 8.5	₩ ₩ |
8.6 to 8.8	₩ |
8.9 to 9.1	||

Table 25.4(b) *Frequency distribution*

Class	Class mid-point	Frequency
7.1 to 7.3	7.2	3
7.4 to 7.6	7.5	5
7.7 to 7.9	7.8	9
8.0 to 8.2	8.1	14
8.3 to 8.5	8.4	11
8.6 to 8.8	8.7	6
8.9 to 9.1	9.0	2

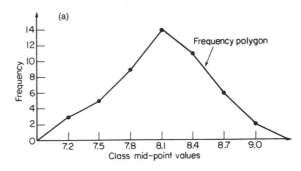

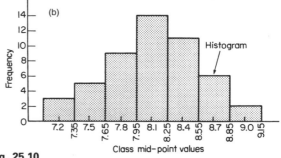

Fig. 25.10

the class frequency. The easiest way to draw a histogram is to mark class mid-point values on the horizontal scale and to draw the rectangles symmetrically about the appropriate class mid-point values and touching one another. A histogram for the data given in *Table 25.4(b)* is shown in *Fig. 25.10(b)*.

Problem 13 The frequency distribution for the masses in kilograms of 50 ingots is:

7.1 to 7.3	3,	7.4 to 7.6	5,	7.7 to 7.9	9,
8.0 to 8.2	14,	8.3 to 8.5	11,	8.6 to 8.8	6,
8.9 to 9.1	2,				

Form a cumulative frequency distribution for these data and draw the corresponding ogive.

A **cumulative frequency distribution** is a table giving values of cumulative frequency for the values of upper class boundaries, and is shown in *Table 25.5*. Columns 1 and 2 show the classes and their frequencies. Column 3 lists the upper class boundary values for the classes given in column 1. Column 4 gives the cumulative frequency values for all frequencies less than the upper class boundary values given in column 3. Thus, for example, for the 7.7 to 7.9 class shown in row 3, the cumulative frequency value is the sum of all frequencies having values of less than 7.95, i.e. $3 + 5 + 9 = 17$, and so on.

The **ogive** for the cumulative frequency distribution given in *Table 25.5* is shown in *Fig. 25.11*. The coordinates corresponding to each upper class boundary/cumulative frequency value are plotted and the coordinates are joined by straight lines. (*Note*: not the best curve drawn through the coordinates as in experimental work.) The ogive is 'anchored' at its start by adding the coordinate (7.05,0).

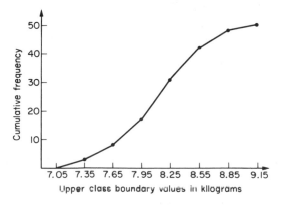

Fig. 25.11

Table 25.5

1 Class	2 Frequency	3 Upper class boundary	4 Cumulative frequency
		Less than	
7.1–7.3	3	7.35	3
7.4–7.6	5	7.65	8
7.7–7.9	9	7.95	17
8.0–8.2	14	8.25	31
8.3–8.5	11	8.55	42
8.6–8.8	6	8.85	48
8.9–9.1	2	9.15	50

Problem 14 The frequency distribution of the diameters of 60 bolts measured in millimetres is as follows:

3.93–3.94	1,	3.95–3.96	5,	3.97–3.98	9,
3.99–4.00	17,	4.01–4.02	15,	4.03–4.04	7,
4.05–4.06	4,	4.07–4.08	2,			

Form a cumulative frequency distribution for these data.

A cumulative frequency distribution is a table listing values of upper class boundaries and cumulative frequencies of less than these upper class boundary values. Such a cumulative frequency distribution for the data given is shown in *Table 25.6*. Columns 1 and 2 show the classes and their frequencies. Column 3 lists the upper class boundary values. In column 4, the sum of the frequencies less than the upper class boundary values given in column 3 are listed. Thus, for the 3.97–3.98 class shown in row 3, the cumulative frequency is the sum of all frequencies less than 3.985, i.e. $1 + 5 + 9 = 15$, and so on.

Table 25.6

1 Class	2 Frequency	3 Upper class boundary	4 Cumulative frequency
3.93–3.94	1	less than 3.945	1
3.95–3.96	5	less than 3.965	6
3.97–3.98	9	less than 3.985	15
3.99–4.00	17	less than 4.005	32
4.01–4.02	15	less than 4.025	47
4.03–4.04	7	less than 4.045	54
4.05–4.06	4	less than 4.065	58
4.07–4.08	2	less than 4.085	60

Problem 15 Draw an ogive (cumulative frequency distribution curve), for the data given in *Table 25.6*.

The ogive for the cumulative frequency distribution given in *Table 25.6* is shown in *Fig. 25.12*. The coordinates corresponding to each upper class boundary/cumulative frequency value are plotted, and the coordinates are joined by straight lines. (*Note*: it is not the best curve drawn through the coordinates as is usual practice when drawing graphs of experimental results.) The ogive is 'anchored' at its start by assuming a zero cumulative frequency for a class one less than the first one given in *Table 25.6*, i.e. (3.925,0).

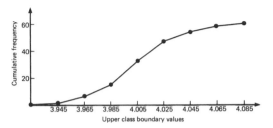

Fig. 25.12

Further problems on presentation of grouped data may be found in section 25.4, Problems 20 to 37, page 233.

25.4 Further problems on presentation of statistical data

Discrete and continuous data

In *Problems 1* and *2*, state whether data relating to the topics given are discrete or continuous.

1 (a) The amount of petrol produced daily, for each of 31 days, by a refinery.
   (b) The amount of coal produced daily by each of 15 miners.
   (c) The number of bottles of milk delivered daily by each of 20 milkmen.
   (d) The size of 10 samples of rivets produced by a machine.
   [(a) continuous (b) continuous (c) discrete (d) continuous]

2 (a) The number of people visiting an exhibition on each of 5 days.
   (b) The time taken by each of 12 athletes to run 100 metres.
   (c) The value of stamps sold in a day by each of 20 post offices.
   (d) The number of defective items produced in each of 10 one-hour periods by a machine.
   [(a) discrete (b) continuous (c) discrete (d) discrete]

Presentation of ungrouped data

3 The number of bottles of milk delivered by each of 5 milkmen in a one-hour period is as follows:

Milkman	A	B	C	D	E
Number of bottles	41	73	62	58	84

Present these data as a pictogram.

$$\left[\begin{array}{l}\text{If one symbol is used to represent 10 bottles, then,}\\ \text{working correct to the nearest 5 bottles, the number}\\ \text{of symbols is A 4, B } 7\tfrac{1}{2}\text{, C 6, D 6 and E } 8\tfrac{1}{2}\end{array}\right]$$

4 The number of vehicles passing a stationary observer on a road in six ten-minute intervals is as shown. Draw a pictogram to represent these data.

Period of Time	1	2	3	4	5	6
Number of Vehicles	35	44	62	68	49	41

$$\left[\begin{array}{l}\text{If one symbol is used to represent 10 vehicles,}\\ \text{working correct to the nearest 5 vehicles gives } 3\tfrac{1}{2},\\ 4\tfrac{1}{2}\text{, 6, 7, 5 and 4 symbols, respectively.}\end{array}\right]$$

5 The number of components produced by a factory in a week is as shown below:

Day	Mon	Tues	Wed	Thur	Fri
Number of Components	1580	2190	1840	2385	1280

Show these data on a pictogram.

$$\left[\begin{array}{l}\text{If one symbol represents 200 components, working}\\ \text{correct to the nearest 100 components gives: Mon. 8,}\\ \text{Tues. 11, Wed. 9, Thur. 12 and Fri. } 6\tfrac{1}{2}\end{array}\right]$$

6 Draw a horizontal bar chart presenting the data given in *Problem 3* above.

$$\left[\begin{array}{l}\text{5 equally spaced horizontal rectangles, the lengths}\\ \text{being proportional to 41, 73, 62, 58 and 84 units,}\\ \text{respectively.}\end{array}\right]$$

7 For the data given in *Problem 4* above, draw a horizontal bar chart.

$$\left[\begin{array}{l}\text{6 equally spaced horizontal rectangles whose}\\ \text{lengths are proportional to 35, 44, 62, 68, 49}\\ \text{and 41, respectively.}\end{array}\right]$$

8 Present the data given in *Problem 5* above on a horizontal bar chart.

$$\left[\begin{array}{l}\text{5 equally spaced horizontal rectangles, whose}\\ \text{lengths are proportional to 1580, 2190, 1840,}\\ \text{2385 and 1280 units, respectively.}\end{array}\right]$$

9 Draw a vertical bar chart depicting the data given in *Problem 3* above.

$$\left[\begin{array}{l}\text{5 equally spaced vertical rectangles, whose heights}\\ \text{are proportional to 41, 73, 62, 58 and 84 units,}\\ \text{respectively.}\end{array}\right]$$

10 For the data given in *Problem 4* above, construct a vertical bar chart.

[6 equally spaced vertical rectangles, whose heights are proportional to 35, 44, 62, 68, 49 and 41 units, respectively.]

11 Depict the data given in *Problem 5* above on a vertical bar chart.

[5 equally spaced vertical rectangles, whose heights are proportional to 1580, 2190, 1840, 2385 and 1280 units, respectively.]

12 A factory produces three different types of components. The percentages of each of these components produced for three, one-month periods are as shown below. Show this information on percentage component bar charts and comment on the changing trend in the percentages of the types of component produced.

Month	1	2	3
Component P	20	35	40
Component Q	45	40	35
Component R	35	25	25

[Three rectangles of equal height, subdivided in the percentages shown in the columns above. P increases by 20% at the expense of Q and R.]

13 An analysis of the expenditure of a small company in thousands of pounds for a two-year period is as shown below.

	Year 1	Year 2
Salaries	40	50
Wages	50	65
Materials	30	60
Rent and rates	15	25

Present these data on percentage component bar charts.

[Two rectangles of equal height, subdivided into the following percentages: year 1 30%, 37%, 22%, 11%; year 2 25%, $32\frac{1}{2}$%, 30%, $12\frac{1}{2}$%.]

14 A company has five distribution centres and the mass of goods in tonnes sent to each centre during four, one-week periods, is as shown.

Week	1	2	3	4
Centre A	147	160	174	158
Centre B	54	63	77	69
Centre C	283	251	237	211
Centre D	97	104	117	144
Centre E	224	218	203	194

Use a percentage component bar chart to present these data and comment on any trends.

[Four rectangles of equal heights, subdivided as follows:
week 1   18%, 7%, 35%, 12%, 28%;
week 2   20%, 8%, 32%, 13%, 27%;
week 3   22%, 10%, 29%, 14%, 25%;
week 4   20%, 9%, 27%, 19% and 25%.
Little change in centres A and B, a reduction of about 5% in C, an increase of about 7% in D and a reduction of about 3% in E.]

15 A factory produces four different types of product. The value of each product produced in a day is as shown.

Product	A	B	C	D
Value of Product (£)	17 000	31 000	44 000	3500

Present these data on a pie diagram.

[A circle of any radius, subdivided into sectors having angles of 64°, 117°, 166° and 13°, respectively.]

16 The employees in a company can be split into the following categories:

managerial 3, supervisory 9, craftsmen 21, semi-skilled 67, others 44

Show these data on a pie diagram.

[A circle of any radius, subdivided into sectors having angles of $7\frac{1}{2}$°, $22\frac{1}{2}$°, $52\frac{1}{2}$°, $167\frac{1}{2}$° and 110°, respectively.]

17 The way in which an apprentice spent his time over a one-month period is as follows:

drawing office 44 hours, production 64 hours, training 12 hours, at college 28 hours.

Use a pie diagram to depict this information.

[A circle of any radius, subdivided into sectors having angles of 107° 156°, 29° and 28°, respectively.]

18 (a) With reference to *Fig. 25.5*, determine the amount spent on labour and materials to produce 1650 units of the product.

(b) If in year 2 of *Fig. 25.4*, 1% corresponds to 2.5 dwellings, how many bungalows are sold in that year?

[(a) £495, (b) 88]

19 (a) If the company sell 23 500 units of the product depicted in *Fig. 25.5*, determine the cost of their overheads per annum.

(b) If 1% of the dwellings represented in year 1 of *Fig. 25.4* corresponds to 2 dwellings, find the total number of houses sold in that year.

[(a) £16 450, (b) 138]

Presentation of grouped data

20 The number of cars passing across a particular road junction is counted daily from Monday to Friday for a ten-week period and the results are shown below. Form a frequency distribution of about 7 classes for these data.

86	87	78	78	82	85	88	71	74	80
77	73	84	79	77	74	81	91	85	82
79	86	72	79	74	76	80	85	83	84
85	88	81	77	90	84	81	84	85	83
83	82	87	81	78	80	82	75	82	80

There is no unique solution, but one solution is: 71–73 3; 74–76 5; 77–79 9; 80–82 14; 83–85 11; 86–88 6; 89–91 2.

21 The mass in kilograms, correct to the nearest one-tenth of a kilogram, of 60 bars of metal are as shown. Form a frequency distribution of about 8 classes for these data.

39.8	40.1	40.3	40.0	40.6	39.7	40.0	40.4	39.6	39.3
39.6	40.7	40.2	39.9	40.3	40.2	40.4	39.9	39.8	40.0
40.2	40.1	40.3	39.7	39.9	40.5	39.9	40.5	40.0	39.9
40.1	40.8	40.0	40.0	40.1	40.2	40.1	40.0	40.2	39.9
39.7	39.8	40.4	39.7	39.9	39.5	40.1	40.1	39.9	40.2
39.5	40.6	40.0	40.1	39.8	39.7	39.5	40.2	39.9	40.3

There is no unique solution, but one solution is: 39.3–39.4 1; 39.5–39.6 5; 39.7–39.8 9; 39.9–40.0 17; 40.1–40.2 15; 40.3–40.4 7; 40.5–40.6 4; 40.7–40.8 2.

22 Draw a histogram for the frequency distribution given in the solution of *Problem 20*.

Rectangles, touching one another, having mid-points of 72, 75, 78, 81, ... and heights proportional to 3, 5, 9, 14, ...

23 Draw a histogram for the frequency distribution given in the solution of *Problem 21*.

Rectangles, touching one another, having mid-points of 39.35, 39.55, 39.75, 39.95, . . . and heights of 1, 5, 9, 17, ...

24 The force in kilonewtons, expressed correct to the nearest kilonewton, required to cause each of 40 blocks of material to fail are as shown. Produce a frequency distribution having about 6 classes for these data. (The frequency distribution should be designed to have unequal class intervals.)

206	244	204	209	201	212	203	219
257	22	281	157	211	126	84	208
213	203	357	202	207	214	318	218
184	207	205	221	217	17	219	203
215	139	210	418	216	296	220	193

There is no unique solution, but one solution is: 1–100 3; 101–200 5; 201–210 13; 211–220 11; 221–320 6; 321–420 2.

25 Form a frequency distribution having about 8 classes for the data given below, which refers to the diameter in millimetres of 60 bolts produced by an automatic process. Draw a histogram and a frequency polygon to depict these data.

3.97	4.02	4.05	4.00	3.98	3.99	4.04	3.97	3.99	3.95
4.01	3.99	4.01	3.96	4.03	4.01	3.99	4.01	4.02	4.04
3.95	4.06	3.98	4.02	4.00	3.97	3.96	4.00	4.07	4.00
3.99	4.02	4.03	3.99	4.01	4.03	4.02	4.01	4.01	4.02
3.98	4.00	3.93	4.05	4.00	3.99	4.00	3.98	4.06	3.97
4.08	4.04	4.01	3.97	3.95	4.02	3.99	4.03	3.99	4.00

There is no unique solution, but one solution is: 3.92–3.94 1; 3.95–3.96 5; 3.97–3.98 9; 3.99–4.00 17; 4.01–4.02 15; 4.03–4.04 7; 4.05–4.06 4; 4.07–4.08 2.

26 The information given below refers to the value of resistance in ohms of a batch of 48 resistors of similar value. Form a frequency distribution for the data, having about 6 classes and draw a frequency polygon and histogram to represent these data diagramatically.

21.0	22.4	22.8	21.5	22.6	21.1	21.6	22.3
22.9	20.5	21.8	22.2	21.0	21.7	22.5	20.7
23.2	22.9	21.7	21.4	22.1	22.2	22.3	21.3
22.1	21.8	22.0	22.7	21.7	21.9	21.1	22.6
21.4	22.4	22.3	20.9	22.8	21.2	22.7	21.6
22.2	21.6	21.3	22.1	21.5	22.0	23.4	21.2

There is no unique solution, but one solution is: 20.5–20.9 3; 21.0–21.4 10; 21.5–21.9 11; 22.0–22.4 13; 22.5–22.9 9; 23.0–23.4 2.

27 The time taken in hours to the failure of 50 specimens of a metal subjected to fatigue failure tests are as shown. Form a frequency distribution, having about 8 classes and unequal class intervals, for these data.

28	22	23	20	12	24	37	28	21	25
21	14	30	23	27	13	23	7	26	19
24	22	26	3	21	24	28	40	27	24
20	25	23	26	47	21	29	26	22	33
27	9	13	35	20	16	20	25	18	22

There is no unique solution, but one solution is: 1–10 3; 11–19 7; 20–22 12; 23–25 14; 26–28 7; 29–38 5; 39–48 2.

28 The number of hours of overtime worked by a group of craftsmen during each of 48 working weeks in a year is as shown below. For these data (a) form a frequency distribution having about seven classes, and then (b)

represent the grouped data diagrammatically by means of (i) a frequency polygon, and (ii) a histogram.

40	39	41	34	37	36	26	47
29	45	41	51	42	46	43	32
48	40	28	52	47	33	52	44
38	53	47	42	51	48	37	30
54	46	35	50	59	43	49	48
41	27	45	46	42	53	25	38

$$\begin{bmatrix} \text{There is no unique solution, but one solution is:} \\ \text{25–29 5; 30–34 4; 35–39 7; 40–44 11; 45–49 12;} \\ \text{50–54 8; 55–59 1.} \end{bmatrix}$$

29 Form a cumulative frequency distribution and hence draw the ogive for the frequency distribution given in the solution to *Problem 26*.

$$\begin{bmatrix} 20.95 \ 3; 21.45 \ 13; 21.95 \ 24; \\ 22.45 \ 37; 22.95 \ 46; 23.45 \ 48. \end{bmatrix}$$

30 For the data given in the solution to *Problem 28*, form a cumulative frequency distribution and hence draw an ogive depicting these data.

$$\begin{bmatrix} 29.5 \ 5; 34.5 \ 9; 39.5 \ 16; 44.5 \ 27; \\ 49.5 \ 39; 54.5 \ 47; 59.5 \ 48. \end{bmatrix}$$

31 Draw a histogram for the frequency distribution given in the solution of *Problem 24*.

$$\begin{bmatrix} \text{Rectangles, touching one another, having mid-points} \\ \text{of 50.5, 150.5, 205.5, 215.5, 270.5 and 370.5. The} \\ \text{heights of the rectangles (frequency per unit class} \\ \text{range) are 0.03, 0.05, 1.3, 1.1, 0.06 and 0.02.} \end{bmatrix}$$

32 Draw a histogram for the frequency distribution given in the solution to *Problem 27*.

$$\begin{bmatrix} \text{Rectangles, touching one another, having mid-points} \\ \text{of 5.5, 15, 21, 24, 27, 33.5 and 43.5. The heights of} \\ \text{the rectangles (frequency per unit class range) are 0.3,} \\ \text{0.78, 4, 4.67, 2.33, 0.5 and 0.2.} \end{bmatrix}$$

33 The frequency distribution for a batch of 48 resistors of similar value, measured in ohms, is:

20.5–20.9 3, 21.0–21.4 10,
21.5–21.9 11, 22.0–22.4 13,
22.5–22.9 9, 23.0–23.4 2.

Form a cumulative frequency distribution for these data.
$$\begin{bmatrix} (20.95 \ 3), (21.45 \ 13), (21.95 \ 24), \\ (22.45 \ 37), (22.95 \ 46), (23.45 \ 48) \end{bmatrix}$$

34 The frequency distribution for the number of hours of overtime worked by a group of craftsmen during each of 48 working weeks in a year is as shown. Form a cumulative frequency distribution for this data.

25–29 5, 30–34 4, 35–39 7, 40–44 11,
45–49 12, 50–54 8, 55–59 1

$$\begin{bmatrix} (29.5 \ 5), (34.5 \ 9), (39.5 \ 16), (44.5 \ 27), \\ (49.5 \ 39), (54.5 \ 47), (59.5 \ 48) \end{bmatrix}$$

35 Draw an ogive for the data given in the solution of *Problem 33*.

36 Draw an ogive for the data given in the solution of *Problem 34*.

37 The diameter in millimetres of a reel of wire is measured in 48 places and the results are as shown.

2.10	2.29	2.32	2.21	2.14	2.22
2.28	2.18	2.17	2.20	2.23	2.13
2.26	2.10	2.21	2.17	2.28	2.15
2.16	2.25	2.23	2.11	2.27	2.34
2.24	2.05	2.29	2.18	2.24	2.16
2.15	2.22	2.14	2.27	2.09	2.21
2.11	2.17	2.22	2.19	2.12	2.20
2.23	2.07	2.13	2.26	2.16	2.12

(a) Form a frequency distribution of diameters having about 6 classes.

(b) Draw a histogram depicting the data.

(c) Form a cumulative frequency distribution.

(d) Draw an ogive for the data.

$$\begin{bmatrix} \text{(a) There is no unique solution, but one solution is:} \\ \quad \text{2.05–2.09 3; 2.10–2.14 10; 2.15–2.19 11;} \\ \quad \text{2.20–2.24 13; 2.25–2.29 9; 2.30–2.34 2.} \\ \text{(b) Rectangles, touching one another, having mid-} \\ \quad \text{points of 2.07, 2.12, . . . and heights of 3, 10, ...} \\ \text{(c) Using the frequency distribution given in the} \\ \quad \text{solution to part (a) gives: 2.095 3; 2.145 13;} \\ \quad \text{2.195 24; 2.245 37; 2.295 46; 2.345 48.} \\ \text{(d) A graph of cumulative frequency against upper} \\ \quad \text{class boundary having the coordinates given in} \\ \quad \text{part (c).} \end{bmatrix}$$

26

Measures of central tendency and dispersion

26.1 Measures of central tendency

A single value, which is representative of a set of values, may be used to give an indication of the general size of the members in a set, the word '**average**' often being used to indicate the single value.

The statistical term used for 'average' is the arithmetic mean or just the **mean**.

Other measures of central tendency may be used and these include the **median** and the **modal** values.

26.2 Mean, median and mode for discrete data

Mean

The **arithmetic mean value** is found by adding together the values of the members of a set and dividing by the number of members in the set. Thus, the mean of the set of numbers: {4, 5, 6, 9} is:

$$\frac{4+5+6+9}{4}, \text{ i.e. } 6$$

In general, the mean of the set: $\{x_1, x_2, x_3, \ldots x_n\}$ is

$$\bar{x} = \frac{x_1 + x_2 + x_3 + \ldots + x_n}{n}, \text{ written as } \frac{\Sigma x}{n}$$

where Σ is the Greek letter 'sigma' and means 'the sum of', and $\bar{x}$ (called x-bar) is used to signify a mean value.

Median

The **median value** often gives a better indication of the general size of a set containing extreme values. The set: {7, 5, 74, 10} has a mean value of 24, which is not really representative of any of the values of the members of the set. The median value is obtained by:

(a) **ranking** the set in ascending order of magnitude, and
(b) selecting the value of the middle member for sets containing an odd number of members, or finding the value of the mean of the two middle members for sets containing an even number of members.

For example, the set: {7, 5, 74, 10} is ranked as {5, 7, 10, 74}, and since it contains an even number of members (four in this case), the mean of 7 and 10 is taken, giving a median value of 8.5. Similarly, the set: {3, 81, 15, 7, 14} is ranked as {3, 7, 14, 15, 81} and the median value is the value of the middle member, i.e. 14.

Mode

The **modal value**, or just the **mode**, is the most commonly occurring value in a set. If two values occur with the same frequency, the set is 'bi-modal'. The set: {5, 6, 8, 2, 5, 4, 6, 5, 3} has a modal value of 5, since the member having a value of 5 occurs three times.

Problem 1   Determine the mean, median and mode for the set:

{2, 3, 7, 5, 5, 13, 1, 7, 4, 8, 3, 4, 3}

The mean value is obtained by adding together the values of the members of the set and dividing by the number of members in the set.

Thus, mean value,

$$\bar{x} = \frac{2+3+7+5+5+13+1+7+4+8+3+4+3}{13}$$

$$= \frac{65}{13} = 5$$

To obtain the median value the set is ranked, that is, placed in ascending order of magnitude, and since the set contains an odd number of members the value of the middle member is the median value. Ranking the set gives:

$$\{1, 2, 3, 3, 3, 4, 4, 5, 5, 7, 7, 8, 13\}$$

The middle term is the seventh member, i.e. 4, thus the median value **is 4**. The modal value is the value of the most commonly occurring member and **is 3**, which occurs three times, all other members only occurring once or twice.

Problem 2   The following set of data refers to the amount of money in £s taken by a news vendor for 6 days. Determine the mean, median and modal values of the set:

$$\{27.90, 34.70, 54.40, 18.92, 47.60, 39.68\}$$

Mean value

$$= \frac{27.90 + 34.70 + 54.40 + 18.92 + 47.60 + 39.68}{6}$$

$$= £37.20$$

The ranked set is: $\{18.92, 27.90, 34.70, 39.68, 47.60, 54.40\}$ Since the set has an even number of members, the mean of the middle two members is taken to give the median value, i.e.

$$\text{median value} = \frac{34.70 + 39.68}{2} = £37.19$$

Since no two members have the same value, this set has **no mode**.

Further problems on mean, median and mode for discrete data may be found in section 26.6, Problems 1 to 4, page 239.

26.3 Mean, median and mode for grouped data

The mean value for a set of grouped data is found by determining the sum of the (frequency × class mid-point values) and dividing by the sum of the frequencies, i.e. mean value

$$\bar{x} = \frac{f_1 x_1 + f_2 x_2 + \ldots + f_n x_n}{f_1 + f_2 + \ldots + f_n} = \frac{\Sigma(fx)}{\Sigma f}$$

where f is the frequency of the class having a mid-point value of x, and so on.

Problem 3   The frequency distribution for the value of resistance in ohms of 48 resistors is as shown. Determine the mean value of resistance.

20.5–20.9   3,   21.0–21.4   10,   21.5–21.9   11
22.0–22.4   13,   22.5–22.9   9,   23.0–23.4   2

The class mid-point/frequency values are:

20.7 3, 21.2 10, 21.7 11, 22.2 13, 22.7 9 and 23.2 2

For grouped data, the mean value is given by:

$$\bar{x} = \frac{\Sigma(fx)}{\Sigma f}$$

where f is the class frequency and x is the class mid-point value. Hence

mean value, $\bar{x}$

$$= \frac{(3 \times 20.7) + (10 \times 21.2) + (11 \times 21.7)}{48} + (13 \times 22.2) + (9 \times 22.7) + (2 \times 23.2)}{48}$$

$$= \frac{1052.1}{48} = 21.919\ldots$$

i.e. **the mean value is 21.9 ohms**, correct to 3 significant figures.

The mean, median and modal values for grouped data may be determined from a **histogram**. In a histogram, frequency values are represented vertically and variable values horizontally. The mean value is given by the value of the variable corresponding to a vertical line drawn through the centroid of the histogram. The median value is obtained by selecting a variable value such that the area of the histogram to the left of a vertical line drawn through the selected variable value is equal to the area of the histogram on the right of the line. The modal value is the variable value obtained by dividing the width of the highest rectangle in the histogram in proportion to the heights of the adjacent rectangles. The method of determining the mean, median and modal values from a histogram is shown in *Problem 4*.

Problem 4 The time taken in minutes to assemble a device is measured 50 times and the results are as shown. Draw a histogram depicting this data and hence determine the mean, median and modal values of the distribution.

14.5–15.5	5,	16.5–17.5	8,	18.5–19.5	16,
20.5–21.5	12,	22.5–23.5	6,	24.5–25.5	3

The histogram is shown in *Fig. 26.1*. The mean value lies at the centroid of the histogram. With reference to any arbitrary axis, say YY shown at a time of 14 minutes, the position of the horizontal value of the centroid can be obtained from the relationship $AM = \Sigma(am)$, where A is the area of the histogram, M is the horizontal distance of the centroid from the axis YY, a is the area of a rectangle of the histogram and m is the distance of the centroid of the rectangle from YY (see Chapter 17). The areas of the individual rectangles are shown circled on the histogram giving a total area of 100 square units. The positions, m, of the centroids of the individual rectangles are 1, 3, 5, . . . units from YY. Thus

$$100M = (10 \times 1) + (16 \times 3) + (32 \times 5) + (24 \times 7)$$
$$+ (12 \times 9) + (6 \times 11)$$

i.e. $M = \dfrac{560}{100} = 5.6$ units from YY

Thus the position of the mean with reference to the time scale is $14 + 5.6$, i.e. **19.6 minutes**.

The median is the value of time corresponding to a vertical line dividing the total area of the histogram into two equal parts. The total area is 100 square units, hence the vertical line must be drawn to give 50 units of area on each side. To achieve this with reference to *Fig. 26.1*, rectangle ABFE must be split so that $50 - (10 + 16)$ units of area lie on one side and $50 - (24 + 12 + 6)$ units of area lie on the other. This shows that the area of ABFE is split so that 24 units

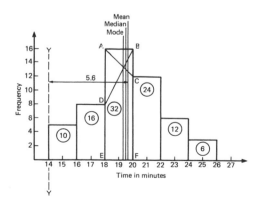

Fig. 26.1

of area lie to the left of the line and 8 units of area lie to the right, i.e. the vertical line must pass through 19.5 minutes. Thus the median value of the distribution is **19.5 minutes.** The mode is obtained by dividing the line AB, which is the height of the highest rectangle, proportionally to the heights of the adjacent rectangles. With reference to *Fig. 26.1*, this is done by joining AC and BD and drawing a vertical line through the point of intersection of these two lines. This gives the mode of the distribution and is **19.3 minutes**.

Further problems on mean, median and mode for grouped data may be found in section 26.6, Problems 5 to 8, page 239.

26.4 Standard deviation

The standard deviation of a set of data gives an indication of the amount of dispersion, or the scatter, of members of the set from the measure of central tendency. Its value is the root-mean-square value of the members of the set and for discrete data is obtained as follows:

(a) determine the measure of central tendency, usually the mean value, $\bar{x}$ (occasionally the median or modal values are specified),

(b) calculate the deviation of each member of the set from the mean, giving
$(x_1 - \bar{x}), (x_2 - \bar{x}), (x_3 - \bar{x}), \ldots,$

(c) determine the squares of these deviations, i.e.
$(x_1 - \bar{x})^2, (x_2 - \bar{x})^2, (x_3 - \bar{x})^2, \ldots,$

(d) find the sum of the squares of the deviations, that is
$(x_1 - \bar{x})^2 + (x_2 - \bar{x})^2 + (x_3 - \bar{x})^2, \ldots,$

(e) divide by the number of members in the set, n, giving
$$\frac{(x_1 - \bar{x})^2 + (x_2 - \bar{x})^2 + (x_3 - \bar{x})^2 + \ldots}{n}$$

(f) determine the square root of (e).

The standard deviation is indicated by σ (the Greek letter small 'sigma') and is written mathematically as:

$$\text{standard deviation, } \sigma = \sqrt{\left\{ \frac{\Sigma(x - \bar{x})^2}{n} \right\}}$$

where x is a member of the set, $\bar{x}$ is the mean value of the set and n is the number of members in the set. The value of standard deviation gives an indication of the distance of the members of a set from the mean value. The set: $\{1, 4, 7, 10, 13\}$ has a mean value of 7 and a standard deviation of about 4.2. The set $\{5, 6, 7, 8, 9\}$ also has a mean value of 7, but the standard deviation is about 1.4. This shows that the members of the second set are mainly much closer to the mean value than the members of the first set. The method of determining the standard deviation for a set of discrete data is shown in *Problem 5*.

Problem 5   Determine the standard deviation from the mean of the set of numbers: {5, 6, 8, 4, 10, 3}, correct to 4 significant figures.

The arithmetic mean, $\bar{x} = \dfrac{\Sigma x}{n} = \dfrac{5 + 6 + 8 + 4 + 10 + 3}{6}$

$$= 6$$

Standard deviation, $\sigma = \sqrt{\left\{\dfrac{\Sigma(x - \bar{x})^2}{n}\right\}}$

The $(x - \bar{x})^2$ values are: $(5 - 6)^2$, $(6 - 6)^2$, $(8 - 6)^2$, $(4 - 6)^2$, $(10 - 6)^2$ and $(3 - 6)^2$.
The sum of the $(x - \bar{x})^2$ values, i.e.

$$\Sigma(x - \bar{x})^2 = 1 + 0 + 4 + 4 + 16 + 9$$

i.e. $\Sigma(x - \bar{x})^2 = 34$

and $\dfrac{\Sigma(x - \bar{x})^2}{n} = \dfrac{34}{6} = 5.\dot{6}$ since there are 6 members in the set. Hence,

standard deviation, $\sigma = \sqrt{\left\{\dfrac{\Sigma(x - \bar{x})^2}{n}\right\}}$

$$= \sqrt{5.\dot{6}} = \textbf{2.380}, \text{ correct to 4}$$
significant figures

For **grouped data**, standard deviation

$$\sigma = \sqrt{\left\{\dfrac{\Sigma\{f(x - \bar{x})^2\}}{\Sigma f}\right\}}$$

where f is the class frequency value, x is the class mid-point value and $\bar{x}$ is the mean value of the grouped data. The method of determining the standard deviation for a set of grouped data is shown in *Problem 6*.

Problem 6   The frequency distribution for the values of resistance in ohms of 48 resistors is as shown. Calculate the standard deviation from the mean of the resistors, correct to 3 significant figures.

20.5–20.9   3,   21.0–21.4   10,   21.5–21.9   11,
22.0–22.4   13,   22.5–22.9   9,   23.0–23.4   2

The standard deviation for grouped data is given by:

$$\sigma = \sqrt{\left\{\dfrac{\Sigma\{f(x - \bar{x})^2\}}{\Sigma f}\right\}}$$

From *Problem 3*, the distribution mean value, $\bar{x} = 21.92$, correct to 4 significant figures. The 'x-values' are the class mid-point values, i.e. 20.7, 21.2, 21.7, ... Thus the $(x - \bar{x})^2$ values are $(20.7 - 21.92)^2$, $(21.2 - 21.92)^2$, $(21.7 - 21.92)^2$, ..., and the $f(x - \bar{x})^2$ values are $3(20.7 - 21.92)^2$, $10(21.2 - 21.92)^2$, $11(21.7 - 21.92)^2$, ... The $\Sigma f(x - \bar{x})^2$ values are $4.465\ 2 + 5.184\ 0 + 0.532\ 4 + 1.019\ 2 + 5.475\ 6 + 3.276\ 8$, i.e. 19.953 2.

$$\dfrac{\Sigma\{f(x - \bar{x})^2\}}{\Sigma f} = \dfrac{19.953\ 2}{48} = 0.415\ 69$$

and **standard deviation, $\sigma = \sqrt{\left\{\dfrac{\Sigma\{f(x - \bar{x})^2\}}{\Sigma f}\right\}}$**

$$= \sqrt{0.415\ 69} = \textbf{0.645}, \text{ correct to}$$
3 significant figures

Further problems on standard deviation may be found in section 26.6, Problems 9 to 14, page 240.

26.5   Quartiles, deciles and percentiles

Other measures of dispersion which are sometimes used are the quartile, decile and percentile values. The **quartile values** of a set of discrete data are obtained by selecting the values of members which divide the set into four equal parts. Thus for the set: {2, 3, 4, 5, 5, 7, 9, 11, 13, 14, 17} there are 11 members and the values of the members dividing the set into four equal parts are 4, 7, and 13. These values are signified by Q_1, Q_2 and Q_3 and called the first, second and third quartile values, respectively. It can be seen that the second quartile value, Q_2, is the value of the middle member and hence is the median value of the set.

For grouped data the ogive may be used to determine the quartile values. In this case, points are selected on the vertical cumulative frequency values of the ogive, such that they divide the total value of cumulative frequency into four equal parts. Horizontal lines are drawn from these values to cut the ogive. The values of the variable corresponding to these cutting points on the ogive give the quartile values (see *Problem 7*).

When a set contains a large number of members, the set can be split into ten parts, each containing an equal number of members. These ten parts are then called **deciles**. For sets containing a very large number of members, the set may be split into one hundred parts, each containing an equal number of members. One of these parts is called a **percentile**.

Problem 7 The frequency distribution given below refers to the overtime worked by a group of craftsmen during each of 48 working weeks in a year.

25–29 5, 30–34 4, 35–39 7, 40–44 11,
45–49 12, 50–54 8, 55–59 1

Draw an ogive for this data and hence determine the quartile values.

The cumulative frequency distribution (i.e. upper class boundary/cumulative frequency values) is:

29.5 5, 34.5 9, 39.5 16, 44.5 27, 49.5 39, 54.5 47, 59.5 48

The ogive is formed by plotting these values on a graph, as shown in *Fig. 26.2*. The total frequency is divided into four equal parts, each having a range of 48/4, i.e. 12. This gives cumulative frequency values of 0 to 12 corresponding to the first quartile, 12 to 24 corresponding to the second quartile, 24 to 36 corresponding to the third quartile and 36 to 48 corresponding to the fourth quartile of the distribution, i.e. the distribution is divided into four equal parts. The quartile values are those of the variable corresponding to cumulative frequency values of 12, 24 and 36, marked Q_1, Q_2 and Q_3 in *Fig. 26.2*. These values, correct to the nearest hour, are **37 hours, 43 hours and 48 hours**, respectively. The Q_2 value is also equal to the median value of the distribution. One measure of the dispersion of a distribution is called the **semi-interquartile range** and is given by $(Q_3 - Q_1)/2$, and is $(48 - 37)/2$ in this case, i.e. $5\frac{1}{2}$ **hours.**

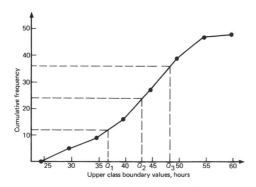

Fig. 26.2

Problem 8 Determine the numbers contained in the (a) 40th to 49th percentile group, and (b) 8th decile group of the set of numbers shown below:

14 22 17 21 30 28 37 7 23 32
24 17 20 22 27 19 26 21 15 29

The set is ranked, giving:

7 14 15 17 17 19 20 21 21 22
22 23 24 26 27 28 29 30 32 37

(a) There are 20 numbers in the set, hence the first 10% will be the two numbers 7 and 14, the second 10% will be 15 and 17, and so on.
Thus the 40th to 49th percentile group will be the numbers **20 and 21**.

(b) The first decile group is obtained by splitting the ranked set into 10 equal groups and selecting the first group, i.e. the numbers 7 and 14. The second decile group are the numbers 15 and 17, and so on.
Thus the 8th decile group contains the numbers **27 and 28**.

Further problems on quartiles, deciles and percentiles may be found in section 26.6, Problems 15 to 19, page 240.

26.6 Further problems on measures of central tendency and dispersion

Mean, median and mode for discrete data

In *Problems 1* to *4*, determine the mean, median and modal values for the sets given.

1 {3, 8, 10, 7, 5, 14, 2, 9, 8}
[mean $7\frac{1}{3}$, median 8, mode 8]

2 {26, 31, 21, 29, 32, 26, 25, 28}
[mean $27\frac{1}{4}$, median 27, mode 26]

3 {4.72, 4.71, 4.74, 4.73, 4.72, 4.71, 4.73, 4.72}
[mean 4.722 5, median 4.72, mode 4.72]

4 {73.8, 126.4, 40.7, 141.7, 28.5, 237.4, 157.9}
[mean 115.2, median 126.4, no mode]

Mean, median and mode for grouped data

5 The frequency distribution given below refers to the heights in centimetres of 100 people. Determine the mean value of the distribution, correct to the nearest millimetre.

150–156   5,   157–163   18,   164–170   20
171–177   27,   178–184   22,   185–191   8

[171.7 cm]

6 Determine the mean value of the grouped data given below, correct to 3 significant figures. The data refers to the mass in kilograms of 70 components.

0.30–0.31   4,   0.32–0.33   10,   0.34–0.35   14,
0.36–0.37   17,   0.38–0.39   12,   0.40–0.41   8,
0.42–0.43   2,   0.44–0.45   2,   0.46–0.47   1

[0.365 kg]

7   The gain of 90 similar transistors is measured and the results are as shown.

83.5–85.5   6,   86.5–88.5   39,   89.5–91.5   27,
92.5–94.5   15,   95.5–97.5   3

By drawing a histogram of this frequency distribution, determine the mean, median and modal values of the distribution.
[mean 89.5, median 89, mode 88.2]

8   The diameters, in centimetres, of 60 holes bored in engine castings are measured and the results are as shown. Draw a histogram depicting these results and hence determine the mean, median and modal values of the distribution.

2.011–2.014   7,   2.016–2.019   16,   2.021–2.024   23,
2.026–2.029   9,   2.031–2.034   5

[mean 2.021 58 cm, median 2.021 52 cm, mode 2.021 67 cm]

Standard deviation

9   Determine the standard deviation from the mean of the set of numbers:
{35, 22, 25, 23, 28, 33, 30} correct to 3 significant figures.
[4.60]

10   The values of capacitances, in microfarads, of ten capacitors selected at random from a large batch of similar capacitors are:

34.3, 25.0, 30.4, 34.6, 29.6, 28.7, 33.4, 32.7, 29.0 and 31.3

Determine the standard deviation from the mean for these capacitors, correct to 3 significant figures.
[2.83 μF]

11   The tensile strength in megapascals for 15 samples of tin were determined and found to be:

34.61, 34.57, 34.40, 34.63, 34.63, 34.51, 34.49, 34.61, 34.52, 34.55, 34.58, 34.53, 34.44, 34.48 and 34.40

Calculate the mean and standard deviation from the mean for these 15 values, correct to 4 significant figures.
[mean 34.53 MPa, standard deviation 0.074 74 MPa]

12   Determine the standard deviation from the mean, correct to 4 significant figures, for the heights of the 100 people given in *Problem 5*.
[9.394 cm]

13   For the frequency distribution of the masses of 70 components given in *Problem 6*, Calculate the standard deviation from the mean, correct to 3 significant figures.
[0.034 6 kg]

14   Calculate the standard deviation from the mean for the data given in *Problem 7*, correct to 3 decimal places.
[2.828]

Quartiles, deciles and percentiles

15   The number of working days lost due to accidents for each of 12 one-monthly periods are as shown. Determine the median and first and third quartile values for this data.

27 37 40 28 23 30 35 24 30 32 31 28

[30, 27.5, 33.5 days]

16   The number of faults occurring on a production line in a nine-week period are as shown below. Determine the median and quartile values for the data.

30 27 25 24 27 37 31 27 35

[27, 26, 33 faults]

17   Determine the quartile values and semi-interquartile range for the frequency distribution given in *Problem 5*.
[Q_1 = 164.5 cm, Q_2 = 172.5 cm, Q_3 = 179 cm, 7.25 cm]

18   Determine the numbers contained in the 5th decile group and in the 70th to 79th percentile groups for the set of numbers:

40 46 28 32 37 42 50 31 48 45
32 38 27 33 40 35 25 42 38 41

[37 and 38; 40 and 41]

19   Determine the numbers in the 6th decile group and in the 80th to 89th percentile group for the set of numbers:

43 47 30 25 15 51 17 21 37 33 44 56 40 49 22
36 44 33 17 35 58 51 35 44 40 31 41 55 50 16

[40, 40, 41; 50, 51, 51]

27

An introduction to normal distribution curves

27.1 Introduction to normal curves

When data are obtained, they can frequently be considered to be a sample (i.e. a few members) drawn at random from a large population (i.e. a set having many members). If the sample number is large it is theoretically possible to choose class intervals which are very small, but which still have a number of members falling within each class. A frequency polygon of these data then has a large number of small line segments, and approximates to a continuous curve. Such a curve is called a **frequency** or a **distribution curve**.

Most distribution curves fall largely into three categories:

(i)   those having a positive skew (see *Fig. 27.1(a)*),
(ii) those having a negative skew (see *Fig. 27.1(b)*), and
(iii) those which are symmetrical (see *Fig. 27.1(c)*).

An extremely important symmetrical distribution curve is called the **normal curve**, which is similar in shape to the distribution curve shown in *Fig. 27.1(c)*. This curve can be described by a mathematical equation and is the basis of much of the work done in more advanced statistics. Many natural occurrences such as the heights or weights of a group of people, the size of components produced by a machine and the life length of certain components, approximate to a normal distribution. The standard deviation of a set of data which approximates to a normal curve can be determined using the techniques introduced in Chapter 26. When a normal curve is drawn, a relationship exists between the area under parts of the curve and the value of the standard deviation of the data. Since a normal curve is symmetrical, the mean value of the data corresponds to the maximum value of the normal distribution curve.

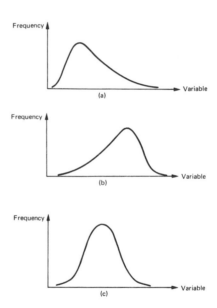

Fig. 27.1

(i)   If two vertical lines are drawn at distances corresponding to one standard deviation on either side of the mean value, then for all normal distribution curves, the area enclosed by the normal curve and the vertical lines at the mean value ± 1 standard deviation is approximately two thirds or $66\frac{2}{3}\%$ of the total area beneath the curve.

(ii) For all normal curves the area enclosed by the curve and vertical lines drawn at distances corresponding to the mean value ± 2 standard deviations is approximately $\frac{19}{20}$ ths or 95% of the total area beneath the curve.

(iii) For all normal curves, the area enclosed by the curve and the vertical lines drawn at distances corresponding to the mean value ±3 standard deviations is approximately $\frac{399}{400}$ or $99\frac{3}{4}$% of the total area beneath the curve.

The area beneath a normal curve is directly proportional to frequency and it follows that for a population which approximates to a normal distribution:

(i)   the values of $\frac{2}{3}$ of the members lie between the mean value ±1 standard deviation,

(ii) the values of $\frac{19}{20}$ of the members lie between the mean value ±2 standard deviations, and

(iii) the values of $\frac{399}{400}$ of the members lie between the mean value ±3 standard deviations.

Problem 1   The mean height of 500 people is 170 cm and the standard deviation is 9 cm. Assuming a normal distribution, determine the number of people likely to have heights of between 152 cm and 197 cm.

The mean value, $\bar{x}$ is 170 cm. Since the standard deviation, σ, is 9 cm, the mean value plus one standard deviation, written as $(\bar{x} + 1\sigma)$, is 170 + 9, i.e. 179 cm. Similarly, $(\bar{x} + 2\sigma)$ is $170 + 2 \times 9 = 188$ cm and $(\bar{x} + 3\sigma)$ is $170 + 3 \times 9 = 197$ cm. Also, $(\bar{x} - 1\sigma)$ is $170 - 9$, i.e. 161 cm and $(\bar{x} - 2\sigma)$ is $170 - 2 \times 9$, i.e. 152 cm. Thus the range of heights required is from $(\bar{x} - 2\sigma)$ to $(\bar{x} + 3\sigma)$. From above, $\frac{19}{20}$ of the area lies between $(\bar{x} \pm 2\sigma)$. Because the normal curve is symmetrical, it follows that half of $\frac{19}{20}$, i.e. $\frac{19}{40}$ of the area lies between $\bar{x}$ and -2σ. Similarly, $\frac{399}{400}$ of the area lies between $(\bar{x} \pm 3\sigma)$, so half of $\frac{399}{400}$, i.e. $\frac{399}{800}$ of the area lies between $\bar{x}$ and $+3\sigma$. Hence, the total area between $(\bar{x} - 2\sigma)$ and $(\bar{x} + 3\sigma)$ is $(\frac{19}{40} + \frac{399}{800})$ of the total area, i.e. 0.973 75 of the total area.

However, from para. 5, the area is proportional to the frequency, hence, 0.973 75 of 500 people, that is **487 people** are likely to have heights of between 152 cm and 197 cm. (In statistics, it is not usual to give results of values determined accurately, since in most instances, it is a probability

which is being determined. Thus the answer to this problem is expressed correct to the nearest whole number.)

Problem 2   For the group of people given in *Problem 1*, determine the number of people likely to have heights of less than 161 cm.

A height of 161 cm corresponds to $170 - 9$ cm, that is, mean minus one standard deviation $(\bar{x} - 1\sigma)$. Since the normal curve is symmetrical, the area to the left of the mean value is half of the total area (see *Fig. 27.2(a)*). The area between $(\bar{x} \pm 1\sigma)$ is $\frac{2}{3}$ of the total area. Hence the area between $\bar{x}$ and -1σ is half of $\frac{2}{3}$, i.e. $\frac{1}{3}$ of the total area (see *Fig. 27.2(b)*). An area of less than -1σ is therefore $\frac{1}{2} - \frac{1}{3}$, i.e. $\frac{1}{6}$ of the total area (see *Fig. 27.2(c)*). However, areas are proportional to frequency, hence $\frac{1}{6}$ of 500, i.e. **83 people**, are likely to have heights of less than 161 cm.

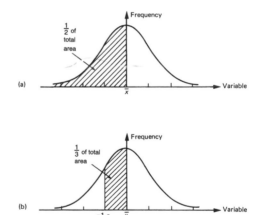

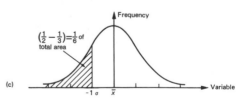

Fig. 27.2

Problem 3   For the group of people given in *Problem 1*, determine the number of people likely to have heights of more than 188 cm.

A height of 188 cm corresponds to 170 + 18, that is, the mean plus two standard deviations ($\bar{x} + 2\sigma$). The area to the right of the mean value is half of the total area under the normal curve (see *Fig. 27.3(a)*). The area between ($\bar{x} \pm 2\sigma$) is $\frac{19}{20}$ of the total area (see para. 4). Hence the area between $\bar{x}$ and $+2\sigma$ is half of $\frac{19}{20}$, i.e. $\frac{19}{40}$ of the total area (see *Fig. 27.3(b)*). An area to the right of the $+2\sigma$ line is therefore $\frac{1}{2} - \frac{19}{40}$, i.e. 0.025 of the total area (see *Fig. 27.3(c)*). Since areas are proportional to frequencies, it is likely that 0.025 of 500 people, i.e. **13 people**, will have heights of more than 188 cm.

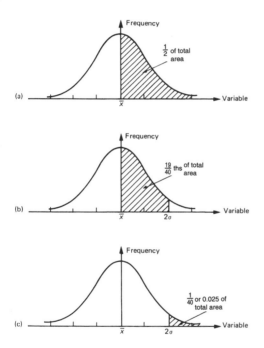

Fig. 27.3

Problem 4   A batch of lemonade bottles has an average content of 753 millilitres and a standard deviation of 1.8 millilitres. If the batch contains 1500 bottles and the contents of the bottles follow a normal distribution, determine:

(a) the number of bottles likely to contain less than 749.4 ml,

(b) the number of bottles likely to contain between 751.2 and 756.6 ml, and

(c) the number of bottles likely to contain more than 758.4 ml.

(a) Since the mean value $\bar{x}$ is 753 and the standard deviation σ is 1.8, 749.4 corresponds to ($\bar{x} - 2\sigma$). The area

between the vertical line drawn through -2σ and the vertical line drawn through the mean value is half of $\frac{19}{20}$ (see *Problem 2*), i.e. $\frac{19}{40}$ of the area to the left of the mean value. Since the normal curve is symmetrical, the total area to the left of the mean value is half the total area, i.e. $\frac{20}{40}$. Thus the area under the curve to the left of the -2σ line is $\left(\frac{20}{40} - \frac{19}{40}\right)$, i.e. $\frac{1}{40}$ or 0.025 of the total area. Area is proportional to frequency, thus the number of bottles having less than 749.4 ml is likely to be 0.025 of 1500, that is, **38 bottles**.

(b) 751.2 corresponds to ($\bar{x} - 1\sigma$) and the area between the mean and -1σ for a normal curve is half of $\frac{2}{3}$, i.e. $\frac{1}{3}$ of the total area. 756.6 corresponds to ($\bar{x} + 2\sigma$) and the area between the mean and $+2\sigma$ is half of $\frac{19}{20}$, i.e. $\frac{19}{40}$ of the total area. Thus, the area between -1σ and $+2\sigma$ is $\left(\frac{1}{3} + \frac{19}{40}\right)$, i.e. $0.808\dot{3}$ of the total area. Thus, the number of bottles likely to have contents of between 751.2 and 756.5 ml is $0.808\dot{3}$ or 1500, i.e. **1213 bottles**.

(c) 758.4 ml corresponds to ($\bar{x} + 3\sigma$) and the area between the mean and $+3\sigma$ for a normal curve is half of $\frac{399}{400}$, i.e. $\frac{399}{800}$ or 0.498 75 of the total area. Thus the area greater than $+3\sigma$ is 0.500 00 − 0.498 75, i.e. 0.001 25 of the total area. Thus the number of bottles likely to have contents of more than 758.4 ml is 0.001 25 × 1500, that is, **2 bottles**.

Further problems on normal curves may be found in section 27.3, Problems 1 to 8, page 244.

27.2 Use of probability paper

It should never be assumed that because data are continuous it automatically follows that it is normally distributed. One of the ways of determining whether data are normally distributed is by using **normal probability paper**, often just called **probability paper**. This is special graph paper which has linear markings on one axis and probability values from 0.01 to 99.99 on the other axis (see *Fig. 27.4*). The divisions on the probability axis are such that a straight line graph results for normally distributed data, when percentage cumulative frequency values are plotted against upper class boundary values. The method used to test the normality of a distribution is shown in *Problems 5* and *6*.

Problem 5 Use normal probability paper to determine whether the data given below are approximately normally distributed.

Class mid-point
value   55 65 75 85 95 105 115 125 135 145
Frequency 16 32 48 68 72 64 44 32 16   8

To test the normality of a distribution the upper class boundary/percentage cumulative frequency values are plotted on probability paper. The upper class boundary values are: 60, 70, 80, ..., 140, 150. The cumulative frequency values are: 16, 48, 96, ..., 392, 400. The percentage cumulative frequency value for a cumulative frequency of 16 is obtained from:

$$\frac{16}{400} \times 100\% = 4\%$$

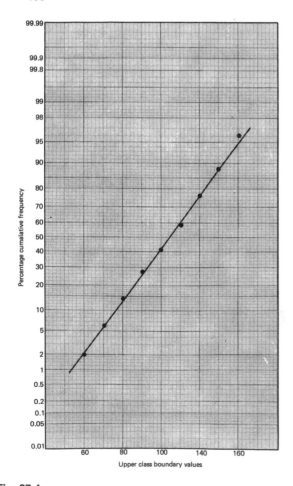

Fig. 27.4

Calculating each of the percentage cumulative frequency values gives: 4, 12, 24, ..., 98, 100. The coordinates of upper class boundary/cumulative frequency values are plotted as shown in *Fig. 27.4*. (When plotting these values, it will always be found that the last coordinates cannot be plotted, since the maximum value on the probability scale is 99.99.) Since all the coordinates lie approximately on a straight line, this indicates that the data given is approximately normally distributed.

Problem 6 The masses of 100 copper ingots are formed into a frequency distribution and the distribution is as follows:

Class mid-point
value (kg)   29.5   30.5   31.5   32.5   33.5
Frequency   3   6   8   19   28

Class mid-point
value (kg)   34.5   35.5   36.5   37.5   38.5
Frequency   19   6   5   4   2

Use probability paper to determine whether this frequency distribution is approximately normal.

The upper class boundary values are 30, 31, 32, 33, ..., 38, 39. The corresponding cumulative frequency values are 3, 9, 17, 36, ..., 98, 100. Since the total frequency is 100, the cumulative frequency values are also the percentage cumulative frequency values. The values of percentage cumulative frequencies and upper class boundaries are plotted as shown in *Fig. 27.5*. It can be seen in *Fig. 27.5* that a group of points at upper class boundary values of 34, 35, and 36 are displaced from a straight line drawn through the remaining coordinates. This shows that the distribution is not normally distributed.

Further problems on the use of probability paper may be found in section 27.3, Problems 9 and 10, page 245.

27.3 Further problems on normal distribution curves

Normal curves

1 A transistor is classed as defective if its gain is less than 69. In a batch of 350 transistors, the mean gain is 75 and the standard deviation is 3. Assuming the value of gain is normally distributed, determine how many are likely to be rejected.
[9]

2 The masses of 800 people are normally distributed, having a mean value of 65 kg and a standard deviation

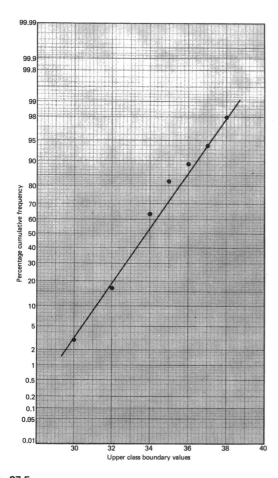

Fig. 27.5

of 5.5 kg. How many people are likely to have masses of less than 48.5 kg?
[1]

3 500 tins of paint have a mean content of 1010 ml, the standard deviation of the contents is 4 ml, and the volume of the contents is normally distributed. Calculate the number of tins which are likely to have a mass of less than (a) 1006 ml, (b) 1002 ml, and (c) 998 ml.
[(a) 83 (b) 13 (c) 1]

4 For the batch of transistors in *Problem 1*, if those having a high gain of more than 81 are rejected, determine the number likely to be rejected due to high gain.
[9]

5 For the 800 people in *Problem 2*, how many are likely to have masses of more than 76 kg?
[20]

6 The diameter of holes produced by a drilling machine bit is 4.05 mm and the standard deviation of the diameters is 0.002 5 mm. For 20 holes drilled using this machine, how many are likely to have diameters of between 4.045 and 4.052 5 mm, assuming the diameters are normally distributed?
[16]

7 The intelligence quotients of 400 children have a mean value of 100 and a standard deviation of 15. Assuming IQs are normally distributed, determine the number of children likely to have IQs of between: (a) 70 and 100, (b) 100 and 130, and (c) 70 and 115.
[(a) 190 (b) 190 (c) 323]

8 The mean mass of tablets produced by a manufacturer is 5 g and the standard deviation of the masses is 0.04 g. In a bottle containing 100 tablets, determine how many are likely to have (a) a mass of less than 4.92 g, (b) a mass of between 4.88 g and 5.04 g, and (c) a mass of more than 5.08 g. Assume the masses of the tablets are normally distributed.
[(a) 3 (b) 83 (c) 3]

Use of probability paper

9 A frequency distribution of 100 measurements is as shown:

Measurement:	16.8	17.0	17.2	17.4	17.6	17.8	18.0
Frequency:	3	8	16	24	24	17	8

Use normal probability paper to determine 'whether this distribution approximates to a normal distribution.
[Yes]

10 Use probability paper to assess whether the frequency distribution given below for the diameters of 60 bolts is approximately normally distributed.
3.93–3.94   1,   3.95–3.96   5,
3.97–3.98   9,   3.99–4.00 17,
4.01–4.02 15,   4.03–4.04   7,
4.05–4.06   4,   4.07–4.08   2
[No]

28

Probability

28.1 Introduction to probability

Probability

The **probability** of something happening is the likelihood or chance of it happening. Values of probability lie between 0 and 1, where 0 represents an absolute impossibility and 1 represents an absolute certainty. The probability of an event happening usually lies somewhere between these two extreme values and is expressed either as a proper or a decimal fraction. Examples of probability are:

that a length of copper wire has zero resistance at 100°C	0
that a fair, six-sided dice will stop with a 3 upwards	$\frac{1}{6}$ or 0.166 7
that a fair coin will land with a head upwards	$\frac{1}{2}$ or 0.5
that a length of copper wire has some resistance at 100°C	1

If p is the probability of an event happening and q is the probability of the same event not happening, then the total probability is $p + q$ and is equal to unity, since it is an absolute certainty that the event either does or does not occur, i.e. $p + q = 1$

Expectation

The **expectation**, E, of an event happening is defined in general terms as the product of the probability p of an event happening and the number of attempts made, n, i.e. $E = pn$.

Thus, since the probability of obtaining a 3 upwards when rolling a fair dice is $\frac{1}{6}$, the expectation of getting a 3 upwards on four throws of the dice is

$$\frac{1}{6} \times 4, \text{ i.e. } \frac{2}{3}$$

Thus expectation is the average occurrence of an event.

Dependent event

A **dependent event** is one in which the probability of one event happening affects the probability of another event happening. Let 5 transistors be taken at random from a batch of 100 transistors for test purposes, and the probability of there being a defective transistor, p_1, be determined. At some later time, let another 5 transistors be taken at random from the 95 remaining transistors in the batch and the probability of there being a defective transistor, p_2, be determined. The value of p_2 is different from p_1 since the batch size has effectively altered from 100 to 95, i.e. the probability p_2 is dependent on probability p_1. Since 5 transistors are drawn, and then another 5 transistors are drawn without replacing the first 5, the second random selection is said to be **without replacement**.

Independent event

An **independent event** is one in which the probability of an event happening does not affect the probability of another event happening. If 5 transistors are taken at random from a batch of transistors and the probability of a defective transistor p_1 is determined and the process is repeated after the original 5 have been replaced in the batch to give p_2, the

p_1 is equal to p_2. Since the 5 transistors are replaced between draws, the second selection is said to be **with replacement**.

The addition law of probability

The addition law of probability is recognized by the word 'or' joining the probabilities. If p_A is the probability of event A happening and p_B is the probability of event B happening, the probability of **event A or event B** happening is given by $p_A + p_B$. Similarly, the probability of events **A or B or C or ... N** happening is given by

$$p_A + p_B + p_C + \ldots + p_N$$

The multiplication law of probability

The multiplication law of probability is recognized by the word '**and**' joining the probabilities. If p_A is the probability of event A happening and p_B is the probability of event B happening, the probability of **event A and event B** happening is given by $p_A \times p_B$. Similarly, the probability of events **A and B and C and ... N** happening is given by

$$p_A \times p_B \times p_C \times \ldots \times p_N$$

28.2 Worked problems on probability

Problem 1 Determine the probabilities of selecting at random (a) a man, and (b) a woman from a crowd containing 20 men and 33 women.

(a) The probability of selecting at random a man, p, is given by the ratio

$$\frac{\text{number of men}}{\text{number in crowd}}, \text{ i.e. } p = \frac{20}{20 + 33} = \frac{20}{53}$$

(b) The probability of selecting at random a women, q, is given by the ratio

$$\frac{\text{number of women}}{\text{number in crowd}}, \text{ i.e. } q = \frac{33}{20 + 33} = \frac{33}{53}$$

Check: the total probability should be equal to 1

$p = \dfrac{20}{53}$ and $q = \dfrac{33}{53}$

The total probability, $p + q = \dfrac{20}{53} + \dfrac{33}{53} = 1$ hence no obvious error has been made.)

Problem 2 Find the expectation of obtaining a 4 upwards with 3 throws of a fair dice.

Expectation is the average occurrence of an event and is defined as the probability times the number of attempts. The probability, p, of obtaining a 4 upwards for one throw of the dice is $\dfrac{1}{6}$. Also, 3 attempts are made, hence $n = 3$ and the expectation, E, is pn, i.e. $\dfrac{1}{6} \times 3 = \dfrac{1}{2}$

Problem 3 Calculate the probabilities of selecting at random:

(a) the winning horse in a race in which 10 horses are running,
(b) the winning horses in both the first and second races if there are 10 horses in each race.

(a) Since only one of the ten horses can win, the probability of selecting at random the winning horse is

$$\frac{\text{number of winners}}{\text{number of horses}}, \text{ i.e. } \frac{1}{10}$$

(b) The probability of selecting the winning horse in the first race is $\dfrac{1}{10}$. The probability of selecting the winning horse in the second race is $\dfrac{1}{10}$. The probability of selecting the winning horses in the first **and** second race is given by the multiplication law of probability, i.e.

$$\text{probability} = \frac{1}{10} \times \frac{1}{10} = \frac{1}{100} \text{ or } \mathbf{0.01}$$

Problem 4 The probability of a component failing in one year due to excessive temperature is $\dfrac{1}{20}$, due to excessive vibration is $\dfrac{1}{25}$ and due to excessive humidity is $\dfrac{1}{50}$. Determine the probabilities that during a one-year period a component:

(a) fails due to excessive temperature and excessive vibration,
(b) fails due to excessive vibration or excessive humidity, and
(c) will not fail because of both excessive temperature and excessive humidity.

Let p_A be the probability of failure due to excessive temperature, then

$$p_A = \frac{1}{20} \text{ and } \bar{p}_A = \frac{19}{20}$$

Let p_B be the probability of failure due to excessive vibration, then

$$p_B = \frac{1}{25} \text{ and } \bar{p}_B = \frac{24}{25}$$

Let p_C be the probability of failure due to excessive humidity, then

$$p_C = \frac{1}{50} \text{ and } \bar{p}_C = \frac{49}{50}$$

(a) The probability of a component failing due to excessive temperature and excessive vibration is given by

$$p_A \times p_B, \text{ i.e. } \frac{1}{20} \times \frac{1}{25} = \frac{1}{500} \text{ or } 0.002$$

(b) The probability of a component failing due to excessive vibration or·excessive humidity is

$$p_B + p_C, \text{ i.e. } \frac{1}{25} + \frac{1}{50} = \frac{3}{50} \text{ or } 0.06$$

(c) The probability that a component will not fail due to excessive temperature and will not fail due to excessive humidity is

$$\bar{p}_A \times \bar{p}_C, \text{ i.e. } \frac{19}{20} \times \frac{49}{50} = \frac{931}{1000} \text{ or } 0.931$$

Problem 5 A batch of 100 capacitors contains 73 which are within the required tolerance values, 17 which are below the required tolerance values, and the remainder are above the required tolerance values. Determine the probabilities that when randomly selecting a capacitor and then a second capacitor:

(a) both are within the required tolerance values when selecting with replacement and (b) the first one drawn is below and the second one drawn is above the required tolerance value, when selection is without replacement.

(a) The probability of selecting a capacitor within the required tolerance values is $\frac{73}{100}$. The first capacitor drawn is now replaced and a second one is drawn from the batch of 100. The probability of this capacitor being within the required tolerance values is also $\frac{73}{100}$.

Thus, the probability of selecting a capacitor within the required tolerance values for both the first **and** the second draw is

$$\frac{73}{100} \times \frac{73}{100} = \frac{5329}{10\ 000} \text{ or } 0.532\ 9$$

(b) The probability of obtaining a capacitor below the required tolerance values on the first draw is $\frac{17}{100}$. There are now only 99 capacitors left in the batch, since the first capacitor is not replaced. The probability of drawing a capacitor above the required tolerance values on the second draw is $\frac{10}{99}$, since there are $(100 - 73 - 17)$ i.e. 10 capacitors above the required tolerance value. Thus, the probability of randomly selecting a capacitor below the required tolerance values and followed by randomly selecting a capacitor above the tolerance values is

$$\frac{17}{100} \times \frac{10}{99} = \frac{170}{9900} = \frac{17}{990} = 0.017\ 2$$

Problem 6 A batch of 40 components contains 5 which are defective. If a component is drawn at random from the batch and tested and then a second component is drawn, determine the probability that neither of the components is defective.

With replacement:
The probability that the component selected on the first draw is satisfactory is $\frac{35}{40}$, i.e. $\frac{7}{8}$. The component is now replaced and a second draw is made. The probability that this component is also satisfactory is $\frac{7}{8}$. Hence, the probability that both the first component drawn **and** the second component drawn are satisfactory is

$$\frac{7}{8} \times \frac{7}{8} = \frac{49}{64} \text{ or } 0.765\ 6$$

Without replacement:
The probability that the first component drawn is satisfactory is $\frac{7}{8}$. There are now only 34 satisfactory components left in the batch and the batch number is 39. Hence, the probability of drawing a satisfactory component on the

second draw is $\dfrac{34}{39}$. Thus the probability that the first component drawn **and** the second component drawn are satisfactory is

$$\frac{7}{8} \times \frac{34}{39} = \frac{238}{312} \text{ or } 0.762\,8$$

Problem 7 A batch of 40 components contains 5 which are defective. If a component is drawn at random from the batch and tested and then a second component is drawn at random, calculate the probability of having one defective component, both with and without replacement.

The probability of having one defective component can be achieved in two ways. If p is the probability of drawing a defective component and q is the probability of drawing a satisfactory component, then the probability of having one defective component is given by drawing a satisfactory component **and** then a defective component **or** by drawing a defective component **and** then a satisfactory one, i.e. by $q \times p + p \times q$.

With replacement:

$$p = \frac{5}{40} = \frac{1}{8} \text{ and } q = \frac{35}{40} = \frac{7}{8}$$

Hence, probability of having a defective component is $\dfrac{1}{8} \times \dfrac{7}{8} + \dfrac{7}{8} \times \dfrac{1}{8}$, i.e.

$$\frac{7}{64} + \frac{7}{64} = \frac{14}{64} = \frac{7}{32} \text{ or } 0.218\,8$$

Without replacement:

$p_1 = \dfrac{1}{8}$ and $q_1 = \dfrac{7}{8}$ on the first of the two draws. The batch number is now 39 for the second draw, thus, $p_2 = \dfrac{5}{39}$ and $q_2 = \dfrac{35}{39}$:

$$p_1 q_2 + q_1 p_2 = \frac{1}{8} \times \frac{35}{39} + \frac{7}{8} \times \frac{5}{39} = \frac{35 + 35}{312}$$

$$= \frac{70}{312} \text{ or } 0.224\,4$$

Problem 8 A box contains 74 brass washers, 86 steel washers and 40 aluminium washers. Three washers are drawn at random from the box without replacement. Determine the probability that all three are steel washers.

Assume, for clarity of explanation, that a washer is drawn at random, then a second, then a third (although this assumption does not affect the results obtained). The total number of washers is 74 + 86 + 40, i.e. 200.

The probability of randomly selecting a steel washer on the first draw is $\dfrac{86}{200}$. There are now 85 steel washers in a batch of 199. The probability of randomly selecting a steel washer on the second draw is $\dfrac{85}{199}$. There are now 84 steel washers in a batch of 198. The probability of randomly selecting a steel washer on the third draw is $\dfrac{84}{198}$. Hence the probability of selecting a steel washer on the first draw and the second draw **and** the third draw is

$$\frac{86}{200} \times \frac{85}{199} \times \frac{84}{198} = \frac{307\,020}{3\,940\,200} = \frac{5117}{65\,670} \text{ or } 0.078$$

Problem 9 For the box of washers given in *Problem 8* above, determine the probability that there are no aluminium washers drawn, when three washers are drawn at random from the box without replacement.

The probability of not drawing an aluminium washer on the first draw is $1 - \left(\dfrac{40}{200}\right)$, i.e. $\dfrac{160}{200}$. There are now 199 washers in the batch of which 159 are not aluminium washers. Hence, the probability of not drawing an aluminium washer on the second draw is $\dfrac{159}{199}$. Similarly, the probability of not drawing an aluminium washer on the third draw is $\dfrac{158}{198}$. Hence the probability of not drawing an aluminium washer on the first **and** second **and** third draws is

$$\frac{160}{200} \times \frac{159}{199} \times \frac{158}{198} = \frac{100\,488}{197\,010} = \frac{50\,244}{98\,505} \text{ or } 0.510$$

Problem 10 For the box of washers in *Problem 8* above, find the probability that there are two brass washers and either a steel or an aluminium washer when three are drawn at random, without replacement.

Two brass washers (A) and one steel washer (B) can be obtained in any of the following ways:

1st draw	*2nd draw*	*3rd draw*
A	A	B
A	B	A
B	A	A

Two brass washers and one aluminium washer (C) can also be obtained in any of the following ways:

1st draw	*2nd draw*	*3rd draw*
A	A	C
A	C	A
C	A	A

Thus there are six possible ways of achieving the combinations specified. If A represents a brass washer, B a steel washer and C an aluminium washer, then the combinations and their probabilities are as shown:

DRAW			*PROBABILITY*
First	Second	Third	
A	A	B	$\frac{74}{200} \times \frac{73}{199} \times \frac{86}{198} = 0.059\,0$
A	B	A	$\frac{74}{200} \times \frac{86}{199} \times \frac{73}{198} = 0.059\,0$
B	A	A	$\frac{86}{200} \times \frac{74}{199} \times \frac{73}{198} = 0.059\,0$
A	A	C	$\frac{74}{200} \times \frac{73}{199} \times \frac{40}{198} = 0.027\,4$
A	C	A	$\frac{74}{200} \times \frac{40}{199} \times \frac{73}{198} = 0.027\,4$
C	A	A	$\frac{40}{200} \times \frac{74}{199} \times \frac{73}{198} = 0.027\,4$

The probability of having the first combination or the second, **or** the third and so on is given by the sum of the probabilities, i.e. by $3 \times 0.059\,0 + 3 \times 0.027\,4$, that is, **0.259 2.**

Further problems on probability may be found in section 28.3, Problems 1 to 12.

28.3 Further problems on probability

1 In a batch of 45 lamps there are 10 faulty lamps. If one lamp is drawn at random, find the probability of it being (a) faulty and (b) satisfactory.

$\left[\text{(a) } \frac{2}{9} \text{ (b) } \frac{7}{9}\right]$

2 A box of fuses are all of the same shape and size and comprises 23 2 A fuses, 47 5 A fuses and 69 13 A fuses. Determine the probability of selecting at random (a) 2 A fuse, (b) a 5 A fuse and (c) a 13 A fuse.

$\left[\text{(a) } \frac{23}{139} \text{ or } 0.165\,5 \text{ (b) } \frac{47}{139} \text{ or } 0.338\,1 \quad \text{(c) } \frac{69}{139} \text{ or } 0.496\,4\right]$

3 (a) Find the probability of having a 2 upwards when throwing a fair 6-sided dice.
(b) Find the probability of having a 5 upwards when throwing a fair 6-sided dice.
(c) Determine the probability of having a 2 and then a 5 on two throws of a fair 6-sided dice.

$\left[\text{(a) } \frac{1}{6} \text{ (b) } \frac{1}{6} \text{ (c) } \frac{1}{36}\right]$

4 The probability of event A happening is $\frac{3}{5}$ and the probability of event B happening is $\frac{2}{3}$. Calculate the probabilities of (a) both A and B happening, (b) only event A happening, i.e. event A happening and event B not happening, (c) only event B happening and (d) either A, or B, or A and B happening.

$\left[\text{(a) } \frac{2}{5} \text{ (b) } \frac{1}{5} \text{ (c) } \frac{4}{15} \text{ (d) } \frac{13}{15}\right]$

5 When testing 1000 soldered joints, 4 failed during a vibration test and 5 failed due to having a high resistance. Determine the probability of a joint failing due to (a) vibration, (b) high resistance, (c) vibration or high resistance and (d) vibration and high resistance.

$\left[\text{(a) } \frac{1}{250} \text{ (b) } \frac{1}{200} \text{ (c) } \frac{9}{1000} \text{ (d) } \frac{1}{50\,000}\right]$

6 The probability that component A will operate satisfactorily for 5 years is $\frac{4}{5}$ and that B will operate satisfactorily over that same period of time is $\frac{3}{4}$. Find the probabilities that in a 5 year period:
(a) both components operate satisfactorily,
(b) only component A will operate satisfactorily, and
(c) only component B will operate satisfactorily.

$\left[\text{(a) } \frac{3}{5} \text{ (b) } \frac{1}{5} \text{ (c) } \frac{3}{20}\right]$

7 In a particular street, 80% of the houses have telephones. If two houses selected at random are visited, calculate the probabilities that (a) they both have a telephone and (b) one has a telephone but the other does not have telephone.

$$\left[(a) \frac{16}{25} \ (b) \frac{8}{25} \right]$$

8 Veroboard pins are packed in packets of 20 by a machine. In a thousand packets, 40 have less than 20 pins. Find the probability that if 2 packets are chosen at random, one will contain less than 20 pins and the other will contain 20 pins or more.

$$\left[\frac{48}{625} \right]$$

9 A batch of 1 kW fire elements contains 16 which are within a power tolerance and 4 which are not. If 3 elements are selected at random from the batch, calculate the probabilities that (a) all three are within the power tolerance and (b) two are within but one is not within the power tolerance.

$$\left[(a) \frac{28}{57} \ (b) \frac{8}{19} \right]$$

10 The statistics on numbers of boys and girls of a group of families each having 3 children are analysed. Assuming equal probability for the birth of a boy or a girl, determine the percentage of the group likely to have (a) two boys and a girl, (b) at least one boy, (c) no girls and (d) at most two girls.

[(a) 12.5% (b) 87.5% (c) 12.5% (d) 87.5%]

11 An amplifier is made up of three transistors, A, B and C. The probabilities of A, B or C being defective are $\frac{1}{20}, \frac{1}{25}$ and $\frac{1}{50}$, respectively. Calculate the percentage of amplifiers produced (a) which work satisfactorily and (b) which have just one defective transistor.

[(a) 89.38% (b) 10.25%]

12 A box contains 14 40 W lamps, 28 60 W lamps and 58 25 W lamps, all the lamps being of the same shape and size. Three lamps are drawn at random from the box, first one, then a second, then a third. Determine the probabilities of: (a) getting one 25 W, one 40 W and one 60 W lamp, with replacement, (b) getting one 25 W, one 40 W and one 60 W lamp without replacement, (c) getting either one 25 W and two 40 W or one 60 W and two 40 W lamps with replacement and (d) getting either one 25 W and two 40 W or one 60 W and two 40 W lamps without replacement.

$$\left[\begin{array}{ll} (a) \ \dfrac{1421}{62\,500} \text{ or } 0.022\,7 & (b) \ \dfrac{2842}{121\,275} \text{ or } 0.023\,4 \\[2ex] (c) \ \dfrac{1421}{125\,000} \text{ or } 0.011\,4 & (d) \ \dfrac{11\,739}{242\,550} \text{ or } 0.048\,4 \end{array} \right]$$

29

Introduction to differentiation

29.1 Introduction to calculus

Calculus is a branch of mathematics involving or leading to calculations dealing with continuously varying functions.

Calculus is a subject which falls into two parts: (i) **differential calculus** (or **differentiation**) and (ii) **integral calculus** (or **integration**).

This chapter deals with differentiation and Chapter 30 deals with integration.

Differentiation is used in calculations involving velocity and acceleration, rates of change and maximum and minimum values of curves.

Integration is used to determine areas under curves, mean and r.m.s. values, volumes, centroids, second moments of area and in the solution of differential equations.

29.2 Functional notation

In an equation such as $y = 3x^2 + 2x - 5$, y is said to be a function of x and may be written as $y = f(x)$. An equation written in the form $f(x) = 3x^2 + 2x - 5$ is termed **functional notation**.

The value of $f(x)$ when $x = 0$ is denoted by $f(0)$, and the value of $f(x)$ when $x = 2$ is denoted by $f(2)$, and so on. Thus when $f(x) = 3x^2 + 2x - 5$, then

$$f(0) = 3(0)^2 + 2(0) - 5 = -5$$
and $f(2) = 3(2)^2 + 2(2) - 5 = 11$
and so on.

> **Problem 1** If $f(x) = 4x^2 - 3x + 2$ find $f(0)$, $f(3)$, $f(-1)$ and $f(3) - f(-1)$.

$f(x)$ $= 4x^2 - 3x + 2$
$f(0)$ $= 4(0)^2 - 3(0) + 2 = \mathbf{2}$
$f(3)$ $= 4(3)^2 - 3(3) + 2 = 36 - 9 + 2 = \mathbf{29}$
$f(-1) = 4(-1)^2 - 3(-1) + 2 = 4 + 3 + 2 = \mathbf{9}$
$f(3) - f(-1) = 29 - 9 = \mathbf{20}$

> **Problem 2** Given that $f(x) = 5x^2 + x - 7$ determine
> (i) $f(2) \div f(1)$, (ii) $f(3 + a)$, (iii) $f(3 + a) - f(3)$ and
> (iv) $\dfrac{f(3 + a) - f(3)}{a}$.

$f(x) = 5x^2 + x - 7$

(i) $f(2) = 5(2)^2 + 2 - 7 = 15$
    $f(1) = 5(1) + 1 - 7 = -1$

$$f(2) \div f(1) = \frac{15}{-1} = \mathbf{-15}$$

(ii) $f(3 + a) = 5(3 + a)^2 + (3 + a) - 7$
    $= 5(9 + 6a + a^2) + (3 + a) - 7$
    $= 45 + 30a + 5a^2 + 3 + a - 7$
    $= \mathbf{41 + 31a + 5a^2}$

(iii) $f(3) = 5(3)^2 + 3 - 7 = 41$
    $f(3 + a) - f(3) = (41 + 31a + 5a^2) - (41)$
    $= \mathbf{31a + 5a^2}$

(iv) $\dfrac{f(3 + a) - f(3)}{a} = \dfrac{31a + 5a^2}{a} = \mathbf{31 + 5a}$

Further problems on functional notation may be found in section 29.8, Problems 1 to 4.

29.3 The gradient of a curve

(a) If a tangent is drawn at a point P on a curve, then the gradient of this tangent is said to be the **gradient of the curve** at P. In *Fig. 29.1*, the gradient of the curve at P is equal to the gradient of the tangent PQ.

(b) For the curve shown in *Fig. 29.2*, let the points A and B have coordinates (x_1, y_1) and (x_2, y_2), respectively. In functional notation $y_1 = f(x_1)$ and $y_2 = f(x_2)$ as shown.

The gradient of the chord AB $= \dfrac{BC}{AC} = \dfrac{BD - CD}{ED}$

$$= \dfrac{f(x_2) - f(x_1)}{(x_2 - x_1)}$$

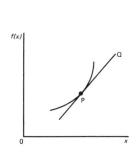

Fig. 29.1

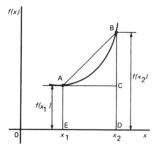

Fig. 29.2

(c) For the curve $f(x) = x^2$ shown in *Fig. 29.3*:

(i) the gradient of chord AB

$$= \dfrac{f(3) - f(1)}{3 - 1} = \dfrac{9 - 1}{2} = 4$$

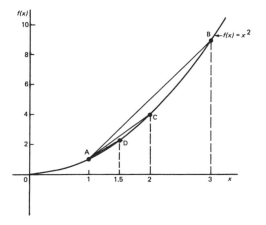

Fig. 29.3

(ii) the gradient of chord AC

$$= \dfrac{f(2) - f(1)}{2 - 1} = \dfrac{4 - 1}{1} = 3$$

(iii) the gradient of chord AD

$$= \dfrac{f(1.5) - f(1)}{1.5 - 1} = \dfrac{2.25 - 1}{0.5} = 2.5$$

(iv) if E is the point on the curve $(1.1, f(1.1))$ then the gradient of chord AE

$$= \dfrac{f(1.1) - f(1)}{1.1 - 1} = \dfrac{1.21 - 1}{0.1} = 2.1$$

(v) if F is the point on the curve $(1.01, f(1.01))$ then the gradient of chord AF

$$= \dfrac{f(1.01) - f(1)}{1.01 - 1} = \dfrac{1.020\ 1 - 1}{0.01} = 2.01$$

Thus as point B moves closer and closer to point A the gradient of the chord approaches nearer and nearer to the value 2. This is called the **limiting value** of the gradient of the chord AB and when B coincides with A the chord becomes the tangent to the curve.

A further problem on the gradient of a curve may be found in section 29.12, Problem 5, page 262.

29.4 Differentiation from first principles

(i) In *Fig. 29.4*, A and B are two points very close together on a curve, δx (delta x) and δy (delta y) representing small increments in the x and y directions, respectively.

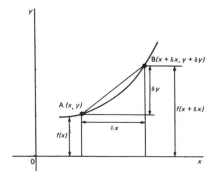

Fig. 29.4

Gradient of chord AB = $\delta y / \delta x$, however

$$\delta y = f(x + \delta x) - f(x)$$

Hence

$$\frac{\delta y}{\delta x} = \frac{f(x + \delta x) - f(x)}{\delta x}$$

As δx approaches zero, $\delta y / \delta x$ approaches a limiting value and the gradient of the chord approaches the gradient of the tangent at A.

(ii) When determining the gradient of a tangent to a curve there are two notations used. The gradient of the curve at A in *Fig. 29.4* can either be written as

$$\lim_{\delta x \to 0} \frac{\delta y}{\delta x} \text{ or } \lim_{\delta x \to 0} \left\{ \frac{f(x + \delta x) - f(x)}{\delta x} \right\}$$

In **Leibniz notation**, $\dfrac{dy}{dx} = \lim\limits_{\delta x \to 0} \dfrac{\delta y}{\delta x}$

In **functional notation**,

$$f'(x) = \lim_{\delta x \to 0} \left\{ \frac{f(x + \delta x) - f(x)}{\delta x} \right\}$$

(iii) dy/dx is the same as $f'(x)$ and is called the **differential coefficient** or the **derivative**. The process of finding the differential coefficient is called **differentiation**. Summarizing, the differential coefficient,

$$\frac{dy}{dx} = f'(x) = \lim_{\delta x \to 0} \frac{\delta y}{\delta x} = \lim_{\delta x \to 0} \left\{ \frac{f(x + \delta x) - f(x)}{\delta x} \right\}$$

Problem 3 Differentiate from first principles $f(x) = x^2$ and determine the value of the gradient of the curve at $x = 2$.

To 'differentiate from first principles' means 'to find $f'(x)$' by using the expression

$$f'(x) = \lim_{\delta x \to 0} \left\{ \frac{f(x + \delta x) - f(x)}{\delta x} \right\}$$

$f(x) = x^2$. Substituting $(x + \delta x)$ for x gives $f(x + \delta x) = (x + \delta x)^2 = x^2 + 2x\delta x + \delta x^2$, hence

$$f'(x) = \lim_{\delta x \to 0} \left\{ \frac{(x^2 + 2x\delta x + \delta x^2) - (x^2)}{\delta x} \right\}$$

$$= \lim_{\delta x \to 0} \left\{ \frac{2x\delta x + \delta x^2}{\delta x} \right\}$$

$$= \lim_{\delta x \to 0} \{2x + \delta x\}$$

As $\delta x \to 0$, $\{2x + \delta x\} \to \{2x + 0\}$. Thus $f'(x) = 2x$, i.e. the differential coefficient of x^2 is $2x$. At $x = 2$, the gradient of the curve, $f'(x) = 2(2) = 4$.

Problem 4 Find the differential coefficient of $y = 5x$.

By definition,

$$\frac{dy}{dx} = f'(x) = \lim_{\delta x \to 0} \left\{ \frac{f(x + \delta x) - f(x)}{\delta x} \right\}$$

The function being differentiated is $y = f(x) = 5x$. Substituting $(x + \delta x)$ for x gives $f(x + \delta x) = 5(x + \delta x) = 5x + 5\delta x$. Hence

$$\frac{dy}{dx} = f'(x) = \lim_{\delta x \to 0} \left\{ \frac{(5x + 5\delta x) - (5x)}{\delta x} \right\}$$

$$= \lim_{\delta x \to 0} \left\{ \frac{5\delta x}{\delta x} \right\} = \lim_{\delta x \to 0} \{5\}$$

Since the term δx does not appear in $\{5\}$ the limiting value as $\delta x \to 0$ of $\{5\}$ is 5. Thus $dy/dx = 5$, i.e. the differential coefficient of $5x$ is 5.

The equation $y = 5x$ represents a straight line of gradient 5 (see Chapter 10). The 'differential coefficient' (i.e. dy/dx or $f'(x)$) means 'the gradient of the curve', and since the slope of the line $y = 5x$ is 5 this result can be obtained by inspection. Hence, in general, if $y = kx$ (where k is a constant) then the slope of the line is k and

$$\frac{dy}{dx} \text{ or } f'(x) = k$$

Problem 5 Find the derivative of $y = 8$.

$y = f(x) = 8$. Since there are no x-values in the original equation, substituting $(x + \delta x)$ for x still gives $f(x + \delta x) = 8$. Hence

$$\frac{dy}{dx} = f'(x) = \lim_{\delta x \to 0} \left\{ \frac{f(x + \delta x) - f(x)}{\delta x} \right\}$$

$$= \lim_{\delta x \to 0} \left\{ \frac{8 - 8}{\delta x} \right\} = 0$$

Thus, when $y = 8$, $\dfrac{dy}{dx} = 0$

The equation $y = 8$ represents a straight horizontal line and the gradient of a horizontal line is zero, hence the result could have been determined by inspection. 'Finding the derivative' means 'finding the gradient', hence, in general, for any horizontal line if $y = k$ (where k is a constant) then $dy/dx = 0$.

Problem 6   Differentiate from first principles $f(x) = 2x^3$.

Substituting $(x + \delta x)$ for x gives

$$f(x + \delta x) = 2(x + \delta x)^3$$
$$= 2(x + \delta x)(x^2 + 2x\delta x + \delta x^2)$$
$$= 2(x^3 + 3x^2\delta x + 3x\delta x^2 + \delta x^3)$$
$$= 2x^3 + 6x^2\delta x + 6x\delta x^2 + 2\delta x^3$$

$$f'(x) = \lim_{\delta x \to 0}\left\{\frac{f(x + \delta x) - f(x)}{\delta x}\right\}$$

$$= \lim_{\delta x \to 0}\left\{\frac{(2x^3 + 6x^2\delta x + 6x\delta x^2 + 2\delta x^3) - (2x^3)}{\delta x}\right\}$$

$$= \lim_{\delta x \to 0}\left\{\frac{(6x^2\delta x + 6x\delta x^2 + 2\delta x^3)}{\delta x}\right\}$$

$$= \lim_{\delta x \to 0}\{6x^2 + 6x\delta x + 2\delta x^2\}$$

Hence $f'(x) = 6x^2$, i.e. the differential coefficient of $2x^3$ is $6x^2$.

Problem 7   Find the differential coefficient of $y = 4x^2 + 5x - 3$ and determine the gradient of the curve at $x = -3$.

$$y = f(x) = 4x^2 + 5x - 3$$
$$f(x + \delta x) = 4(x + \delta x)^2 + 5(x + \delta x) - 3$$
$$= 4(x^2 + 2x\delta x + \delta x^2) + 5x + 5\delta x - 3$$
$$= 4x^2 + 8x\delta x + 4\delta x^2 + 5x + 5\delta x - 3$$

$$\frac{dy}{dx} = f'(x)$$

$$= \lim_{\delta x \to 0}\left\{\frac{f(x + \delta x) - f(x)}{\delta x}\right\}$$

$$= \lim_{\delta x \to 0}\frac{(4x^2 + 8x\delta x + 4\delta x^2 + 5x + 5\delta x - 3)}{\delta x} \frac{- (4x^2 + 5x - 3)}{\delta x}$$

$$= \lim_{\delta x \to 0}\left\{\frac{8x\delta x + 4\delta x^2 + 5\delta x}{\delta x}\right\}$$

$$= \lim_{\delta x \to 0}\{8x + 4\delta x + 5\}$$

i.e. $\dfrac{dy}{dx} = f'(x) = 8x + 5$

At $x = -3$, the gradient of the curve $= dy/dx = f'(x) = 8(-3) + 5 = -19$.

Further problems on differentiation from first principles may be found in section 29.12, Problems 6 to 19, page 262.

29.5   Differentiation of $y = ax^n$ by the general rule

From differentiation by first principles, a general rule for differentiating ax^n emerges where a and n are any constants. This rule is:

if $y = ax^n$ then $\dfrac{dy}{dx} = anx^{n-1}$

or, if $f(x) = ax^n$ then $f'(x) = anx^{n-1}$

(Each of the results obtained in *Problems 3* to *7* may be deduced by using this general rule.)

When differentiating, results can be expressed in a number of ways. For example:

(i)   if $y = 3x^2$ then $dy/dx = 6x$,
(ii)   if $f(x) = 3x^2$ then $f'(x) = 6x$,
(iii)   the differential coefficient of $3x^2$ is $6x$,
(iv)   the derivative of $3x^2$ is $6x$, and
(v) $d/dx(3x^2) = 6x$.

Problem 8   Using the general rule, differentiate the following with respect to x:

(a) $y = 5x^7$ (b) $y = 3\sqrt{x}$ (c) $y = 4/x^2$.

(a)   Comparing $y = 5x^7$ with $y = ax^n$ shows that $a = 5$ and $n = 7$. Using the general rule,

$$\frac{dy}{dx} = anx^{n-1} = (5)(7)x^{7-1} = 35x^6$$

(b) $y = 3\sqrt{x} = 3x^{1/2}$. Hence $a = 3$ and $n = \dfrac{1}{2}$.

$$\frac{dy}{dx} = anx^{n-1} = (3)(\tfrac{1}{2})x^{1/2-1} = \frac{3}{2}x^{-1/2} = \frac{3}{2x^{1/2}} = \frac{3}{2\sqrt{x}}$$

(c) $y = 4/x^2 = 4x^{-2}$. Hence $a = 4$ and $n = -2$.

$$\frac{dy}{dx} = anx^{n-1} = (4)(-2)x^{-2-1} = -8x^{-3} = -\frac{8}{x^3}$$

Problem 9   Find the differential coefficient of

$$y = \frac{2}{5}x^3 - \frac{4}{x^3} + 4\sqrt{x^5} + 7$$

$$y = \frac{2}{5}x^3 - \frac{4}{x^3} + 4\sqrt{x^5} + 7$$

i.e.

$$y = \frac{2}{5}x^3 - 4x^{-3} + 4x^{5/2} + 7$$

$$\frac{dy}{dx} = \left(\frac{2}{5}\right)(3)x^{3-1} - (4)(-3)x^{-3-1} + (4)\left(\frac{5}{2}\right)x^{(5/2)-1} + 0$$

$$= \frac{6}{5}x^2 + 12x^{-4} + 10x^{3/2}$$

i.e.

$$\frac{dy}{dx} = \frac{6}{5}x^2 + \frac{12}{x^4} + 10\sqrt{x^3}$$

Problem 10   If $f(t) = 5t + \dfrac{1}{\sqrt{t^3}}$ find $f'(t)$.

$$f(t) = 5t + \frac{1}{\sqrt{t^3}} = 5t + \frac{1}{t^{3/2}} = 5t^1 + t^{-3/2}$$

Hence $f'(t) = (5)(1)t^{1-1} + \left(-\dfrac{3}{2}\right)t^{(-3/2)-1}$

$$= 5t^0 - \frac{3}{2}t^{-5/2}$$

i.e. $f'(t) = 5 - \dfrac{3}{2t^{5/2}} = 5 - \dfrac{3}{2\sqrt{t^5}}$

Problem 11   Differentiate $y = \dfrac{(x + 2)^2}{x}$ with respect to x.

$$y = \frac{(x + 2)^2}{x} = \frac{x^2 + 4x + 4}{x} = \frac{x^2}{x} + \frac{4x}{x} + \frac{4}{x}$$

i.e. $y = x + 4 + 4x^{-1}$

Hence $\dfrac{dy}{dx} = 1 + 0 + (4)(-1)x^{-1-1}$

$$= 1 - 4x^{-2} = 1 - \frac{4}{x^2}$$

Further problems on differentiation of $y = ax^n$ by the general rule may be found in section 29.12, Problems 20 to 31, page 262.

29.6   Differentiation of sin $a\theta$ and cos $a\theta$

Figure 29.5(a) shows a graph of $y = \sin \theta$. The gradient is continually changing as the curve moves from 0 to A to B to C to D. The gradient, given by $dy/d\theta$, may be plotted in a corresponding position below $y = \sin \theta$, as shown in *Fig. 29.5(b)*.

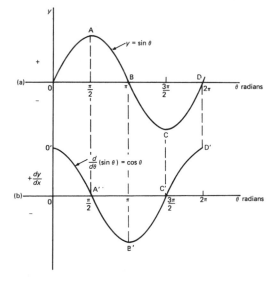

Fig. 29.5

(i)   At 0, the gradient is positive and is at its steepest. Hence 0' is a maximum positive value.

(ii)   Between 0 and A the gradient is positive but is decreasing in value until at A the gradient is zero, shown as A'.

(iii)   Between A and B the gradient is negative but is increasing in value until at B the gradient is at its steepest. Hence B' is a maximum negative value.

(iv) If the gradient of $y = \sin\theta$ is further investigated between B and C and C and D then the resulting graph of $dy/d\theta$ is seen to be a cosine wave.

Hence the rate of change of $\sin\theta$ is $\cos\theta$, i.e.

$$\text{if } y = \sin\theta \text{ then } \frac{dy}{d\theta} = \cos\theta$$

It may also be shown that:

$$\text{if } y = \sin a\theta, \quad \frac{dy}{d\theta} = a\cos a\theta \text{ (where } a \text{ is a constant)}$$

(v) If a similar exercise is followed for $y = \cos\theta$ then the graph of $dy/d\theta$ is found to be a graph of $\sin\theta$, but displaced by π radians. It is, in fact, a graph of $-\sin\theta$. Thus

$$\text{if } y = \cos\theta, \quad \frac{dy}{d\theta} = -\sin\theta$$

It may also be shown that:

$$\text{if } y = \cos a\theta, \quad \frac{dy}{d\theta} = -a\sin a\theta \text{ (where } a \text{ is a constant)}$$

Problem 12   Differentiate the following with respect to the variable: (a) $y = 2\sin 5\theta$ (b) $f(t) = 3\cos 2t$.

(a) $y = 2\sin 5\theta$

$$\frac{dy}{d\theta} = (2)(5)\cos 5\theta = \mathbf{10\cos\theta}$$

(b) $f(t) = 3\cos 2t$
  $f'(t) = (3)(-2)\sin 2t = \mathbf{-6\sin 2t}$

Problem 13   Find the differential coefficient of $y = 2\sin 2x - 3\cos 4x$.

  $y = 7\sin 2x - 3\cos 4x$

$$\frac{dy}{dx} = (7)(2)\cos 2x - (3)(-4)\sin 4x$$

$$= \mathbf{14\cos 2x + 12\sin 4x}$$

Problem 14   An alternating voltage is given by $v = 100\sin 200t$ volts, where t is the time in seconds. Calculate the rate of change of voltage when (a) $t = 0.005$ s and (b) $t = 0.01$ s.

$v = 100\sin 200t$ volts. The rate of change of v is given by dv/dt.

$$\frac{dv}{dt} = (100)(200)\cos 200t = 20\,000\cos 200t$$

(a) When $t = 0.005$ s,

$$\frac{dv}{dt} = 20\,000\cos(200)(0.005) = 20\,000\cos 1$$

$\cos 1$ means 'the cosine of 1 radian' (make sure your calculator is on radians – not degrees). Hence

$$\frac{dv}{dt} = \mathbf{10\,806 \text{ volts per second}}$$

(b) When $t = 0.01$ s,

$$\frac{dv}{dt} = 20\,000\cos(200)(0.01)$$

$$= 20\,000\cos 2$$

$$= \mathbf{-8323 \text{ volts per second}}$$

Further problems on the differentiation of $\sin a\theta$ and $\cos a\theta$ may be found in section 29.12, Problems 32 to 37, page 263.

29.7 Differentiation of e^{ax} and ln ax

If a graph of $y = e^x$ is plotted, as shown in Chapter 9, and the gradient of the curve determined at intervals, and plotted in a corresponding position below $y = e^x$, it is found that the shape of the gradient curve is also e^x, i.e.

$$\text{if } y = e^x, \text{ then } \frac{dy}{dx} = e^x$$

It may also be shown that

$$\text{if } y = e^{ax}, \text{ then } \frac{dy}{dx} = ae^{ax}$$

Therefore if $y = 2e^{6x}$, then

$$\frac{dy}{dx} = (2)(6e^{6x}) = \mathbf{12e^{6x}}$$

If a graph of $y = \ln x$ is plotted and the gradient of the curve determined at intervals, and plotted in a corresponding position below $y = \ln x$, it is found that the shape of the gradient curve produced is that of $1/x$, i.e.

if $y = \ln x$, then $\dfrac{dy}{dx} = \dfrac{1}{x}$

It may also be shown that

if $y = \ln ax$, then $\dfrac{dy}{dx} = \dfrac{1}{x}$

(Note that in the latter expression a does not appear in the dy/dx term.)

Thus if $y = \ln 4x$, then $\dfrac{dy}{dx} = \dfrac{1}{x}$

Problem 15   Differentiate the following with respect to the variable:

(a) $y = 3e^{2x}$ (b) $f(t) = \dfrac{4}{3e^{5t}}$

(a) If $y = 3e^{2x}$ then

$\dfrac{dy}{dx} = (3)(2e^{2x}) = \mathbf{6e^{2x}}$

(b) If $f(t) = \dfrac{4}{3e^{5t}} = \dfrac{4}{3}e^{-5t}$, then

$f'(t) = \left(\dfrac{4}{3}\right)(-5e^{-5t}) = \dfrac{-20}{3}e^{-5t} = \dfrac{\mathbf{-20}}{\mathbf{3e^{5t}}}$

Problem 16   Differentiate $y = 5 \ln 3x$.

If $y = 5 \ln 3x$, $\dfrac{dy}{dx} = (5)\left(\dfrac{1}{x}\right) = \dfrac{5}{x}$

Further problems on the differentiation of e^{ax} and $\ln ax$ may be found in section 29.12, Problems 38 to 41, page 263.

29.8 Summary of standard derivatives

The five differential coefficients used are summarized in Table 29.1.

Table 29.1

y or $f(x)$	$\dfrac{dy}{dx}$ or $f'(x)$
ax^n	$an\,x^{n-1}$
$\sin ax$	$a \cos ax$
$\cos ax$	$-a \sin ax$
e^{ax}	ae^{ax}
$\ln ax$	$\dfrac{1}{x}$

Problem 17   Find the gradient of the curve $y = 3x^2 - 7x + 2$ at the point $(1,-2)$.

If $y = 3x^2 - 7x + 2$, then

$\text{gradient} = \dfrac{dy}{dx} = 6x - 7$

At the point $(1,-2)$, $x = 1$, hence **gradient** $= 6(1) - 7 = $ **-1**.

Problem 18   If $y = \dfrac{3}{x^2} - 2 \sin 4x + \dfrac{2}{e^x} + \ln 5x$ find dy/dx.

$y = \dfrac{3}{x^2} - 2 \sin 4x + \dfrac{2}{e^x} + \ln 5x$

$= 3x^{-2} - 2 \sin 4x + 2e^{-x} + \ln 5x$

$\dfrac{dy}{dx} = (3)(-2x^{-3}) - 2(4 \cos 4x) + (2)(-e^{-x}) + \dfrac{1}{x}$

$= \dfrac{-6}{x^3} - 8 \cos 4x - \dfrac{2}{e^x} + \dfrac{1}{x}$

Further problems on standard derivatives may be found in section 29.12, Problems 42 and 43, page 263.

29.9 Successive differentiation

When a function $y = f(x)$ is differentiated with respect to x the differential coefficient is written as dy/dx or $f'(x)$. If the expression is differentiated again, the second differential coefficient is obtained and is written as dy^2/dx^2 (pronounced dee two y by dee x squared) or $f''(x)$ (pronounced f double-dash x). By successive differentiation further higher derivatives such as dy^3/dx^3 and dy^4/dx^4 may be obtained.

Thus if $y = 5x^4$,

$$\frac{dy}{dx} = 20x^3, \frac{d^2y}{dx^2} = 60x^2, \frac{d^3y}{dx^3} = 120x,$$

$$\frac{d^4y}{dx^4} = 120 \text{ and } \frac{d^5y}{dx^5} = 0$$

Problem 19 If $f(x) = 4x^5 - 2x^3 + x - 3$, find $f''(x)$.

$$f(x) = 4x^5 - 2x^3 + x - 3$$
$$f'(x) = 20x^4 - 6x^2 + 1$$

$$f''(x) = 80x^3 - 12x = 4x(20x^2 - 3)$$

Problem 20 Given $y = \frac{2}{3}x^3 - \frac{4}{x^2} + \frac{1}{2x} - \sqrt{x}$ determine dy^2/dx^2.

$$y = \frac{2}{3}x^3 - \frac{4}{x^2} + \frac{1}{2x} - \sqrt{x} = \frac{2}{3}x^3 - 4x^{-2} + \frac{1}{2}x^{-1} - x^{1/2}$$

$$\frac{dy}{dx} = \left(\frac{2}{3}\right)(3)x^2 - 4(-2)x^{-3} + \left(\frac{1}{2}\right)(-1)x^{-2} - \left(\frac{1}{2}\right)x^{-1/2}$$

i.e.

$$\frac{dy}{dx} = 2x^2 + 8x^{-3} - \frac{1}{2}x^{-2} - \frac{1}{2}x^{-1/2}$$

$$\frac{d^2y}{dx^2} = 4x - 24x^{-4} + x^{-3} + \frac{1}{4}x^{-3/2}$$

$$= 4x - \frac{24}{x^4} + \frac{1}{x^3} + \frac{1}{4\sqrt{x^3}}$$

Further problems on successive differentiation may be found in section 29.12, Problems 44 to 46, page 263.

29.10 Rates of change

(i) If a quantity y depends on and varies with a quantity x then the rate of change of y with respect to x is dy/dx. Thus, for example, the rate of change of pressure p with height h is dp/dh.

(ii) A rate of change with respect to time is usually just called 'the rate of change', the 'with respect to time' being assumed. Thus, for example, a rate of change of

voltage, v is dv/dt and a rate of change of temperature, θ, is $d\theta/dt$, and so on.

Problem 21 The length l metres of a certain metal rod at temperature $t°C$ is given by $l = 1 + 0.000\,03t + 0.000\,000\,3t^2$. Determine the rate of change of length, in mm/°C, when the temperature is (a) 100°C and (b) 250°C.

The rate of change of length means dl/dt. Since length $l = 1 + 0.000\,003t + 0.000\,000\,3t^2$, then

$$\frac{dl}{dt} = 0.000\,03 + 0.000\,000\,6t$$

(a) When $t = 100°C, \frac{dl}{dt} = 0.000\,03 + (0.000\,000\,6)(100)$

$$= 0.000\,09 \text{ m/°C} = \textbf{0.09 mm/°C}$$

(b) When $t = 250°C, \frac{dl}{dt} = 0.000\,03 + (0.000\,000\,6)(250)$

$$= 0.000\,18 \text{ m/°C} = \textbf{0.18 mm/°C}$$

Problem 22 The luminous intensity I candelas of a lamp at varying voltage V is given by $I = 5 \times 10^{-4}V^2$. Determine the voltage at which the light is increasing at a rate of 0.4 candelas per volt.

The rate of change of light with respect to voltage is given by dI/dV. Since $I = 5 \times 10^{-4}V^2$,

$$\frac{dI}{dV} = (5 \times 10^{-4})(2)V = 10 \times 10^{-4}V = 10^{-3}V$$

When the light is increasing at 0.4 candelas per volt then $+0.4 = 10^{-3}V$, from which

$$\text{voltage } V = \frac{0.4}{10^{-3}} = 0.4 \times 10^{+3} = \textbf{400 volts}$$

Problem 23 Newtons law of cooling is given by $\theta = \theta_0 e^{-kt}$, where the excess of temperature at zero time is $\theta_0°C$ and at time t seconds is $\theta°C$. Determine the rate of change of temperature after 50 s, given that $\theta_0 = 15°C$ and $k = -0.02$.

The rate of change of temperature is $d\theta/dt$. Since $\theta = \theta_0 e^{-kt}$

then $\frac{d\theta}{dt} = (\theta_0)(-k)e^{-kt} = -k\theta_0 e^{-kt}$

When $\theta_0 = 15$, $k = -0.02$ and $t = 50$, then

$$\frac{d\theta}{dt} = -(-0.02)(15)\,e^{-(-0.02)(50)}$$

$$= 0.3e' = \textbf{0.815°C/s}$$

Problem 24   The pressure p of the atmosphere at height h above ground level is given by $p = p_0 e^{-h/c}$, where p_0 is the pressure at ground level and c is a constant. Determine the rate of change of pressure with height when $p_0 = 10^5$ pascals and $c = 6 \times 10^4$ at 1500 m.

The rate of change of pressure with height is dp/dh. Since $p = p_0 e^{-h/c}$ then

$$\frac{dp}{dh} = (p_0)\,-\frac{1}{c}e^{-h/c} = -\frac{p_0}{c}e^{-h/c}$$

When $p_0 = 10^5$, $c = 6 \times 10^4$ and $h = 1500$, then

$$\frac{dp}{dh} = \frac{10^5}{6 \times 10^4}\,e^{(-1500/6 \times 10^4)}$$

$$= -\frac{5}{3}e^{-0.025} = \textbf{-1.63 Pa/m}$$

Further problems on rates of change may be found in section 29.12, Problems 47 to 51, page 263.

29.11   Maximum and minimum points

In *Fig. 29.6*, the gradient (or rate of change) of the curve changes from positive between 0 and P to negative between P and Q and then positive again between Q and R. At point P the gradient is zero and, as x increases, the gradient of the curve changes from positive just before P to negative just after. Such a point is called a **maximum point** and appears as the 'crest of a wave'. At point Q, the gradient is also zero and, as x increases, the gradient of the curve changes from negative just before Q to positive just after. Such a point is called a **minimum value**, and appears as the 'bottom of a valley'. Points such as P and Q are given the general name of **turning points**, or **stationary points**.

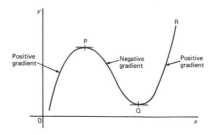

Fig. 29.6

Procedure for finding and distinguishing between stationary points

(i)   Given $y = f(x)$, determine dy/dx (i.e. $f'(x)$).

(ii)   Let $dy/dx = 0$ and solve for the values of x.

(iii)   Substitute the values of x into the original equation, $y = f(x)$, to find the corresponding y-ordinate values. This establishes the coordinates of the stationary points.

To determine the nature of the stationary points:

Either

(iv)   Determine the sign of the gradient of the curve just before and just after the stationary points. If the sign change for the gradient of the curve is:
   (a) **positive to negative** – the point is a **maximum** one;
   (b) **negative to positive** – the point is a **minimum** one.

or

(v)   Find d^2y/dx^2 and substitute into it the values of x found in (ii). If the result is:
   (a) **positive** – the point is a **minimum** one;
   (b) **negative** – the point is a **maximum** one.

Consider the equation $y = x^2 - 2x + 3$. Gradient of the curve,

$$\frac{dy}{dx} = 2x - 2$$

At the turning point, the gradient is zero, hence $2x - 2 = 0$, from which $2x = 2$ and $x = 1$. When $x = 1$, $y = (1)^2 - 2(1) + 3 = 2$, hence at the coordinates (1,2) a turning point occurs. To determine the nature of the turning point:

Method 1
Consider the gradient of the curve at a value of x just less than 1, say 0.9.
At $x = 0.9$, gradient $= 2x - 2 = 2(0.9) - 2 = -0.2$
Now consider the gradient of the curve at a value of x just greater than 1, say 1.1.

At $x = 1.1$, gradient $= 2x - 2 = 2(1.1) - 2 = 0.2$
Hence the gradient has changed from negative just before the turning point at $x = 1$, to positive just after. This indicates a **minimum value**.

Method 2
If the gradient of the curve, $dy/dx = 2x - 2$, then $d^2y/dx^2 = 2$, which is positive, hence the turning point is a **minimum**.
   A graph of $y = x^2 - 2x + 3$ with the minimum point at $(1,2)$ is shown in *Fig. 29.7*.

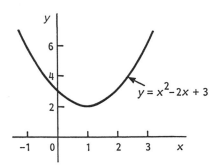

Fig. 29.7

Problem 25   Locate the turning point on the curve $y = 2x^2 - 4x$ and determine its nature by examining the sign of the gradient on either side.

Following the above procedure:

(i)   Since $y = 2x^2 - 4x$, $dy/dx = 4x - 4$.
(ii) At a turning point, $dy/dx = 0$. Hence $4x - 4 = 0$, from which, $x = 1$.
(iii) When $x = 1$, $y = 2(1)^2 - 4(1) = -2$.
      Hence the coordinates of the turning point are (1,−2).
(iv) If x is slightly less than 1, say 0.9, then $dy/dx = 4(0.9) - 4 = -0.4$, i.e. negative. If x is slightly greater than 1, say 1.1, then $dy/dx = 4(1.1) - 4 = 0.4$, i.e. positive. Since the gradient of the curve is negative just before the turning point and positive just after **(1,−2) is a minimum point**.

Problem 26   Find the maximum and minimum values of the curve $y = x^3 - 3x + 5$ by (a) examining the gradient on either side of the turning points, and (b) determining the sign of the second derivative.

Since $y = x^3 - 3x + 5$ then $dy/dx = 3x^2 - 3$. For a maximum or minimum value $dy/dx = 0$. Hence $3x^2 - 3 = 0$,

from which, $3x^2 = 3$ and $x = \pm 1$
When $x = 1$, $y = (1)^3 - 3(1) + 5 = 3$
When $x = -1$, $y = (-1)^3 - 3(-1) + 5 = 7$
Hence (1,3) and (−1,7) are the coordinates of the turning points.

(a) Considering the point (1,3):
    If x is slightly less than 1, say 0.9, then $dy/dx = 3(0.9)^2 - 3$, which is negative.
    If x is slightly more than 1, say 1.1, then $dy/dx = 3(1.1)^2 - 3$, which is positive.
    Since the gradient changes from negative to positive, **the point (1,3) is a minimum point**.

    Considering the point (−1,7):
    If x is slightly less than −1, say −1.1, then $dy/dx = 3(-1.1)^2 - 3$, which is positive.
    If x is slightly more than −1, say −0.9, then $dy/dx = 3(-0.9)^2 - 3$, which is negative.
    Since the gradient changes from positive to negative, **the point (−1,7) is a maximum point**.
(b) Since $dy/dx = 3x^2 - 3$, then $dy^2/dx^2 = 6x$.
    When $x = 1$, $\dfrac{dy^2}{dx^2}$ is positive, hence (1,3) is a **minimum value**
    When $x = -1$, $\dfrac{dy^2}{dx^2}$ is negative, hence (−1,7) is a **maximum value**
    Thus the maximum value is 7 and the minimum value is 3.

It can be seen that the second differential method of determining the nature of the turning points is, in this case, quicker than investigating the gradient.

Problem 27   Locate the turning points on the following curves and determine whether they are maximum or minimum points:

(a) $y = 4\theta + e^{-\theta}$ (b) $y = 3(\ln \theta - \theta)$.

(a) Since $y = 4\theta + e^{-\theta}$ then $dy/d\theta = 4 - e^{-\theta} = 0$ for a maximum or minimum value. Hence $4 = e^{-\theta}$, $\dfrac{1}{4} = e^{\theta}$, giving $\theta = \ln\left(\dfrac{1}{4}\right) = -1.386\,3$ (see Chapter 9).
    When $\theta = 1.386\,3$,

$$y = 4(-1.386\,3) + e^{-(-1.386\,3)} = -5.545\,2 + 4.000\,0$$
$$= -1.545\,2$$

Thus (−1.386 3,−1.545 2) are the coordinates of the turning point.

$\dfrac{d^2y}{d\theta^2} = e^{-\theta}$ When $\theta = -1.386\ 3$, $\dfrac{d^2y}{d\theta^2} = e^{+1.386\ 3} = 4.0$,

which is positive, hence $(-1.386\ 3, -1.545\ 2)$ is a **minimum point**.

(b) Since $y = 3(\ln \theta - \theta) = 3 \ln \theta - 3\theta$, then $dy/d\theta = (3/\theta) - 3 = 0$ for a maximum or minimum value. Hence $3/\theta = 3$, $3 = 3\theta$ and $\theta = 1$.
When $\theta = 1$, $y = 3 (\ln 1 - 1) = 3(0 - 1) = -3$, hence $(1, -3)$ are the coordinates of the turning point.

$\dfrac{d^2y}{d\theta^2} = -\dfrac{3}{\theta^2}$ When $\theta = 1$, $\dfrac{d^2y}{d\theta^2} = -3$

which is negative, hence $(1, -3)$ **is a maximum value**.

Further problems on maximum and minimum points may be found in section 29.12, Problems 52 to 59, page 264.

29.12 Further problems on the introduction to differentiation

Functional notation

1 If $f(x) = 6x^2 - 2x + 1$ find $f(0)$, $f(1)$, $f(2)$, $f(-1)$ and $f(-3)$.
   [1, 5, 21, 9, 61]
2 If $f(x) = 2x^2 + 5x - 7$ find $f(1)$, $f(2)$, $f(-1)$ and $f(2) - f(-1)$.
   [0, 11, -10, 21]
3 Given $f(x) = 3x^3 + 2x^2 - 3x + 2$ prove that $f(1) = \dfrac{1}{7}f(2)$.
4 If $f(x) = -x^2 + 3x + 6$ find $f(2)$, $f(2 + a)$, $f(2 + a) - f(2)$ and $\dfrac{f(2 + a) - f(2)}{a}$.
   [8; $-a^2 - a + 8$; $-a^2 - a$; $-a - 1$]

The gradient of a curve

5 Plot the curve $f(x) = 4x^2 - 1$ for values of x from $x = -1$ to $x = +4$. Label the coordinates $(3, f(3))$ and $(1, f(1))$ as J and K, respectively. Join points J and K to form the chord JK. Determine the gradient of chord JK. By moving J nearer and nearer to K determine the gradient of the tangent of the curve at K.
   [16; 8]

Differentiation from first principles

In *Problems 6* to *17*, differentiate from first principles.

6 $y = x$
   [1]
7 $y = 7x$
   [7]
8 $y = 4x^2$
   [8x]
9 $y = 5x^3$
   [15x^2]
10 $y = -2x^2 + 3x - 12$
   [-4x + 3]
11 $y = 23$
   [0]
12 $f(x) = 9x$
   [9]
13 $f(x) = \dfrac{2x}{3}$
   $\left[\dfrac{2}{3}\right]$
14 $f(x) = 9x^2$
   [18x]
15 $f(x) = -7x^3$
   [-21x^2]
16 $f(x) = x^2 + 15x - 4$
   [2x + 15]
17 $f(x) = 4$
   [0]
18 Determine $\dfrac{d}{dx}(4x^3)$ from first principles.
   [12x^2]
19 Find $\dfrac{d}{dx}(3x^2 + 5)$ from first principles.
   [6x]

Differentiation of y = ax^n by the general rule

20 Using the general rule for ax^n check the results of *Problems 6* to *19*.

In *Problems 21* to *28*, determine the differential coefficients with respect to the variable:

21 $y = 7x^4$
   [28x^3]
22 $y = \sqrt{x}$
   $\left[\dfrac{1}{2\sqrt{x}}\right]$

23 $y = \sqrt{t^3}$

$$\left[\frac{3}{2}\sqrt{t}\right]$$

24 $y = 6 + \dfrac{1}{x^3}$

$$\left[\frac{-3}{x^4}\right]$$

25 $y = 3x - \dfrac{1}{\sqrt{x}} + \dfrac{1}{x}$

$$\left[3 + \frac{1}{2\sqrt{x^3}} - \frac{1}{x^2}\right]$$

26 $y = \dfrac{5}{x^2} - \dfrac{1}{\sqrt{x^7}} + 2$

$$\left[-\frac{10}{x^3} + \frac{7}{2\sqrt{x^9}}\right]$$

27 $y = 3(t - 2)^2$
[$6t - 12$]

28 $y = (x + 1)^3$
[$3x^2 + 6x + 3$]

29 Differentiate $f(x) = 6x^2 - 3x + 5$ and find the gradient of the curve at (a) $x = -1$ and (b) $x = 2$.
[$12x - 3$; (a) -15; (b) 21]

30 Find the differential coefficient of $y = 2x^3 + 3x^2 - 4x - 1$ and determine the gradient of the curve at $x = 2$.
[$6x^2 + 6x - 4$; 32]

31 Determine the derivative of $y = -2x^3 + 4x + 7$ and determine the gradient of the curve at $x = -1.5$.
[$-6x^2 + 4$; -9.5]

Differentiation of sin *aθ* and cos *aθ*

32 Show graphically that the rate of change of $\sin \theta$ is $\cos \theta$.

33 Show graphically that the rate of change of $\cos \theta$ is $- \sin \theta$.

34 Differentiate with respect to x:
(a) $y = 4 \sin 3x$ (b) $y = 2 \cos 6x$.
[(a) $12 \cos 3x$ (b) $-12 \sin 6x$]

35 Given $f(\theta) = 2 \sin 3\theta - 5 \cos 2\theta$, find $f'(\theta)$.
[$6 \cos 3\theta + 10 \sin 2\theta$]

36 An alternating current is given by $i = 5 \sin 100t$ amperes, where t is the time in seconds. Determine the rate of change of current when $t = 0.01$ seconds.
[4.21 A/s]

37 $v = 50 \sin 40t$ volts represents an alternating voltage where t is the time in seconds. At a time of 20×10^{-3} seconds, find the rate of change of voltage.
[35.87 V/s]

Differentiation of e^{ax} and ln *ax*

38 Differentiate with respect to x (a) $y = 5e^{3x}$ (b) $y = \dfrac{2}{7e^{2x}}$.

$$\left[(a)\ 15e^{3x}\ (b)\ \frac{-4}{7e^{2x}}\right]$$

39 Given $f(\theta) = 5 \ln 2\theta - 4 \ln 3\theta$ determine $f'(\theta)$.

$$\left[\frac{5}{\theta} - \frac{4}{\theta} = \frac{1}{\theta}\right]$$

40 If $f(t) = 4 \ln t + 2$ evaluate $f'(t)$ when $t = 0.25$.
[16]

41 Evaluate dy/dx when $x = 1$, given $y = 3e^{4x} - (5/2e^{3x}) + 6 \ln 5x$. Give the answer correct to 3 significant figures.
[664]

Differentiation of standard derivatives

42 Find the gradient of the curve $y = 2x^4 + 3x^3 - x + 4$ at the points $(0,4)$ and $(1,8)$.
[$-1,16$]

43 Differentiate

$$y = \frac{2}{x^2} + 2 \ln 2x - 2(\cos 5x + 3 \sin 2x) - \frac{2}{e^{3x}}$$

$$\left[\frac{-4}{x^3} + \frac{2}{x} + 10 \sin 5x - 12 \cos 2x + \frac{6}{e^{3x}}\right]$$

Successive differentiation

44 If $y = 3x^4 + 2x^3 - 3x + 2$ find (a) d^2y/dx^2 (b) d^3y/dx^3.
[(a) $36x^2 + 12x$ (b) $72x + 12$]

45 (a) Given $f(t) = \dfrac{2}{5}t^2 - \dfrac{1}{t^3} + \dfrac{3}{t} - \sqrt{t} + 1$ determine $f''(t)$.

(b) Evaluate $f''(t)$ when $t = 1$.

$$\left[(a)\ \frac{4}{5} - \frac{12}{t^5} + \frac{6}{t^3} + \frac{1}{4\sqrt{t^3}}\ (b)\ -4.95\right]$$

46 Find the second differential coefficient with respect to the variable of the following:
(a) $3 \sin 2t + \cos t$ (b) $2 \ln 4\theta$.
[(a) $-(12 \sin 2t + \cos t)$ (b) $-2/\theta^2$]

Rates of change

47 The length, l metres, of a rod of metal at temperature $\theta°C$ is given by $l = 1 + 2 \times 10^{-4}\theta + 4 \times 10^{-6}\theta^2$. Determine the rate of change of l, in cm/°C, when the temperature is (a) 100°C (b) 400°C.
[(a) 0.1 cm/°C (b) 0.34 cm/°C]

48 An alternating current, i amperes, is given by $i = $

$100 \sin 2\pi ft$, where f is the frequency in hertz and t the time in seconds. Determine the rate of change of current when $t = 12$ ms, given that $f = 50$ Hz.
[−25 420 A/s]

49 The luminous intensity, I candelas, of a lamp is given by $I = 8 \times 10^{-4}V^2$, where V is the voltage. Find (a) the rate of change of luminous intensity with voltage when $V = 100$ volts, and (b) the voltage at which the light is increasing at a rate of 0.5 candelas per volt.
[(a) 0.16 cd/V (b) 312.5 V]

50 The voltage across the plates of a capacitor at any time t seconds is given by $v = Ve^{-t/CR}$, where V, C and R are constants. Given $V = 200$ volts, $C = 0.1 \times 10^{-6}$ farads and $R = 2 \times 10^6$ ohms find (a) the initial rate of change of voltage, and (b) the rate of change of voltage after 0.2 s.
[(a) −1000 V/s (b) 367.9 V/s]

51 Newton's law of cooling is given by $\theta = \theta_o e^{-kt}$, where the excess of temperature at zero time is $\theta_o\,°C$ and at time t seconds is $\theta\,°C$. Given $\theta_o = 15°C$ and $k = -0.02$, find the time when the rate of change of temperature is $1°C/s$.
[60.2 s]

Maximum and minimum points

52 Define (a) a maximum point and (b) a minimum point.

53 Sketch the following curves and state for each:
(a) the value, and the coordinates, of the first maximum point, and

(b) the value, and the coordinates, of the first minimum point.
(i) $y = 4 \sin 2\theta$ (ii) $y = 3 \cos 3\theta$

$$\left[(i)\ (a)\ 4\ \text{at}\ \left(\frac{\pi}{4},4\right)\ (b)\ -4\ \text{at}\ \left(\frac{3\pi}{4},-4\right) \right.$$

$$\left. (ii)\ (a)\ 3\ \text{at}\ (0,3)\ (b)\ -3\ \text{at}\ \left(\frac{\pi}{3},-3\right) \right]$$

In *Problems 54* to *58*, find the turning points and distinguish between them.

54 $y = 3x^2 - 4x + 2$
$$\left[\text{Minimum at}\ \left(\frac{2}{3},\frac{2}{3}\right) \right]$$

55 $x = \theta(6 - \theta)$
[Maximum at (3,9)]

56 $y = 5x - 2 \ln x$
[Minimum at (0.400 0,3.832 6)]

57 $y = 2x - e^x$
[Maximum at (0.693 1,−0.613 7)]

58 $x = 8t + \dfrac{1}{2t^2}$
$$\left[\text{Minimum at}\ \left(\frac{1}{2},6\right) \right]$$

59 The speed, v, of a car (in m/s) is related to time t s by the equation $v = 3 + 12t - 3t^2$. Determine the maximum speed of the car in km/h.
[54 km/h]

30

Introduction to integration

30.1 Integration

The process of integration reverses the process of differentiation. In differentiation, if $f(x) = 2x^2$ then $f'(x) = 4x$. Thus the integral of $4x$ is $2x^2$, i.e. integration is the process of moving from $f'(x)$ to $f(x)$. By similar reasoning, the integral of $2t$ is t^2.

Integration is a process of summation or adding parts together and an elongated S, shown as $\int$, is used to replace the words 'the integral of'. Hence, from above, $\int 4x = 2x^2$ and $\int 2t$ is t^2.

In differentiation, the differential coefficient dy/dx indicates that a function of x is being differentiated with respect to x, the dx indicating that it is 'with respect to x'. In integration the variable of integration is shown by adding d (the variable) after the function to be integrated.

Thus $\int 4x\, dx$ means 'the integral of $4x$ with respect to x', and $\int 2t\, dt$ means 'the integral of $2t$ with respect to t'.

As stated above, the differential coefficient of $2x^2$ is $4x$, hence $\int 4x\, dx = 2x^2$. However, the differential coefficient of $2x^2 + 7$ is also $4x$. Hence $\int 4x\, dx$ is also equal to $2x^2 + 7$. To allow for the possible presence of a constant, whenever the process of integration is performed, a constant 'c' is added to the result. Thus

$$\int 4x\, dx = 2x^2 + c \text{ and } \int 2t\, dt = t^2 + c$$

'c' is called the **arbitrary constant of integration**.

30.2 The general solution of $\int ax^n\, dx$

The general solution of integrals of the form $\int ax^n\, dx$, where a and n are constants and $n \neq -1$, is given by:

$$\int ax^n\, dx = \frac{ax^{n+1}}{n+1} + c$$

Using this rule gives:

$$\int 3x^4\, dx = \frac{3x^{4+1}}{4+1} + c = \frac{3}{5}x^5 + c$$

and $\int \frac{4}{9}t^3\, dt = \frac{4}{9}\left(\frac{t^{3+1}}{3+1}\right) + c = \frac{4}{9}\left(\frac{t^4}{4}\right) + c = \frac{1}{9}t^4 + c$

Both of these results may be checked by differentiation.

30.3 Standard integrals

From Chapter 29, $\dfrac{d}{dx}(\sin ax) = a\cos ax$

Since integration is the reverse process of differentiation it follows that $\int a\cos ax\, dx = \sin ax + c$

or $\int \cos ax\, dx = \dfrac{1}{a}\sin ax + c$

By similar reasoning

$$\int \sin ax\, dx = -\frac{1}{a}\cos ax + c$$

$$\int e^{ax}\,dx = \frac{1}{a}e^{ax} + c \quad \text{and} \quad \int\frac{1}{x}\,dx = \ln x + c$$

Thus $\int ax^n\,dx = \dfrac{ax^{n+1}}{n+1} + c$ **except when** $n = -1$

When $n = -1$ then $\int x^{-1}\,dx = \int\frac{1}{x}\,dx = \ln x + c$

Summary of standard integrals

Table 30.1

(i) $\int ax^n\,dx = \dfrac{ax^{n+1}}{n+1} + c$ (except when $n = -1$)

(ii) $\int\cos ax\,dx = \dfrac{1}{a}\sin ax + c$

(iii) $\int\sin ax\,dx = -\dfrac{1}{a}\cos ax + c$

(iv) $\int e^{ax}\,dx = \dfrac{1}{a}e^{ax} + c$

(v) $\int\dfrac{1}{x}\,dx = \ln x + c$

Problem 1 Determine (a) $\int 3x^2\,dx$ (b) $\int 2t^3\,dt$.

The general rule is $\int ax^n\,dx = \dfrac{ax^{n+1}}{n+1} + c$

(a) When $a = 3$ and $n = 2$ then

$$\int 3x^2\,dx = \frac{3x^{2+1}}{2+1} + c = \frac{3x^3}{3} + c = x^3 + c$$

(b) When $a = 2$ and $n = 3$ then

$$\int 2t^3\,dt = \frac{2t^{3+1}}{3+1} + c = \frac{2}{4}t^4 + c = \frac{1}{2}t^4 + c$$

Each of these results may be checked by differentiating them.

Problem 2 Determine (a) $\int 8\,dx$ (b) $\int\frac{2}{3}x\,dx$.

(a) $\int 8\,dx$ is the same as $\int 8x^0\,dx$, and, using the general rule when $a = 8$ and $n = 0$ gives

$$\int 8x^0\,dx = \frac{8x^{0+1}}{0+1} + c = 8x + c$$

In general, if k is a constant then $\int k\,dx = kx + c$.

(b) When $a = \dfrac{2}{3}$ and $n = 1$ then

$$\int\frac{2}{3}x\,dx = \left(\frac{2}{3}\right)\frac{x^{1+1}}{(1+1)} + c = \left(\frac{2}{3}\right)\frac{x^2}{2} + c = \frac{1}{3}x^2 + c$$

Problem 3 Determine $\int\left(2 + \frac{5}{7}x - 6x^2\right)dx$.

$\int\left(2 + \frac{5}{7}x - 6x^2\right)dx$ may be written as $\int 2\,dx + \int\frac{5}{7}x\,dx - \int 6x^2\,dx$, i.e. each term is integrated separately. (This splitting up of terms only applies for addition and subtraction.) Hence

$$\int\left(2 + \frac{5}{7}x - 6x^2\right)dx$$

$$= \frac{2x^{0+1}}{0+1} + \left(\frac{5}{7}\right)\frac{x^{1+1}}{(1+1)} - \frac{6x^{2+1}}{(2+1)} + c$$

$$= 2x + \left(\frac{5}{7}\right)\frac{x^2}{2} - 6\frac{x^3}{3} + c = 2x + \frac{5}{14}x^2 - 2x^3 + c$$

Note that when an integral contains more than one term there is no need to have an arbitrary constant for each; just a single constant at the end is sufficient.

Problem 4 Determine (a) $\int\sqrt{x}\,dx$ (b) $\int\frac{3}{x^2}\,dx$.

When n is fractional or negative the general rule for integrals of the form $\int ax^n\,dx$ can still be applied.

(a) $\int\sqrt{x}\,dx = \int x^{1/2}\,dx$. Using the general rule, where $a = 1$ and $n = \dfrac{1}{2}$ gives:

$$\int x^{1/2}\,dx = \frac{(1)x^{(1/2)+1}}{\frac{1}{2}+1} + c = \frac{x^{3/2}}{\frac{3}{2}} + c$$

$$= \frac{2}{3}x^{3/2} + c = \frac{2}{3}\sqrt{x^3} + c$$

(b) $\int\frac{3}{x^2}\,dx = \int 3x^{-2}\,dx$.

Using the general rule, where $a = 3$ and $n = -2$ gives:

$$\int 3x^{-2}\,dx = \frac{3x^{-2+1}}{-2+1} + c = \frac{3x^{-1}}{-1} + c$$

$$= -3x^{-1} + c = -\frac{3}{x} + c$$

Problem 5 Determine (a) $\int \left(\dfrac{x^3 - 2x}{3x}\right) dx$

(b) $\int (1 - x)^2\,dx$

(a) $\int \left(\dfrac{x^3 - 2x}{3x}\right) dx = \int \left(\dfrac{x^3}{3x} - \dfrac{2x}{3x}\right) dx = \int \left(\dfrac{x^2}{3} - \dfrac{2}{3}\right) dx$

$$= \left(\frac{1}{3}\right)\frac{x^{2+1}}{(2+1)} - \frac{2}{3}x + c$$

$$= \frac{1}{9}x^3 - \frac{2}{3}x + c$$

(b) $\int (1 - x)^2\,dx = \int (1 - 2x + x^2)\,dx$

$$= \frac{(1)x^{0+1}}{0+1} - \frac{(2)x^{1+1}}{(1+1)} + \frac{x^{2+1}}{2+1} + c$$

$$= x - x^2 + \frac{1}{3}x^3 + c$$

This problem shows that functions often have to be rearranged into the standard form of $\int ax^n\,dx$ before it is possible to integrate them.

Problem 6 Determine (a) $\int 5 \cos 3x\,dx$

(b) $\int 3 \sin 2x\,dx$.

(a) $\int 5 \cos 3x\,dx = 5 \int \cos 3x\,dx$

$$= (5)\frac{1}{3} \sin 3x + c \text{ (from Table 30.1(ii))}$$

$$= \frac{5}{3} \sin 3x + c$$

(b) $\int 3 \sin 2x\,dx = 3 \int \sin 2x\,dx$

$$= 3\left(-\frac{1}{2} \cos 2x\right) + c \text{ (from Table 30.1(iii))}$$

$$= -\frac{3}{2} \cos 2x + c$$

Problem 7 Determine (a) $\int 5e^{3x}\,dx$ (b) $\int \dfrac{5}{e^{2x}}\,dx$

(a) $\int 5e^{3x}\,dx = 5 \int e^{3x}\,dx = (5)\dfrac{1}{3}e^{ax} + c$

(from Table 30.1(iv))

$$= \frac{5}{3}e^{ax} + c$$

(b) $\int \dfrac{6}{e^{2x}}\,dx = \int 6e^{-2x}\,dx = 6\int e^{-2x}\,dx$

$$= (6)\frac{1}{-2}e^{-2x} + c \text{ (from Table 30.1(v))}$$

$$= -3e^{-2x} + c \text{ or } \frac{-3}{e^{2x}} + c$$

Further problems on integration may be found in section 30.6, Problems 1 to 8, page 271.

30.4 Limits and definite integrals

Integrals containing an arbitrary constant c in their results are called **indefinite integrals** since their precise value cannot be determined without further information.

Definite integrals are those in which limits are applied.

If an expression is written as $[x]_a^b$, b is called the upper limit and a the lower limit.

The operation of applying the limits is defined as $[x]_a^b = (b) - (a)$.

The increase in the value of the integral x^2 as x increases from 1 to 3 is written as $\int_1^3 x^2\,dx$.
Applying the limits gives:

$$\int_1^3 x^2\,dx = \left[\frac{x^3}{3} + c\right]_1^3$$

$$= \left\{\frac{(3)^3}{3} + c\right\} - \left\{\frac{(1)^3}{3} + c\right\}$$

$$= (9 + c) - \left(\frac{1}{3} + c\right) = 8\frac{2}{3}$$

Note that the 'c' term always cancels out when limits are applied and it need not be shown with definite integrals.

Problem 8 Evaluate (a) $\int_1^2 4x\,dx$ (b) $\int_{-2}^3 (5 - x^2)\,dx$.

(a) $\int_1^2 4x\,dx = \left[\dfrac{4x^2}{2}\right]_1^2 = [2x^2]_1^2 = \{2(2)^2\} - \{2(1)^2\}$

$$= 8 - 2 = 6$$

(b) $\int_{-2}^3 (5 - x^2)\,dx = \left[5x - \dfrac{x^3}{3}\right]_{-2}^3$

$$= \left\{5(3) - \frac{(3)^3}{3}\right\} - \left\{5(-2) - \frac{(-2)^3}{3}\right\}$$

$$= \{15 - 9\} - \left\{-10 - \frac{-8}{3}\right\} = 6 + 10 - \frac{8}{3}$$

$$= 13\frac{1}{3}$$

Problem 9 Evaluate (a) $\int_0^2 x(3 + 2x)dx$

(b) $\int_{-1}^1 \frac{(x^4 - 5x^2 + x)}{x}\,dx$

(a) $\int_0^2 x(3 + 2x)\,dx = \int_0^2 (3x + 2x^2)\,dx = \left[\frac{3x^2}{2} + \frac{2x^3}{3}\right]_0^2$

$$= \left\{\frac{3(2)^2}{2} + \frac{2(2)^3}{3}\right\} - \{0 + 0\} = 6 + \frac{16}{3}$$

$$= 11\frac{1}{3}$$

(b) $\int_{-1}^1 \frac{(x^4 - 5x^2 + x)}{x}\,dx$

$$= \int_{-1}^1 (x^3 - 5x + 1)\,dx = \left[\frac{x^4}{4} - \frac{5x^2}{2} + x\right]_{-1}^1$$

$$= \left\{\frac{1}{4} - \frac{5}{2} + 1\right\} - \left\{\frac{(-1)^4}{4} - \frac{5(-1)^2}{2} + (-1)\right\}$$

$$= \left\{\frac{1}{4} - \frac{5}{2} + 1\right\} - \left\{\frac{1}{4} - \frac{5}{2} - 1\right\} = 2$$

Problem 10 Evaluate (a) $\int_0^{\pi/3} 2 \sin 3x\,dx$

(b) $\int_1^2 \cos 2x\,dx$ (correct to 4 significant figures)

(a) $\int_0^{\pi/3} 2 \sin 3x\,dx = \left[\frac{-2}{3} \cos 3x\right]_0^{\pi/3} = \frac{-2}{3} [\cos 3x]_0^{\pi/3}$

$$= \frac{-2}{3}\left[\cos 3\frac{\pi}{3} - \cos 0\right]$$

$$= \frac{-2}{3}[\cos \pi - \cos 0]$$

$$= \frac{-2}{3}[-1 - 1] \text{ (note } \cos \pi \text{ means the cosine of } \pi \text{ radians)}$$

$$= \frac{-2}{3}(-2) = \frac{4}{3} \text{ or } 1\frac{1}{3}$$

(b) $\int_1^2 \cos 2x\,dx = \left[\frac{1}{2} \sin 2x\right]_1^2 = \frac{1}{2} [\sin 2x]_1^2$

$$= \frac{1}{2}[\sin 4 - \sin 2]$$

(where sin 4 means the sine of 4 **radians**)

$$= \frac{1}{2}[(-0.756\,80\ldots) - (0.909\,297\ldots)]$$

$$= -\mathbf{0.833\,0}, \text{ correct to 4 significant figures}$$

Problem 11 Evaluate $\int_1^2 3e^{2t}\,dt$, correct to 2 decimal places.

$$\int_1^2 3e^{2t}\,dt = \left[\frac{3}{2}e^{2t}\right]_1^2 = \frac{3}{2}[e^{2t}]_1^2 = \frac{3}{2}[e^4 - e^2]$$

$$= \frac{3}{2}[54.598\ldots - 7.389\ldots]$$

$$= \mathbf{70.81}, \text{ correct to 2 decimal places}$$

Further problems on definite integrals may be found in section 30.6, Problems 9 to 14, page 271.

30.5   Areas under and between curves using integration

The area shown shaded in *Fig. 30.1* may be determined using approximate methods such as the trapezoidal rule, the mid-ordinate rule or Simpson's rule (see Chapter 16), or precisely by using integration.
The **shaded area** in *Fig. 30.1* is given by:

$$\int_a^b y\,dx = \int_a^b f(x)\,dx$$

There are several instances in engineering and science where the area beneath a curve needs to be accurately determined. For example, the area between limits of: velocity/time graph gives distance travelled; force/distance graph gives work done; voltage/current graph gives power, and so on.

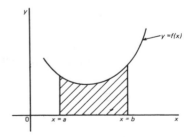

Fig. 30.1

Should a curve drop below the x-axis, then y ($= f(x)$) becomes negative and $\int f(x)\,dx$ is negative. When determining such areas by integration, a negative sign is placed before the integral. For the curve shown in *Fig. 30.2*, the total shaded area is given by (area E + area F + area G).

By integration,

total shaded area $= \int_a^b f(x)\,dx - \int_b^c f(x)\,dx + \int_c^d f(x)\,dx.$

(Note that this is **not** the same as $\int_a^d f(x)\,dx$.)

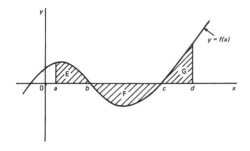

Fig. 30.2

It is usually necessary to sketch a curve in order to check whether it crosses the x-axis.

Problem 12 Determine the area enclosed by $y = 2x + 3$, the x-axis and ordinates $x = 0$ and $x = 3$.

$y = 2x + 3$ is a straight line graph as shown in *Fig. 30.3*, where the area enclosed by $y = 2x + 3$, the x-axis and ordinates $x = 0$ and $x = 3$ is shown shaded. By integration,

$$\text{shaded area} = \int_0^3 y\,dx = \int_0^3 (2x + 3)\,dx$$

$$= \left[\frac{2x^2}{2} + 3x \right]_0^3 = (9 + 9) - (0 + 0)$$

$$= \textbf{18 square units}$$

This answer may be checked since the shaded area is a trapezium. Area of trapezium

$$= \frac{1}{2}(\text{sum of parallel sides})(\text{perpendicular distance between parallel sides})$$

$$= \frac{1}{2}(3 + 9)(3) = \textbf{18 square units}$$

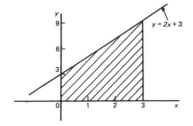

Fig. 30.3

Problem 13 Sketch the graph of $y = 2x^2 + 3$ from $x = -2$ to $x = 3$. Find, by integration, the area enclosed by the curve, the x-axis and ordinates $x = -1$ and $x = 2$.

Since $2x^2 + 3$ is a quadratic expression, the curve $y = 2x^2 + 3$ is a parabola, cutting the y-axis at $y = 3$, as shown in *Fig. 30.4*. The area required is shown shaded. By integration,

$$\text{shaded area} = \int_{-1}^2 y\,dx = \int_{-1}^2 (2x^2 + 3)\,dx$$

$$= \left[\frac{2x^3}{3} + 3x \right]_{-1}^2$$

$$= \left\{ \frac{2(2)^3}{3} + 3(2) \right\} - \left\{ \frac{2(-1)^3}{3} + 3(-1) \right\}$$

$$= \left(\frac{16}{3} + 6 \right) - \left(-\frac{2}{3} - 3 \right)$$

$$= \left(11\frac{1}{3} \right) - \left(-3\frac{2}{3} \right) = \textbf{15 square units}$$

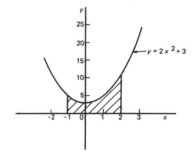

Fig. 30.4

Problem 14 Sketch the graph $y = x^3 - 2x^2 - 5x + 6$ between $x = -2$ and $x = 3$, and determine the area enclosed by the curve and the x-axis.

A table of values is produced and the graph sketched, as shown in *Fig. 30.5*, where the area enclosed by the curve and the x-axis is shown shaded.

x	-2	-1	0	1	2	3
x^3	-8	-1	0	1	8	27
$-2x^2$	-8	-2	0	-2	-8	-18
$-5x$	10	5	0	-5	-10	-15
$+6$	6	6	6	6	6	6
y	0	8	6	0	-4	0

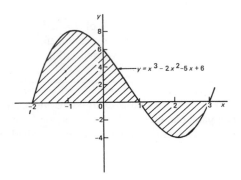

Fig. 30.5

The shaded area = $\int_{-2}^{1} y\, dx - \int_{1}^{3} y\, dx$, the minus sign before

the second integral being necessary since the enclosed area is below the *x*-axis. Hence

$$\text{shaded area} = \int_{-2}^{1} (x^3 - 2x^2 - 5x + 6)\, dx$$

$$- \int_{1}^{3} (x^3 - 2x^2 - 5x + 6)\, dx$$

$$\int_{-2}^{1} (x^3 - 2x^2 - 5x + 6)\, dx$$

$$= \left[\frac{x^4}{4} - \frac{2x^3}{3} - \frac{5x^2}{2} + 6x \right]_{-2}^{1}$$

$$= \left\{ \frac{1}{4} - \frac{2}{3} - \frac{5}{2} + 6 \right\}$$

$$- \left\{ \frac{(-2)^4}{4} - \frac{2(-2)^3}{3} - \frac{5(-2)^2}{2} + 6(-2) \right\}$$

$$= \left(3\frac{1}{12} \right) - \left(-12\frac{2}{3} \right) = -15\frac{3}{4} \text{ square units}$$

$$\int_{1}^{3} (x^3 - 2x^2 - 5x + 6)\, dx$$

$$= \left[\frac{x^4}{4} - \frac{2x^3}{3} - \frac{5x^2}{2} + 6x \right]_{1}^{3}$$

$$= \left\{ \frac{81}{4} - 18 - \frac{45}{2} + 18 \right\} - \left\{ \frac{1}{4} - \frac{2}{3} - \frac{5}{2} + 6 \right\}$$

$$= \left(-2\frac{1}{4} \right) - \left(3\frac{1}{12} \right) = -5\frac{1}{3} \text{ square units}$$

Hence,

$$\mathbf{shaded\ area} = \left(15\frac{3}{4} \right) - \left(-5\frac{1}{3} \right) = 21\frac{1}{12} \textbf{ square units}$$

Problem 15   Determine the area enclosed by the curve $y = \cos\theta$, the ordinates $\theta = 0$ and $\theta = \pi/2$ and the θ-axis.

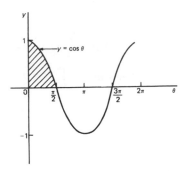

Fig. 30.6

The graph of $y = \cos\theta$ is shown in *Fig. 30.6* and the required area is shown shaded.

$$\text{shaded area} = \int_{0}^{\pi/2} \cos\theta\, d\theta = [\sin\theta]_{0}^{\pi/2}$$

$$= \sin\frac{\pi}{2} - \sin 0 = \mathbf{1\ square\ unit}$$

Problem 16   Determine the area enclosed by the curve $y = 3x^2 + 6$, the *x*-axis and ordinates $x = 1$ and $x = 4$, (a) by integration, (b) by the trapezoidal rule, (c) by the mid-ordinate rule, and (d) by Simpson's rule.

The curve $y = 3x^2 + 6$ is shown plotted in *Fig. 30.7*.

(a) **By integration,**

$$\text{shaded area} = \int_{1}^{4} y\, dx = \int_{1}^{4} (3x^2 + 6)\, dx$$

x	0	1.0	1.5	2.0	2.5	3.0	3.5	4.0
$y = 3x^2 + 6$	6.0	9.0	12.75	18.0	24.75	33.0	42.75	54.0

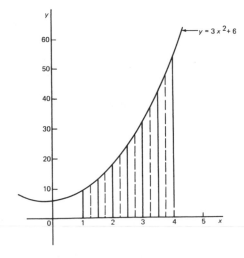

Fig. 30.7

$$= [x^3 + 6x]_1^4 = \{(4)^3 + 6(4)\} - \{(1)^3 + 6(1)\}$$
$$= (64 + 24) - (1 + 6) = \textbf{81 square units}$$

(b) **By the trapezoidal rule** (see Chapter 16, section 16.1, para. (b)):

$$\text{Area} = (\text{width of interval})\left\{\frac{1}{2}(\text{first} + \text{last ordinates})\right.$$

$$\left. + \text{ sum of remaining ordinates}\right\}$$

Selecting 6 intervals, each of width 0.5, gives:

$$\text{Area} = (0.5)\left\{\frac{1}{2}(9 + 54) + 12.75 + 18 + 24.75\right.$$

$$\left. + 33 + 42.75\right\}$$

$$= \textbf{81.375 square units}$$

(c) **By the mid-ordinate rule** (see Chapter 16, section 16.1, para. (c)):

$$\text{Area} = (\text{width of interval})(\text{sum of mid-ordinates})$$

Selecting 6 intervals, each of width 0.5, gives the mid-ordinates as shown by the broken line in *Fig. 30.7*. Thus

$$\text{Area} = (0.5)(11 + 15 + 21 + 28.5 + 37.5 + 48)$$

$$= \textbf{80.5 square units}$$

(d) **By Simpson's rule** (see Chapter 16, section 16.1, para. (d)):

$$\text{Area} = \frac{1}{3}(\text{width of interval}) \times$$

$$\left\{\left(\begin{array}{c}\text{first} + \text{last} \\ \text{ordinate}\end{array}\right) + 4\left(\begin{array}{c}\text{sum of even} \\ \text{ordinates}\end{array}\right)\right.$$

$$\left. + 2\left(\begin{array}{c}\text{sum of remaining} \\ \text{odd ordinates}\end{array}\right)\right\}$$

Selecting 6 intervals, each of width 0.5, gives:

$$\text{Area} = \frac{1}{3}(0.5)[(9 + 54) + 4(12.75 + 24.75 + 42.75)$$

$$+ 2(18 + 33)]$$

$$= \textbf{81 square units}$$

Integration gives the correct result for the area under a curve. Simpson's rule is seen to be the most accurate of the three approximate methods in this case.

Further problems on areas under and between curves using integration may be found in section 30.6, Problems 15 to 24, page 272.

30.6 Further problems on introduction to integration

Integration

Determine the indefinite integrals in *Problems 1* to *8*.

1 (a) $\int 5\, dx$ (b) $\int 7x\, dx$

$$\left[(a)\ 5x + c\ (b)\ \frac{7x^2}{2} + c\right]$$

2 (a) $\int \frac{2}{3}x^2\, dx$ (b) $\int \frac{4}{5}x^3\, dx$

$$\left[(a)\ \frac{2}{9}x^3 + c\ (b)\ \frac{1}{5}x^4 + c\right]$$

3 (a) $\int (3 + 2x - 4x^2)\, dx$ (b) $3\int (x - 5x^2)\, dx$

$$\left[(a)\ 3x + x^2 - \frac{4x^3}{3} + c\ (b)\ 3\left(\frac{x^2}{2} - \frac{5x^3}{3}\right) + c\right]$$

4 (a) $\int \frac{3x^2 - 5x}{2x}\, dx$ (b) $\int (2 + x)^2\, dx$

$$\left[(a)\ \frac{3}{4}x^2 - \frac{5}{2}x + c\ (b)\ 4x + 2x^2 + \frac{x^3}{3} + c\right]$$

5 (a) $\int \frac{2}{x^3}\, dx$ (b) $\int \frac{9}{4x^2}\, dx$

$$\left[(a)\ \frac{-1}{x^2} + c\ (b)\ \frac{-9}{4x} + c\right]$$

6 (a) $\int 7\sqrt{x^3}\, dx$ (b) $\int \frac{4}{\sqrt{x}}\, dx$

$$\left[(a)\ \frac{14}{5}\sqrt{x^5} + c\ (b)\ 8\sqrt{x} + c\right]$$

7 (a) $\int 4 \sin 3x\, dx$ (b) $2\int \cos 5x\, dx$

$$\left[(a)\ \frac{-4}{3}\cos 3x + c\ (b)\ \frac{2}{5}\sin 5x + c\right]$$

8 (a) $\int 5\, e^{2t}\, dt$ (b) $\int \frac{-3}{e^{4x}}\, dx$

$$\left[(a)\ \frac{5}{2}\, e^{2t} + c\ (b)\ \frac{3}{4e^{2x}} + c\right]$$

Definite integrals

Evaluate the definite integrals in *Problems 9* to *14* (where necessary correct to 4 significant figures).

9 (a) $\int_1^4 3x^2\, dx$ (b) $\int_{-1}^2 2x\, dx$

[(a) 80 (b) 3]

10 (a) $\int_0^2 (5x - 2x^2)\, dx$ (b) $\int_1^3 (x^2 - 4x + 3)\, dx$

$$\left[\text{(a) } 4\frac{2}{3} \text{ (b) } -1\frac{1}{3} \right]$$

11 (a) $\int_0^\pi 3 \cos\theta\, d\theta$ (b) $\int_0^{\pi/2} 5 \cos\theta\, d\theta$

[(a) 0 (b) 5]

12 (a) $\int_1^2 (3 - x)^2\, dx$ (b) $\int_{-2}^1 \dfrac{2x + x^2}{x}\, dx$

$$\left[\text{(a) } 2\frac{1}{3} \text{ (b) } 4\frac{1}{2} \right]$$

13 (a) $\int_0^{\pi/4} 4 \cos 2x\, dx$ (b) $\int_0^1 4 \sin 4t\, dt$

[(a) 2 (b) 1.654]

14 (a) $\int_1^2 \dfrac{1}{4} e^{3t}\, dt$ (b) $\int_{-1}^3 \dfrac{1}{5e^{2x}}\, dx$

[(a) 31.95 (b) 0.738 7]

Area under and between curves using integration

15 Show by integration that the area of a rectangle formed by the line $y = 4$, the ordinate $x = 1$ and $x = 6$ and the x-axis is 20 square units.

16 Show by integration that the area of the triangle formed by the line $y = 3x$, the ordinates $x = 0$ and $x = 5$ and the x-axis is 37.5 square units.

17 Sketch the curve $y = x^2 + 5$ between $x = -1$ and $x = 4$. Find by integration the area enclosed by the curve the x-axis and ordinates $x = 1$ and $x = 3$. Use an approximate method to find the area and compare your result with that obtained by integration.

$$\left[18\frac{2}{3} \text{ square units} \right]$$

18 Determine the area enclosed by $y = 2x^3$, the x-axis and ordinates $x = 0$ and $x = 2$.
[8 square units]

19 Determine the area enclosed between the curve $y = 6 - x - x^2$ and the x-axis.

$$\left[20\frac{5}{6} \text{ square units} \right]$$

20 Sketch the curve $y = x(x - 1)(x - 3)$ and use Simpson's rule to find the area enclosed by the curve and the x-axis. Compare your answer with the true area obtained by integration.

$$\left[3\frac{1}{12} \text{ square units} \right]$$

21 Find the area enclosed by the curve $y = x^2 + x - 6$, the x-axis and ordinates $x = -2$ and $x = 1$.

$$\left[16\frac{1}{2} \text{ square units} \right]$$

22 Sketch the curve $y = x^3 - 2x^2 - 3x$ between $x = -2$ and $x = 4$. Determine the area enclosed by the curve and the x-axis.

$$\left[24\frac{1}{3} \text{ square units} \right]$$

23 A vehicle has a velocity $v = (3 + 4t)$ m/s after t seconds. How far does it travel in the first 3 seconds? Determine also the distance travelled in the fourth second.

(Distance travelled $= \int_{t_1}^{t_2} v\, dt$)

[27m, 17m]

24 The force F newtons acting on a body at a distance x metres from a fixed point is given by $F = 2x + 3x^2$. Determine the work done when the body moves from the position where $x = 1$ m to that where $x = 4$ m.

(Work done $= \int_{x_1}^{x_2} F\, dx$)

[78 N m]

Index